Methodologies Used in Remote Sensing Data Analysis and Remote Sensors for Precision Agriculture

Methodologies Used in Remote Sensing Data Analysis and Remote Sensors for Precision Agriculture

Editors

Jiyul Chang
Sigfredo Fuentes

MDPI • Basel • Beijing • Wuhan • Barcelona • Belgrade • Manchester • Tokyo • Cluj • Tianjin

Editors
Jiyul Chang
South Dakota State University
USA

Sigfredo Fuentes
The University of Melbourne
Australia

Editorial Office
MDPI
St. Alban-Anlage 66
4052 Basel, Switzerland

This is a reprint of articles from the Special Issue published online in the open access journal *Sensors* (ISSN 1424-8220) (available at: https://www.mdpi.com/journal/sensors/special_issues/remote_data).

For citation purposes, cite each article independently as indicated on the article page online and as indicated below:

LastName, A.A.; LastName, B.B.; LastName, C.C. Article Title. *Journal Name* **Year**, *Volume Number*, Page Range.

ISBN 978-3-0365-6614-6 (Hbk)
ISBN 978-3-0365-6615-3 (PDF)

Contents

About the Editors

Jiyul Chang

Jiyul Chang who is a Senior Lecturer in Agronomy, Horticulture, and Plant science Department, South Dakota State University. Dr. Chang has conducted many precision agriculture projects which include geospatial analysis of field spatiotemporal variations using GIS, remote sensing data analysis for crop and nutrient management, and management zone map/prescription map development using various images (satellites and drones). Dr. Chang has also taught precision agriculture courses (mapping, remote sensing, agricultural GIS software). Currently, Dr. Chang is carrying out projects on nitrogen management and crop management using drone images and the 7Gen FnA iLEARN-USDA project.

Sigfredo Fuentes

Sigfredo Fuentes is an Associate Professor in Digital Agriculture, Food and Wine Sciences. Professor Fuentes' scientific interests include climate change impacts on agriculture, development of new computational tools for plant physiology, food, and wine science, new and emerging sensor technology, proximal, short- and long-range remote sensing using robots and UAVs, machine learning, and artificial intelligence.

Editorial

Methodologies Used in Remote Sensing Data Analysis and Remote Sensors for Precision Agriculture

Sigfredo Fuentes [1] and Jiyul Chang [2],*

[1] Digital Agriculture Food and Wine, School of Agriculture and Food, Faculty of Veterinary and Agricultural Science, The University of Melbourne, Parkville, VIC 3010, Australia

[2] Department of Agronomy, Horticulture & Plant Science, South Dakota State University, Brookings, SD 57007, USA

* Correspondence: jiyul.chang@sdstate.edu

Citation: Fuentes, S.; Chang, J. Methodologies Used in Remote Sensing Data Analysis and Remote Sensors for Precision Agriculture. *Sensors* **2022**, *22*, 7898. https://doi.org/10.3390/s22207898

Received: 4 October 2022
Accepted: 9 October 2022
Published: 17 October 2022

Publisher's Note: MDPI stays neutral with regard to jurisdictional claims in published maps and institutional affiliations.

When adopting remote sensing techniques in precision agriculture, there are two main areas to consider: data acquisition and data analysis methodologies. Imagery and remote sensor data collected using different platforms provide a variety of information volumes and formats. For example, recent research in precision agriculture has used multispectral images from different platforms, such as satellites, airborne, and, most recently, drones. These images have been used for various analyses, from the detection of pests and diseases, growth and water status of crops, to yield estimations. However, accurately detecting specific biotic or abiotic stresses requires a narrow range of spectral information to be analyzed for each application. In data analysis, the volume and complexity of data formats obtained using the latest technologies in remote sensing (e.g., a cube of data for hyperspectral imagery) demands complex data processing systems and data analysis using multiple inputs to estimate specific categorical or numerical targets. New and emerging methodologies within artificial intelligence, such as machine learning and deep learning, have enabled us to deal with these increasing data volumes and complex analyses.

This Special Issue (SI) mainly focused on (i) advanced methodologies for remotely sensed data collected by different types of sensors and platforms for precision agriculture and (ii) the implementation of various sensors for specific targets in precision agriculture. High-quality research was published in this SI from researchers from various countries, including China, the USA, Slovenia, Spain, Germany, Brazil, Australia, and Singapore. The SI's studies have been ordered following the application within the soil–plant–atmosphere continuum starting with the soil salinity precision monitoring using unmanned aerial vehicles (UAV) and multispectral imagery [1]; the evaluation of optical sensors for the diagnosis of nitrogen content for wheat plants [2]; the detection of root-knot nematode infestation in potato plants using hyperspectral imagery [3]; detection of powdery mildew using hyperspectral, thermal, and RGB imagery [4]; leaf area index estimations for wheat using hyperspectral reflectance data [5]; vineyard canopy characteristics and vigor assessment using UAV and satellite imagery [6]; estimation of crop vegetation parameters using satellite and UAV spectral remote sensing [7]; above-ground biomass estimation of oat plants using UAV remote sensing and machine learning [8]; wheat lodging estimation using multispectral UAV imagery and deep learning [9]; yield estimation for guinea grass using UAV remote sensing [10]; and wheat yield prediction from satellite imagery, meteorological data, and machine learning modeling [11].

Different sensor technologies, such as SPAD, Dualex 4, and RapidSCAN, were implemented to assess the accuracy of estimating nitrogen levels in winter wheat, with Dualex 4 being the sensor with the best performance [2].

Different machine and deep learning analytical methods were employed to analyze imagery and the numerical data from various research studies. For soils, among the methodologies used were partial least square (PLS) back propagation neural networks

(BPNN), support vector machines (SVM), and random forests (RF) to construct retrieval models to estimate soil salinity using regression models [1]. The latter was the most accurate method resulting in determination coefficients of $R^2 = 0.724$ for the modeling stage and 0.745 for validation. For roots and disease estimation, hyperspectral imagery for disease detection on potato tubers (diseased and non-diseased), and machine learning modeling using SVM classifiers plus dimensionality reduction methods with accuracies over 60% [3]. Other multisource vegetation indices extracted from hyperspectral, thermal, and RGB imaging have been used coupled with RF and SVM regression algorithms to target a powdery mildew index on wheat. The former machine learning methodology resulted in higher and more stable performances and $R^2 > 0.86$ [4].

In terms of canopy architecture, hyperspectral reflectance data from winter wheat was used as inputs for a combination of algorithms at different phenological stages to estimate LAI as targets. The best performance was obtained in flowering and filling stages with $0.87 < R^2 < 0.71$ for modeling and $0.84 < R^2 < 0.77$ for validation, respectively [5]. Other canopy-related parameters for vineyards, such as the normalized differential vegetation index (NDVI) obtained from UAV and satellite multispectral data using simple linear regression from individual plants and clusters of plants according to the spatial footprint of imagery. The NDVI was then related to the tree row volume resulting in moderate R^2 for vigor estimation [6]. Other multispectral/hyperspectral parameters from satellite–UAV data comparisons were performed to estimate crop vegetation parameters, such as LAI, leaf chlorophyll concentration, and canopy water content, with no clear superiority for either remote-sensed data on the estimations [2]. For above-ground biomass estimation of oats, UAV-based remote sensing multispectral imagery and derived vegetation indices (VIs) were coupled with PLS, SVM, and artificial neural networks (ANN) and RF algorithms. These studies' results showed various low to moderate accuracies in predicting above-ground biomass [8]. The highest accuracy was obtained by combining RGB + digital surface model (DSM) with 89% [9]. Deep learning based on convolutional neural networks (CNN) with different architectures to analyze RGB from a UAV platform was used to estimate dry matter yield for guinea grass resulting in correlation coefficients of $0.79 < R < 0.62$ [10]. Furthermore, RF algorithms were also used for wheat yield prediction based on satellite-based NDVI combined with meteorological data in Australia, resulting in $0.89 < R2 < 0.42$ for different locations [11].

After climatic anomalies, plants can suffer from lodging, such as wheat, and the damage estimation can be helpful in decision-making. Multispectral imagery from a UAV platform was used to estimate lodging in wheat coupled with a lightweight network model method based on RGB-DSM with 89% accuracy in the lodging estimation.

It has been shown in this SI that remote sensing coupled with artificial intelligence and machine learning are powerful tools to estimate parameters from soil salinity, plant biotic and abiotic stresses/damage, canopy architecture characteristics, and yield estimation.

Institutional Review Board Statement: Not applicable.

Informed Consent Statement: Not applicable.

Data Availability Statement: Not applicable.

Conflicts of Interest: The authors declare no conflict of interest.

References

1. Yu, X.; Chang, C.; Song, J.; Zhuge, Y.; Wang, A. Precise Monitoring of Soil Salinity in China's Yellow River Delta Using UAV-Borne Multispectral Imagery and a Soil Salinity Retrieval Index. *Sensors* **2022**, *22*, 546. [CrossRef] [PubMed]
2. Jiang, J.; Wang, C.; Wang, H.; Fu, Z.; Cao, Q.; Tian, Y.; Zhu, Y.; Cao, W.; Liu, X. Evaluation of Three Portable Optical Sensors for Non-Destructive Diagnosis of Nitrogen Status in Winter Wheat. *Sensors* **2021**, *21*, 5579. [CrossRef] [PubMed]
3. Lapajne, J.; Knapič, M.; Žibrat, U. Comparison of Selected Dimensionality Reduction Methods for Detection of Root-Knot Nematode Infestations in Potato Tubers Using Hyperspectral Imaging. *Sensors* **2022**, *22*, 367. [CrossRef] [PubMed]
4. Feng, Z.; Song, L.; Duan, J.; He, L.; Zhang, Y.; Wei, Y.; Feng, W. Monitoring Wheat Powdery Mildew Based on Hyperspectral, Thermal Infrared, and RGB Image Data Fusion. *Sensors* **2022**, *22*, 31. [CrossRef] [PubMed]

5. Li, C.; Wang, Y.; Ma, C.; Ding, F.; Li, Y.; Chen, W.; Li, J.; Xiao, Z. Hyperspectral Estimation of Winter Wheat Leaf Area Index Based on Continuous Wavelet Transform and Fractional Order Differentiation. *Sensors* **2021**, *21*, 8497. [CrossRef] [PubMed]
6. Campos, J.; García-Ruíz, F.; Gil, E. Assessment of Vineyard Canopy Characteristics from Vigour Maps Obtained Using UAV and Satellite Imagery. *Sensors* **2021**, *21*, 2363. [CrossRef] [PubMed]
7. Wijesingha, J.; Dayananda, S.; Wachendorf, M.; Astor, T. Comparison of Spaceborne and UAV-Borne Remote Sensing Spectral Data for Estimating Monsoon Crop Vegetation Parameters. *Sensors* **2021**, *21*, 2886. [CrossRef] [PubMed]
8. Sharma, P.; Leigh, L.; Chang, J.; Maimaitijiang, M.; Caffé, M. Above-Ground Biomass Estimation in Oats Using UAV Remote Sensing and Machine Learning. *Sensors* **2022**, *22*, 601. [CrossRef] [PubMed]
9. Yang, B.; Zhu, Y.; Zhou, S. Accurate Wheat Lodging Extraction from Multi-Channel UAV Images Using a Lightweight Network Model. *Sensors* **2021**, *21*, 6826. [CrossRef] [PubMed]
10. De Oliveira, G.S.; Marcato Junior, J.; Polidoro, C.; Osco, L.P.; Siqueira, H.; Rodrigues, L.; Jank, L.; Barrios, S.; Valle, C.; Simeão, R.; et al. Convolutional Neural Networks to Estimate Dry Matter Yield in a Guineagrass Breeding Program Using UAV Remote Sensing. *Sensors* **2021**, *21*, 3971. [CrossRef] [PubMed]
11. Pang, A.; Chang, M.W.L.; Chen, Y. Evaluation of Random Forests (RF) for Regional and Local-Scale Wheat Yield Prediction in Southeast Australia. *Sensors* **2022**, *22*, 717. [CrossRef] [PubMed]

sensors

MDPI

Article

Precise Monitoring of Soil Salinity in China's Yellow River Delta Using UAV-Borne Multispectral Imagery and a Soil Salinity Retrieval Index

Xinyang Yu [1,2], Chunyan Chang [1], Jiaxuan Song [1], Yuping Zhuge [1,*] and Ailing Wang [1]

1 College of Resources and Environment, Shandong Agricultural University, Tai'an 271018, China; yuxy.12b@igsnrr.ac.cn (X.Y.); chyan0103@sdau.edu.cn (C.C.); SJxuan020903@163.com (J.S.); ailingwang@sdau.edu.cn (A.W.)
2 Tropical Research and Education Center/Department of Agricultural and Biological Engineering, Institute of Food and Agricultural Sciences, University of Florida, Homestead, FL 33031, USA
* Correspondence: zhugeyp@sdau.edu.cn

Abstract: Monitoring salinity information of salinized soil efficiently and precisely using the unmanned aerial vehicle (UAV) is critical for the rational use and sustainable development of arable land resources. The sensitive parameter and a precise retrieval method of soil salinity, however, remain unknown. This study strived to explore the sensitive parameter and construct an optimal method for retrieving soil salinity. The UAV-borne multispectral image in China's Yellow River Delta was acquired to extract band reflectance, compute vegetation indexes and soil salinity indexes. Soil samples collected from 120 different study sites were used for laboratory salt content measurements. Grey correlation analysis and Pearson correlation coefficient methods were employed to screen sensitive band reflectance and indexes. A new soil salinity retrieval index (SSRI) was then proposed based on the screened sensitive reflectance. The Partial Least Squares Regression (PLSR), Multivariable Linear Regression (MLR), Back Propagation Neural Network (BPNN), Support Vector Machine (SVM), and Random Forest (RF) methods were employed to construct retrieval models based on the sensitive indexes. The results found that green, red, and near-infrared (NIR) bands were sensitive to soil salinity, which can be used to build SSRI. The SSRI-based RF method was the optimal method for accurately retrieving the soil salinity. Its modeling determination coefficient (R^2) and Root Mean Square Error ($RMSE$) were 0.724 and 1.764, respectively; and the validation R^2, $RMSE$, and Residual Predictive Deviation (RPD) were 0.745, 1.879, and 2.211.

Keywords: soil salinity sensitive parameter; random forest; support vector machine; optimal retrieval model; remote sensing

Citation: Yu, X.; Chang, C.; Song, J.; Zhuge, Y.; Wang, A. Precise Monitoring of Soil Salinity in China's Yellow River Delta Using UAV-Borne Multispectral Imagery and a Soil Salinity Retrieval Index. *Sensors* 2022, 22, 546. https://doi.org/10.3390/s22020546

Academic Editors: Jiyul Chang and Sigfredo Fuentes

Received: 6 November 2021
Accepted: 8 January 2022
Published: 11 January 2022

Publisher's Note: MDPI stays neutral with regard to jurisdictional claims in published maps and institutional affiliations.

1. Introduction

Soil is a vital component of the ecosystem. It plays a crucial role in the structure and operation of the land ecosystem [1,2]. However, the degradation of soil resources has emerged as one of the world's most pressing ecological concerns. Soil salinization has already become a significant symptom of soil degradation that affects 10% of the world's agricultural land [3,4]. The search for a reliable monitoring index and precise regression method for soil salinity is essential to globally assess soil salinization and its severe implications for agriculture and food security.

Ecological parameter measurement and airborne/satellite remote sensing (RS) monitoring technologies are two commonly utilized soil salinity assessment methods. Traditional methods rely on field surveys and electrical conductivity measurements, which are accurate but time and labor-intensive [5,6], and do not allow for monitoring of the spatial distribution pattern of soil salinity content. Multi- and hyperspectral satellite RS technology has been used in soil salinity monitoring since the 1990s [7,8]. Azabdaftari et al. (2016), for instance, computed vegetation indexes in the Adana region of Turkey using Landsat multispectral images from four different times [9]. Morgan et al. (2018) forecasted soil salinity in Cairo, Egypt using Sentinel-2 multispectral data [10]. Hyperspectral images such as EO-1 and HJ-1A were also employed as data sources to accurately detect soil salinity [11,12]. Different from the satellite RS means, the Unmanned Aerial Vehicle (UAV)-borne spectral sensors are highly maneuverable and have been used to monitor soil salinity since the 2010s. Hu et al. (2019) used electromagnetic induction equipment and a hyperspectral camera mounted on a UAV platform to evaluate and estimate field-scale soil salinity [13]. Ivushkin (2019) looked into the use of UAVs to measure salt stress in quinoa plants [14]. Wang et al. (2019) extracted the salt content of extremely salty soil in China's Yellow River Estuary and compared the retrieval findings with the inverse distance weighted interpolation results to achieve more accurate saline soil extraction [15]. To boost the spectral resolution to retrieve soil salinity, Ma (2020) combined Sentinel-2A and UAV multispectral images to increase the spectral resolution to inverse regional soil salinity [16]. Satellite RS imagery-based soil salinity studies have indicated that the index in the visible to infrared spectrum may better measure soil salinity, which can increase the accuracy of soil salinity retrieval [17–19]. The majority of vegetation indexes can indirectly indicate soil salinity [20]. However, few studies focused on the detection of UAV band information sensitive to soil salinity, which is essential for the construction of a reliable soil salinity monitoring index to help efficiently predict the soil salinity conditions.

For the soil salinity regression method, several approaches such as partial least square (PLS), BP Neural Network (BPNN), Support Vector Machines (SVM), and random forest (RF) were introduced and applied [15,21]. For instance, Ma (2018) increased the accuracy of soil salinization retrieval by combining numerous mathematical changes on soil surface reflectance with regression analysis of collected soil data [12]. Machine learning algorithms were used by Yao et al. (2019) to infer agriculture soil salt concentration from UAV multispectral RS images [22]. The determination coefficients for validation were more than 0.69. To improve regional retrieval precision, Chen et al. (2021) presented a differentiated fusion method for calculating satellite and ground spectral variables of soil salinity based on sample differences [23]. Spectral parameters and correlation salinity indexes have been converted and filtered to retrieve soil salinity. In resource management and allocation, the river delta region has a high degree of social-ecological interdependence and competition. In China, the Yellow River Delta (YRD) features shallow groundwater levels (0–2 m), significant salinity, and surface salinity. Soil salinization affects over 70% of YRD's land, making the region's biological ecosystem severely vulnerable [24]. Soil salinization has long been a major source of soil degradation in the YRD, limiting local agricultural productivity. Precise monitoring of soil salinity is essential to assess soil salinization. However, screening and design of sensitive parameters, as well as a suitable retrieval method, is, nevertheless, unknown.

This study thus strived to explore the sensitive parameter and construct an optimal method for soil salinity retrieval. The Yellow River Delta (YRD) in China was selected as the study area to experiment. UAV RS image and ground truth data collected during the spring season were used as the data source. Sensitive bands and spectral parameters of soil salinity were identified using grey correlation analysis and Pearson correlation coefficient approaches. PLSR, MLR, BPNN, SVM, and RF modeling methods were used to create soil salt retrieval models based on reflectance, vegetation index, and salinity index. The accuracies were evaluated quantitatively to find the optimal retrieval model. This study is expected to serve as a guide for the selection of sensitive criteria and the optimal soil salinity prediction algorithms, which can be used in other regions to retrieve soil salinity efficiently.

2. Materials and Methods

2.1. Study Area

The study was conducted in a representative arable region of Kenli District, YRD (37°35′6″~37°35′14′ N, 118°20′31″~118°20′46″ E). The climate of the study area is a temperate continental monsoon climate, which is dry and windy in spring. With a potential evapotranspiration–precipitation ratio of 7.6, potential evapotranspiration considerably outnumbers precipitation in spring, resulting in limited vegetation covering in the study area and severe salt deposition in the soil. The groundwater table is also shallow and mineralized. Arable and abandoned lands are the most common land uses, and coastal (tidal) salty soil with a light texture and high capillary action is the most common soil type. Hydrogeological conditions in the study area may contribute to soil salinization [25].

2.2. Image Acquisition and Preprocessing

The spring season in the study area is the period of high evapotranspiration and accumulation of soil salinity, which is crucial for the development of winter wheat. On 19 March 2021, a field survey was conducted to collect soil samples and obtain UAV images (Figure 1). The DJI Matrice 600 Pro (SZ DJI Technology Co., Ltd. Shenzhen, China) and the Parrot Sequoia agriculture multispectral camera, which includes Green (G), Red (R), Red Edge (REG), and Near-infrared (NIR) bands, are part of the UAV image acquisition system (Table 1). During the UAV image acquisition period, the UAV's flying height was set to 50 m, and the spatial resolution was set to 5 cm. Each flight trace had a 60% overlap ratio. After that, the UAV image and the associated GPS data were loaded into the Pix4D Mapper for preprocessing, which included geometric correction, radiometric calibration, and orthorectification.

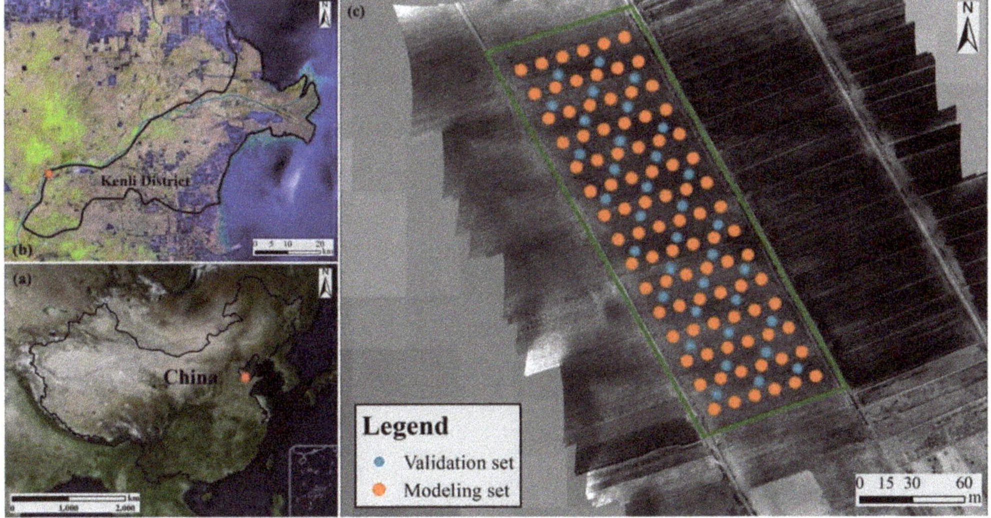

Figure 1. Location of the study area. (**a**) Location of the Kenli District in China; (**b**) test area in the Kenli district; (**c**) UAV image covering the test area.

Table 1. Band information of multispectral camera sensor.

ID	Band	Abbreviation	Center Wavelength (nm)	Bandwidth (nm)
1	Green	G	550	40
2	Red	R	660	40
3	Red edge	REG	735	10
4	Near-infrared	NIR	790	40

2.3. Soil Sampling and Laboratory Procedures

One hundred and twenty sample sites and 40 ground control points were evenly distributed in the test area. An EC110 portable salinity meter equipped with a 2225FST series probe (in which the temperature correction for the electrical conductivity had already been completed) (Spectrum Technologies Inc., Dallas/Fort Worth, TX, USA) was used to make five measurements at and near each sampling site, with a range of no more than 5 cm × 5 cm. Using the five-point sampling approach, samples from 0–10 cm soil surface layer were taken at each survey location and put into separate sealed plastic bags. Meanwhile, the hand-held differential GPS (Trimble GEO 7X, Trimble Inc., Sunnyvale, CA, USA) was used to record the longitude and latitude coordinates of each sampling location, while the camera captured and recorded the surrounding environmental information.

Soil samples were treated in the laboratory for natural air drying at room temperature. Coarse fragments such as stones were discarded. All the soil samples were then physically milled, thoroughly mixed, sieved to obtain the fraction less than 2 mm (fine earth fraction), and packaged in separate bags for salt content analysis. The soil samples were processed into the soil solution at a soil-to-water ratio of 1:5 [26,27]. The soil conductivity value was measured using an EC110 conductivity meter, and 30 extracts were chosen at random to compute the matching soil total salt concentration [28]. Equation (1) depicts the conversion connection between soil total salt concentration and extraction solution conductivity in the studied region [4].

$$S_t = 2.180 \times EC_{1:5} + 0.727 \tag{1}$$

where S_t is the total salt content of the soil (g/kg), and $EC_{1:5}$ is the conductivity of soil extract (mS/cm) with a soil–water ratio of 1:5. $EC_{1:5}$ is used to calculate the total salt content of different soil samples without measuring the ion composition, as shown in Equation (1). For each treatment, the measurement was performed five times.

2.4. Construction of Soil Salinity Retrieval Index

The sensitive reflectance will be used to build a new soil salinity retrieval index. Before that, the correlation coefficient technique and grey correlation analysis between band reflectance and soil salinity content were primarily computed to screen sensitive band reflectance. The grey correlation analysis technique is a statistical analysis approach using several factors. It is used to calculate the degree of correlation among components based on the similarity or dissimilarity of development patterns among factors, i.e., the grey correlation degree [29]. The Pearson correlation coefficient measures the degree of linear association between two distance variables. The Pearson correlation analysis is a type of factor correlation analysis that is appropriate for continuous variables [30]. Besides, the band diagnostic index (P_i) was employed in this study to further improve the accuracy and reliability of screening sensitive band reflectance. The calculation equation of the band diagnostic index is shown below.

$$P_i = R_i \times \sigma_i \tag{2}$$

where R_i is the correlation coefficient between the reflectance value on each band and the soil salinity, and σ_i is the standard deviation of reflectance value of band i [31].

2.5. Validation

To examine the performance of the new proposed index, six vegetation indexes, six salinity indexes, and one brightness index were used as comparisons to conduct screening, model construction, and validation process. The vegetation index is calculated using the standard multispectral RS bands R and NIR, and it includes the Normalized Difference Vegetation Index (NDVI), Difference Vegetation Index (DVI), Soil Adjusted Vegetation Index (SAVI), and Ratio Vegetation Index (RVI). Based on the band operation of the NDVI, the Green Normalized Difference Vegetation Index (GNDVI) and the Red Normalized Difference Vegetation Index (NDVI$_{REG}$) were calculated and classed as VI indexes. The salinity index stands for the soil salinity index. It is represented by six algebras (SI-T, SI1, SI2, SI3, NDSI, and SRSI), with Soil Remote Sensing Index (SRSI) being the transformation and synthesis index of the Soil Salinity Index SI1 and the vegetation index NDVI (Table 2). The brightness index (BI) is determined using the R and NIR bands.

Table 2. Spectral indexes and equations. G represents the reflectance of the green band, R denotes the reflectance of the red band, REG is the reflectance of the red edge band, and NIR is the reflectance of the near-infrared band.

Index Type	Spectral Index	Equation	Reference
VI	Normalized Difference Vegetation Index (NDVI)	$\frac{NIR-R}{NIR+R}$	[19]
	Difference Vegetation Index (DVI)	$NIR - R$	[19]
	Soil Adjusted Vegetation Index (SAVI)	$\frac{(1+L)\times(NIR-R)}{NIR+R+L}, L = 0.5$	[32]
	Ratio Vegetation Index (RVI)	$\frac{NIR}{R}$	[32]
	Green Normalized Difference Vegetation Index (GNDVI)	$\frac{NIR-G}{NIR+G}$	[33]
	Red Normalized Vegetation Difference Index (NDVI$_{REG}$)	$\frac{NIR-REG}{NIR-REG}$	[33]
SI	Salinity Index (SI-T)	$\frac{R}{NIR}\times100$	[34]
	Salinity Index 1 (SI1)	$\sqrt{G \times R}$	[34]
	Salinity Index 2 (SI2)	$\sqrt{G^2+R^2+NIR^2}$	[35]
	Salinity Index 3 (SI3)	$\sqrt{G^2+R^2}$	[35]
	Normalized Difference Salinity Index (NDSI)	$\frac{R-NIR}{R+NIR}$	[36]
	Soil Remote Sensing Index (SRSI)	$\sqrt{(NDVI-1)^2+SI1^2}$	[32]
BI	Brightness index (BI)	$\sqrt{R^2+NIR^2}$	[36]

The soil salinity retrieval model and comparison techniques were constructed using QGIS, SPSS, and Matlab. Based on the newly constructed index and sensitive VI and SI, the retrieval models of soil salinity were built using the Partial Least Squares Regression (PLSR [37]), Multivariable Linear Regression (MLR [38]), Back Propagation Neural Network (BPNN [39]), Support Vector Machine (SVM [40]), and Random Forest (RF [41]) methods. The determination coefficient (R^2), root mean square error ($RMSE$), and residual predictive deviation (RPD) were employed to evaluate the regression results. R^2 represents the consistency with which the model was established and validated. If R^2 is near to one, the model is more robust and has a better fitting degree. The $RMSE$ is used to evaluate the model's prediction performance. The lower the $RMSE$, the better the model's prediction ability. The RPD is the ratio of the measured value's standard deviation to the predicted error. When RPD is less than 1.4, the model cannot predict measured values; $1.4 \leq RPD < 2$ indicates that the model can roughly predict those values, and RPD more than or equal to 2.0 shows that the model has exceptional prediction ability. Models with high R^2 and RPD values perform better in terms of prediction and stability [42].

3. Results

3.1. Statistical Analysis of Soil Samples

The soil salt concentration varied from 0.264 to 20.651 g/kg throughout the test area, with an average of 7.583 g/kg and a standard deviation of 5.766 g/kg (Table 3). The salinity of the soil in the test area was typically high. Modeling set's soil salinity varied from 0.277 to 20.675 g/kg, with an average of 7.575 g/kg and a standard deviation of 5.735 g/kg. Validation set's soil salinity varied from 0.258 to 20.250 g/kg, with an average of 7.627 g/kg and a standard deviation of 5.864 g/kg. The mean and standard deviation of the modeling and validation sets are comparable to the statistical findings of all sample sets, which may decrease model creation and validation deviation in the latter stage and has modeling reliability.

3.2. Selection of Sensitive Bands

The correlation findings of UAV image reflectance showed that grey correlation coefficients have larger absolute values than Pearson correlation coefficients for the four-band reflectance and soil salinity content. Grey correlation coefficients between G, R, NIR, and salinity content were 0.567, 0.569, and 0.612, respectively, and were all significant at the

0.01 level (Table 4). Relative correlation coefficients were 0.532 ($p < 0.01$), 0.522 ($p < 0.01$), and 0.557 ($p < 0.01$) for G, R, NIR, and salinity content, which showed the same pattern as that of the grey correlation. Among the four bands, NIR had the greatest correlation coefficient.

Table 3. Statistics of soil salinity content.

Sample Set	Minimum (g/kg)	Maximum (g/kg)	Average (g/kg)	SD (g/kg)	Sample Size
All	0.264	20.651	7.583	5.766	120
Modeling set	0.277	20.675	7.575	5.735	90
Validation set	0.258	20.250	7.627	5.864	30

Table 4. Correlation analysis of sensitive reflectance with soil salinity.

Reflectance	Grey Correlation Coefficient	Pearson Correlation Coefficient
G	0.567 **	0.532 **
R	0.569 **	0.522 **
REG	0.550 *	S0.509 *
NIR	0.612 **	0.557 **

* Significant at 0.05 level, ** significant at 0.01 level.

To further improve the accuracy and reliability of screening sensitive band reflectance, the diagnostic index P_i of G, R, REG, and NIR were computed. We can find that G, B, and NIR bands were higher than that of REG (Table 5), which further indicated that the soil reflectance of green, red, and near-infrared bands of UAV multispectral image were sensitive to soil salt information, which can be used to construct a sensitive soil salinity retrieval index.

Table 5. Diagnostic index of UAV image reflectance.

	G	R	REG	NIR
R_i	0.567 **	0.569 **	0.550 *	0.612 **
σ_i	0.791	0.761	0.470	0.732
P_i	0.472	0.456	0.273	0.435

* Significant at 0.05 level, ** significant at 0.01 level.

3.3. Construction of Soil Salinity Retrieval Model

This study compared various combinations of the three-*soil* salinity sensitive bands (R, G, and NIR), e.g., addition, subtraction, and division (Table 6), and analyzed the relationship between these transformation indexes and soil salinity information. Finally, we devised a new index, namely the Soil Salinity Retrieval Index (SSRI, Equation (3)) to detect soil salinity by relying on the three sensitive bands.

$$SSRI = \frac{NIR}{\sqrt{R * G}} \tag{3}$$

where G, R, and NIR is the green, red, and near-infrared band reflectance of the UAV image, respectively.

Table 6. Equation combinations of the G, R, NIR.

ID	Algebra Operation
1	R+G+NIR
2	R-G-NIR, G-R-NIR, NIR-R-G
3	$\sqrt[3]{R * G * NIR}$
4	$\frac{R}{\sqrt{G*NIR}}$, $\frac{G}{\sqrt{R*NIR}}$, $\frac{NIR}{\sqrt{R*G}}$

3.4. Correlation Analysis

The correlations of proposed SSRI, VIs, and SIs with soil salinity content were shown in Table 7. Among the 14 indexes, SSRI showed the higher grey correlation and Pearson correlation coefficients, 0.689 and 0.632, respectively. NDVI and DVI were the only two VIs that demonstrated a significant association ($p < 0.01$), with NDVI having the strongest correlation (0.619, 0.602). SI, SI-T, SI3, NDSI, and SRSI had a significant association with soil salinity ($p < 0.01$), with SRSI having the highest value of correlation (Table 7). Therefore, NDVI, SRSI, and SSRI were utilized to build soil salinity retrieval models.

Table 7. Correlation analysis of sensitive spectral index with soil salinity.

Spectral Index	Grey Correlation Coefficient	Pearson Correlation Coefficient
SSRI	0.689 **	0.632 **
NDVI	0.619 **	0.602 **
DVI	0.601 **	0.557 **
SRVI	0.512 *	0.476 *
RVI	0.517 *	0.458 *
GNDVI	0.557 **	0.514 *
NDVI$_{REG}$	0.507 *	0.454
Salinity Index (SI-T)	0.607 **	0.559 **
Salinity Index 1 (SI1)	0.556 **	0.514 *
Salinity Index 2 (SI2)	−0.390	−0.200
Salinity Index 3 (SI3)	0.637 **	0.601**
NDSI	0.535 *	0.474*
SRSI	0.677 **	0.615**
Brightness Index (BI)	0.235	0.229

* Significant at 0.05 level, ** significant at 0.01 level.

3.5. Retrieval Accuracy

The RF, BPNN, SVM, PLSR, and MLR were used to create retrieval models of soil salinity based on the NDVI image. The results showed that the NDVI-based RF model showed the highest modeling and validation accuracies (R^2 = 0.625 and 0.633) among the five methods and then was BPNN, SVM, PLSR, and MLR in order of modeling and validation accuracies (Table 8). However, only the *RPD* of the RF model topped 1.4, which is the rough sample prediction threshold. Therefore, in the test area, NDVI is not suited for accurate soil salinity retrieval.

Table 8. Accuracy statistics of the NDVI based retrieval model.

Modeling Method	Modeling Accuracy		Validation Accuracy		
	R^2	*RMSE*	R^2	*RMSE*	*RPD*
RF	0.625	2.977	0.633	2.789	1.425
BPNN	0.601	3.375	0.610	3.090	1.397
SVM	0.584	3.547	0.591	3.274	1.363
PLSR	0.557	3.645	0.566	3.455	1.321
MLR	0.492	3.988	0.488	4.714	0.670

Table 9 displayed the statistically accurate findings of the five modeling approaches using SRSI. According to the statistical data, the accuracy of modeling and validation of the five modeling approaches is in the following order: RF > BPNN > PLSR > SVM > MLR. Except for the MLR model, the modeling and validation accuracy of the other four models are all more than 0.6. The vegetation index has the potential to extract soil salinity with acceptable accuracies.

In the test area, the R^2 values of RF, BPNN, SVM, PLSR, and MLR based on SSRI (Table 10) showed stronger fitting impacts than the retrieval model based on NDVI and

SRSI (Tables 8 and 9). Furthermore, the modeling and validation accuracies of the five techniques (RF, BPNN, SVM, PLSR, and MLR) were all higher than 0.6, and the *RPD* of the RF model is more than 2.2 (Table 10), which indicates that the RF has adequate soil salinity retrieval capacity.

Table 9. Accuracy statistical results of SRSI retrieval model.

Modeling Method	Modeling Accuracy		Validation Accuracy		
	R^2	*RMSE*	R^2	*RMSE*	*RPD*
RF	0.667	2.554	0.679	2.443	1.878
BPNN	0.641	2.631	0.653	2.781	1.750
SVM	0.619	3.205	0.621	3.029	1.549
PLSR	0.633	2.980	0.639	2.991	1.583
MLR	0.537	3.652	0.526	3.631	0.998

Table 10. Accuracy statistical results of soil salinity index retrieval model based on SSRI.

Modeling Method	Modeling Accuracy		Validation Accuracy		
	R^2	*RMSE*	R^2	*RMSE*	*RPD*
RF	0.724	1.764	0.745	1.879	2.211
BPNN	0.699	1.989	0.682	2.376	2.043
SVM	0.665	2.554	0.658	3.002	1.675
PLSR	0.671	2.275	0.689	2.897	1.748
MLR	0.639	3.091	0.622	2.994	1.464

The comparison of the modeling and validation accuracies (Tables 8–10) indicated that the retrieval models based on the proposed SSRI were more accurate than those based on vegetation index and soil salinity index. The soil salt retrieval modeling and validation accuracy were all greater than 0.638, and the *RPD* values were all greater than 1.463. Besides, among the five prediction modeling approaches, the order of modeling and validation accuracy was RF, BPNN, PLSR, SVM, and MLR. The modeling and validation accuracies of the RF modeling approach in various models were all greater than 0.6, and *RPD* values were above 1.424 (Tables 8–10). Among them, the R^2 and *RMSE* of the modeling set using the SSRI-based RF method were 0.724 and 1.746; and the R^2, *RMSE*, and *RPD* of the validation set were 0.745, 1.879, and 2.211 (Figure 2), which were the highest. The optimal retrieval model of soil salinity in the test area is the SSRI-based RF method.

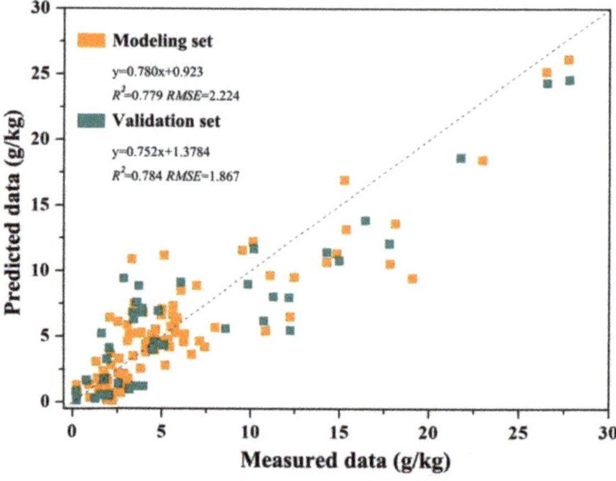

Figure 2. Scatter plot of the optimal retrieval model (SSRI-based RF method) of soil salinity based on UAV imagery.

4. Discussions

The sensitive parameter and optimal retrieval method for soil salinity monitoring using UAV multispectral imagery were investigated in this study. The proposed soil salinity retrieval index (SSRI) based RF method was found to show the best accuracy in predicting soil salinity. The modeling R^2 and $RMSE$ were 0.724 and 1.764, respectively; and the validation R^2, $RMSE$, and RPD were 0.745, 1.879, and 2.211, respectively, which were the highest among all the models built using the five prediction approaches based on SSRI, vegetation index, and salinity index.

Compared to existing soil salinity retrieval studies using UAV imagery, this study screened sensitive band information and combined them to form a feasible index to help retrieve soil salinity. The retrieval values of soil salinity in the whole test area using the SSRI-based RF model (Figure 3) ranged from 0.323 to 21.210 g/kg, with an average value of 6.871 g/kg, which was close to the descriptive statistical results of the soil samples (Table 3). The test area can be divided into five grades based on the saline soil grading standard (Wang et al., 2019), namely extremely saline soil (salt content greater than 10.0 g/kg), severely saline soil (salt content 6.0–10.0 g/kg), moderately saline soil (salt content 4.0–6.0 g/kg), slightly saline soil (salt content 2.0–4.0 g/kg), and non-saline soil (Figure 3). According to the area statistical figures, the extremely saline soil occupied the lowest share of 5.3 percent of the five grades. Severely and moderately saline soil zones accounted for 15.5 and 13.6 percent of the overall test area, respectively. The proposal of slightly saline soil was 65.4 percent, the highest of the five categories. This pattern of soil salinity distribution is consistent with the observation in Figure 2, i.e., more than half of the sample locations were in the slightly saline region. The non-saline region encompassed 10.2 percent of the left test area. The geographical analysis demonstrated that soil salinization is widespread in the test area, with the majority of test sites belonging to the saline soil grade.

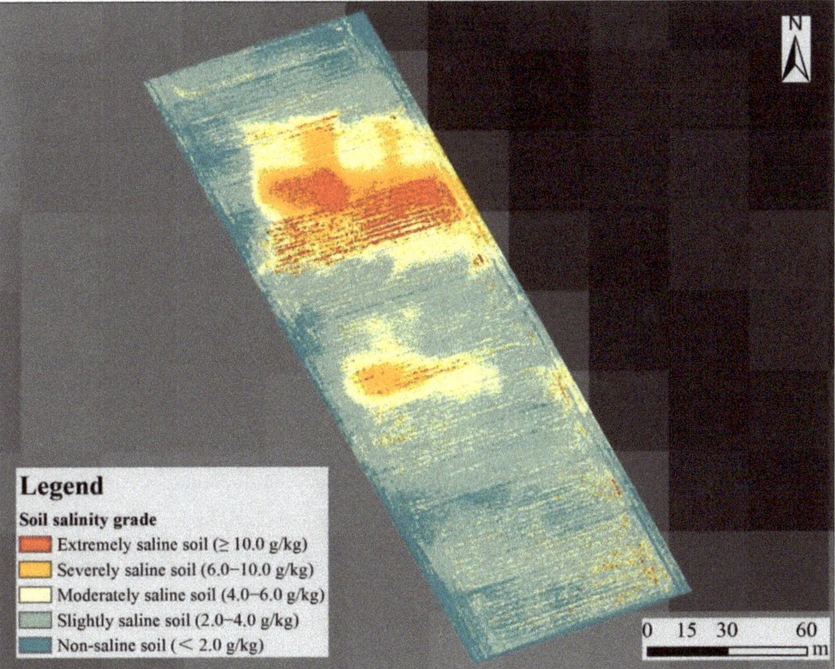

Figure 3. Retrieval map of soil salinity using the SRSI based RF method.

Visible and NIR bands displayed significant correlation links with soil salinity according to the results of two spectral screening analysis methodologies. The main minerals involved in the salinization of the soil of the YRD are rock salt and gypsum, with the main anions being Cl^- and SO_4^{2-} and the main cations being Na^+ and Ca^{2+} [11,43]. Previous research found that although NaCl has no spectral characteristics in the visible and near-infrared bands, NaCl is correlated with gypsum [44]. Gypsum possesses absorption qualities in the visible and near-infrared bands, which can help reveal soil salinity spectral information. Xu et al. (2018) found that gypsum has molecular vibration absorption spectrum features in the NIR band, visible and NIR band can collect SO_4^{2-} spectral information [45]. Furthermore, studies have shown that salinized soil has higher reflectance in the visible and NIR bands than non-salinized soil [15,46]. Hence, spectral information of salinized soil retrieved from RS data can be used to estimate soil salinity in visible and near-infrared bands.

This study explored the sensitive parameters and optimal method to retrieve soil salinity, while soil samples were collected in the surface layer of soil (0–10 cm). For agriculture and food security, more attention should be paid to the indirect approach to a salinization assessment of root-zone (0–100 cm) [47]. Besides, the soil sample collection and measurement were conducted in one site. The proposed SSRI and the findings need more examination to test the reliability in further research. Furthermore, UAV multispectral image and the SSRI-based RF method can efficiently predict soil salinity with acceptable accuracy, whereas the UAV's battery duration time prevents it from being used in large regional-scale soil salinity assessment. Recently, studies have fused satellite RS data with UAV images to derive regional-scale soil salinity, which is useful for estimating soil salinity across wide areas. However, it should be noted the variations in band wavelengths, meteorological conditions at the time of acquisition, and sensor compatibility between aviation and aerospace platforms are distinctly different. How to eliminate these uncertainties is a direction where further endeavors should be made in.

5. Conclusions

This study explored the sensitive parameter and optimal method for the accurate retrieval and spatial distribution of soil salinity. The sensitive band of soil salinity was discovered to be the band G, R, and NIR, a soil salinity retrieval index (SSRI) was proposed accordingly to retrieve soil salinity. SSRI-based RF method was the optimal combination that can accurately retrieve the soil salinity. Further study will be conducted in other salinized regions to examine the findings of this study.

Author Contributions: Conceptualization, Y.Z.; methodology, C.C.; software, J.S.; validation, X.Y.; investigation, A.W.; writing—original draft, X.Y.; writing—review and editing, Y.Z. and A.W. All authors have read and agreed to the published version of the manuscript.

Funding: This research was funded by the Major Science and Technology Projects in Shandong Province, grant number 2019JZZY010723, and Natural Foundation of Shandong Province, grant number ZR2019MD014. The funders have no role in study design, data collection, and analysis, decision to publish, or preparation of the manuscript.

Institutional Review Board Statement: Not applicable.

Informed Consent Statement: Not applicable.

Data Availability Statement: Data used in this study are available under request.

Acknowledgments: We would like to thank the kind help of the editor and the reviewers to improve the manuscript.

Conflicts of Interest: The authors declare no conflict of interest.

References

1. Green, S.M.; Dungait, J.; Tu, C.; Buss, H.L.; Sanderson, N.; Hawkes, S.J.; Xing, K.; Yue, F.; Hussey, V.L.; Peng, J.; et al. Soil functions and ecosystem services research in the Chinese karst critical zone. *Chem. Geol.* **2019**, *527*, 119107. [CrossRef]
2. Norris, C.E.; Quideau, S.A.; Landhusser, S.M.; Drozdowski, B.; Hogg, K.E.; Oh, S.W. Assessing structural and functional indicators of soil nitrogen availability in reclaimed forest ecosystems using 15 n-labeled aspen litter. *Can. J. Soil Sci.* **2018**, *98*, 357–368. [CrossRef]
3. Ghassemi, F.; Jakeman, A.J.; Nix, H.A. *Salinisation of Land and Water Resources: Human Causes, Extent, Management and Case Studies*; CAB international: Canberra, Australia, 1995; pp. 1–3.
4. Wang, Z. Spatial and Temporal Variability of Soil Moisture and Salinity, Affecting Factors and Forecasting Model in the Typical Area of the Yellow River Delta. 2017. Available online: https://d.wanfangdata.com.cn/thesis/D01212536 (accessed on 8 February 2018).
5. Fourati, H.T.; Bouaziz, M.; Benzina, M.; Bouaziz, S. Detection of terrain indices related to soil salinity and mapping salt-affected soils using remote sensing and geostatistical techniques. *Environ. Monit. Assess* **2017**, *189*, 177. [CrossRef] [PubMed]
6. Koganti, T.; Narjary, B.; Zare, E.; Pathan, A.L.; Huang, J.; Triantafilis, J. Quantitative mapping of soil salinity using the dualem-21s instrument and em inversion software. *Land Degrad. Dev.* **2018**, *29*, 1768–1781. [CrossRef]
7. Jiang, H.; Shu, H. Optical remote-sensing data based research on detecting soil salinity at different depth in an arid-area oasis, Xinjiang, China. *Earth Sci. Inform.* **2018**, *12*, 43–56. [CrossRef]
8. Wang, D. *Quantitative Inversion of Water and Salt in Coastal Saline Soil in the Yellow River Delta*; Shandong Agricultural University: Tai'an, China, 2020.
9. Azabdaftari, A.; Sunar, F. Soil salinity mapping using multitemporal Landsat data. *ISPRS—Int. Arch. Photogramm. Remote Sens. Spat. Inf. Sci.* **2016**, *7*, 3–9. [CrossRef]
10. Morgan, R.S.; Abd, E.-H.M.; Rahim, S. Soil salinity mapping utilizing sentinel-2 and neural networks. *Indian J. Agric. Res.* **2018**, *52*, 524–529. [CrossRef]
11. Weng, Y.L.; Gong, P. Soil salinity measurements on the Yellow River Delta. *J. Nanjing Univ. Nat. Sci.* **2006**, *42*, 602–610.
12. Ma, C. Retrieval of soil salt content based on Sentinel-1 dual-polarization radar image. *Trans. Chin. Soc. Agric. Eng.* **2018**, *34*, 153–158.
13. Hu, J.; Peng, J.; Zhou, Y.; Xu, D.; Shi, Z. Quantitative estimation of soil salinity using uav-borne hyperspectral and satellite multispectral images. *Remote Sens.* **2019**, *11*, 736. [CrossRef]
14. Ivushkin, K.; Bartholomeus, H.; Bregt, A.K.; Pulatov, A.; Franceschini, M.H.; Kramer, H.; van Loo, E.N.; Roman, V.J.; Finkers, R. Uav based soil salinity assessment of cropland. *Geoderma* **2018**, *338*, 502–512. [CrossRef]
15. Wang, D.; Chen, H.; Wang, G.; Cong, J.; Wang, X.; Wei, X. Salinity Inversion of Severe Saline Soil in the Yellow River Estuary Based on UAV Multi-Spectra. *Chin. Agric. Sci.* **2019**, *52*, 1698–1709.
16. Ma, Y.; Chen, H.Y.; Zhao, G.X.; Wang, Z.R.; Wang, D.Y. Spectral index fusion for salinized soil salinity inversion using Sentinel-2A and UAV images in a coastal area. *IEEE Access.* **2020**, *8*, 159595–159608. [CrossRef]
17. Aldabaa, A.; Weindorf, D.C.; Chakraborty, S.; Sharma, A.; Li, B. Combination of proximal and remote sensing methods for rapid soil salinity quantification. *Geoderma* **2015**, *239–240*, 34–46. [CrossRef]
18. Rao, B.; Sharma, R.; Ravi, S.; Das, S.N.; Dwivedi, R.S.; Thammappa, S.S.; Venkataratnam, L. Spectral behaviour of salt-affected soils. *Int. J. Remote Sens.* **1995**, *16*, 2125–2136. [CrossRef]
19. Shrestha, R.P. Relating soil electrical conductivity to remote sensing and other soil properties for assessing soil salinity in northeast Thailand. *Land Degrad. Dev.* **2006**, *17*, 677–689. [CrossRef]
20. Ramos, T.; Castanheira, N.; Oliveira, A.; Paz, A.; Gonalves, M. Soil salinity assessment using vegetation indices derived from Sentinel-2 multispectral data. application to Lezíria Grande, Portugal. *Agric. Water Manag.* **2020**, *241*, 106387. [CrossRef]
21. Masoud, A. Predicting salt abundance in slightly saline soils from Landsat ETM+ imagery using spectral mixture analysis and soil spectrometry. *Geoderma* **2014**, *217–218*, 45–56. [CrossRef]
22. Yao, Z.; Chen, J.; Zhang, Z.; Tan, C.; Wei, G.; Wang, X. Effect of plastic film mulching on the accuracy of soil salinity retrieval by UAV multispectral remote sensing. *Trans. Chin. Soc. Agric. Eng.* **2019**, *35*, 89–97.
23. Chen, H.; Ma, Y.; Zhu, A.; Wang, Z.; Zhao, G.; Wei, Y. Soil salinity inversion based on differentiated fusion of satellite image and ground spectra. *Int. J. Appl. Earth Obs. Geoinf.* **2021**, *101*, 102360. [CrossRef]
24. Jiang, C.; Pan, S.; Chen, S. Recent morphological changes of the Yellow River submerged delta: Causes and environmental implications. *Geomorphology* **2017**, *293*, 93–107. [CrossRef]
25. Fan, X.; Pedroli, B.; Liu, G.; Liu, Q.; Liu, H.; Shu, L. Soil salinity development in the yellow river delta in relation to groundwater dynamics. *Land Degrad. Dev.* **2011**, *23*, 175–189. [CrossRef]
26. He, Y.; Desutter, T.; Prunty, L.; Hopkins, D.; Jia, X.; Wysocki, D.A. Evaluation of 1:5 soil to water extract electrical conductivity methods. *Geoderma* **2012**, *185–186*, 12–17. [CrossRef]
27. FAO. *Standard Operating Procedure for Saturated Soil Paste Extract*; Global Soil Laboratory Network (GLOSOLAN): Rome, Italy, 2021.
28. Lu, R. *Soil Agrochemical Analysis Method*; China Agricultural Science and Technology Press: Beijing, China, 2002.
29. Yeh, Y.L.; Chen, T.C. Application of grey correlation analysis for evaluating the artificial lake site in Pingtung plain, Taiwan. *Can. J. Civ. Eng.* **2011**, *31*, 56–64. [CrossRef]
30. Li, Y. *Research on Soil Salinity in the Yellow River Delta Based on Remote Sensing*; Chang'an University: Xi'an, China, 2018.

31.	Piao, S.; Fang, J.; Ji, W.; Guo, Q.; Ke, J.; Tao, S. Variation in a satellite-based vegetation index in relation to climate in china. *J. Veg. Sci.* **2004**, *15*, 219–226. [CrossRef]

32.	Alhammadi, M.S.; Glenn, E.P. Detecting date palm trees health and vegetation greenness change on the eastern coast of the United Arab Emirates using SAVI. *Int. J. Remote Sens.* **2008**, *29*, 1745–1765. [CrossRef]

33.	Abderrazak, B.; Ali, E.B.; Rachid, B.; Hassan, R. Sentinel-MSI vnir and swir bands sensitivity analysis for soil salinity discrimination in an arid landscape. *Remote Sens.* **2018**, *10*, 855.

34.	Allbed, A.; Kumar, L.; Aldakheel, Y. Assessing soil salinity using soil salinity and vegetation indices derived from ikonos high-spatial resolution imageries: Applications in a date palm dominated region. *Geoderma* **2014**, *230–231*, 1–8. [CrossRef]

35.	Yahiaoui, I.; Douaoui, A.; Zhang, Q.; Ziane, A. Soil salinity prediction in the lower cheliff plain (algeria) based on remote sensing and topographic feature analysis. *J. Arid. Land* **2015**, *7*, 794–805. [CrossRef]

36.	Khan, N.M.; Rastoskuev, V.V. Mapping salt-affected soils using remote sensing indicators—A simple approach with the use of GIS IDRISI. *Ratio* **2001**, *11*, 5–9.

37.	Geladi, P.; Kowalski, B. Partial Least-Squares Regression: A Tutorial. *Analytica Chimica Acta* **1986**, *185*, 1–17. [CrossRef]

38.	Zelterman, D. *Multivariable Linear Regression*; Springer International Publishing: Berlin, Germany, 2015.

39.	Hecht-Nielsen, R. Theory of the Backpropagation Neural Network. *Neural Netw.* **1988**, *1*, 593–605. [CrossRef]

40.	Amari, S.; Wu, S. Improving support vector machine classifiers by modifying kernel functions. *Neural Netw.* **1999**, *12*, 783–789. [CrossRef]

41.	Breiman, L. Random forest. *Mach. Learn.* **2001**, *45*, 5–32. [CrossRef]

42.	Terhoeven-Urselmans, T.; Schmidt, H.; Joergensen, R.G.; Ludwig, B. Usefulness of near-infrared spectroscopy to determine biological and chemical soil properties: Importance of sample pre-treatment. *Soil Biol. Biochem.* **2008**, *40*, 1178–1188. [CrossRef]

43.	An, L.S.; Zhao, Q.S.; Ye, S.Y.; Liu, G.Q.; Ding, X.G. Water-salt interactions factors and vegetation effects in the groundwater ecosystem in Yellow River Delta. *Adv. Water Sci.* **2011**, *22*, 689–695.

44.	Goldshleger, N.; Ben-Dor, E.; Lugassi, R.; Eshel, G. Soil degradation monitoring by remote sensing: Examples with three degradation processes. *Soil Sci. Soc. Am. J.* **2010**, *74*, 1433–1445. [CrossRef]

45.	Xu, W. *Spectral Discriminant Analysis of Martian Simulated Minerals and Brines*; Shandong University: Weihai, China, 2018.

46.	Fan, X.; Liu, Y.; Tao, J.; Weng, Y. Soil salinity retrieval from advanced multi-spectral sensor with partial least square regression. *Remote Sens.* **2015**, *7*, 488–511. [CrossRef]

47.	Scudiero, E.; Skaggs, T.; Anderson, R.; Corwin, D. Soil degradation in farmlands of California's San Joaquin Valley resulting from drought-induced land-use changes. In Proceedings of the EGU General Assembly Conference Abstracts, Vienna, Austria, 23–28 April 2016; EPSC2016-728.

Article

Evaluation of Three Portable Optical Sensors for Non-Destructive Diagnosis of Nitrogen Status in Winter Wheat

Jie Jiang [1,2,3,4,5,†], Cuicun Wang [1,2,3,4,5,†], Hui Wang [1,2,3,4,5], Zhaopeng Fu [1,2,3,4,5], Qiang Cao [1,2,3,4,5], Yongchao Tian [1,2,3,4,5], Yan Zhu [1,2,3,4,5], Weixing Cao [1,2,3,4,5] and Xiaojun Liu [1,2,3,4,5,*]

[1] National Engineering and Technology Center for Information Agriculture, Nanjing Agricultural University, Nanjing 210095, China; 2019201009@njau.edu.cn (J.J.); 2019201008@njau.edu.cn (C.W.); 2019101173@njau.edu.cn (H.W.); 2018801234@njau.edu.cn (Z.F.); qiangcao@njau.edu.cn (Q.C.); yctian@njau.edu.cn (Y.T.); yanzhu@njau.edu.cn (Y.Z.); caow@njau.edu.cn (W.C.)

[2] MOE Engineering Research Center of Smart Agricultural, Nanjing Agricultural University, Nanjing 210095, China

[3] MARA Key Laboratory for Crop System Analysis and Decision Making, Nanjing Agricultural University, Nanjing 210095, China

[4] Jiangsu Key Laboratory for Information Agriculture, Nanjing Agricultural University, Nanjing 210095, China

[5] Jiangsu Collaborative Innovation Center for Modern Crop Production, Nanjing Agricultural University, Nanjing 210095, China

* Correspondence: liuxj@njau.edu.cn; Tel.: +86-25-8439-6804; Fax: +86-25-8439-6672

† The first two authors contributed equally to this work.

Citation: Jiang, J.; Wang, C.; Wang, H.; Fu, Z.; Cao, Q.; Tian, Y.; Zhu, Y.; Cao, W.; Liu, X. Evaluation of Three Portable Optical Sensors for Non-Destructive Diagnosis of Nitrogen Status in Winter Wheat. *Sensors* **2021**, *21*, 5579. https://doi.org/10.3390/s21165579

Academic Editors: Jiyul Chang and Sigfredo Fuentes

Received: 13 July 2021
Accepted: 17 August 2021
Published: 19 August 2021

Abstract: The accurate estimation and timely diagnosis of crop nitrogen (N) status can facilitate in-season fertilizer management. In order to evaluate the performance of three leaf and canopy optical sensors in non-destructively diagnosing winter wheat N status, three experiments using seven wheat cultivars and multi-N-treatments (0–360 kg N ha^{-1}) were conducted in the Jiangsu province of China from 2015 to 2018. Two leaf sensors (SPAD 502, Dualex 4 Scientific+) and one canopy sensor (RapidSCAN CS-45) were used to obtain leaf and canopy spectral data, respectively, during the main growth period. Five N indicators (leaf N concentration (LNC), leaf N accumulation (LNA), plant N concentration (PNC), plant N accumulation (PNA), and N nutrition index (NNI)) were measured synchronously. The relationships between the six sensor-based indices (leaf level: SPAD, Chl, Flav, NBI, canopy level: NDRE, NDVI) and five N parameters were established at each growth stages. The results showed that the Dualex-based NBI performed relatively well among four leaf-sensor indices, while NDRE of RS sensor achieved a best performance due to larger sampling area of canopy sensor for five N indicators estimation across different growth stages. The areal agreement of the NNI diagnosis models ranged from 0.54 to 0.71 for SPAD, 0.66 to 0.84 for NBI, and 0.72 to 0.86 for NDRE, and the kappa coefficient ranged from 0.30 to 0.52 for SPAD, 0.42 to 0.72 for NBI, and 0.53 to 0.75 for NDRE across all growth stages. Overall, these results reveal the potential of sensor-based diagnosis models for the rapid and non-destructive diagnosis of N status.

Keywords: nitrogen indicator; nitrogen nutrition diagnosis; optical sensor; spectral index

1. Introduction

Nitrogen (N) is an essential nutrient that improves crop growth and grain yield. The excessive application of N fertilizers can lead to low N use efficiency, resulting in environmental pollution and a loss of grain quality [1,2]. Precision N management could be used to optimize N application by considering the temporal and spatial variability of crop N status in practical production [3,4]. However, this promising strategy requires the development and application of real-time and non-destructive technologies for in-season crop N nutrition diagnosis.

Leaf and plant N concentrations (LNC, PNC) have been used as vital parameters of crop N status [5]. However, N concentrations are dependent on the plant biomass, such that two different plants with the same N concentration but differ in plant biomass. The critical N concentration dilution curve (CNDC) reflects the power–function relationship between crop critical N concentrations and plant biomass [6]. Based on the CNDC, the N nutrition index (NNI) could be calculated to effectively diagnose crop N nutrition status [7]. For example, NNI values greater than one indicate excessive N status, while values less than one correspond to N deficiency. Previous studies indicated that the NNI diagnosis model has been successfully used for characterizing corn (R^2 = 0.33–0.68), wheat (R^2 = 0.73–0.86), and pepper (R^2 = 0.19–0.84) N status throughout crop growth stages [8–10]. However, the calculation of NNI requires complicated chemical analysis to determine PNC. Plant biomass measurements using destructive sampling are also time-consuming and unsuitable for in-season N management [11,12].

The application of proximal and remote sensing technology can provide an efficient method for real time crop N status estimations [13]. Optical transmission measurements with a handheld SPAD-502 chlorophyll meter (SPAD meter) have been widely used for crop N nutrition assessments due to their portability, fast responses, and affordable cost [14,15]. However, SPAD are easily influenced by the crop growth stage and cultivar leaf structure, with over-fertilized plants being challenging to detect due to chlorophyll saturation. The Dualex 4 Scientific+ sensor (Dualex) is a portable leaf fluorescence sensor that measures Chl values through leaf transmittance at 710 and 850 nm and epidermal flavonoid (Flav) levels through the assessment of chlorophyll fluorescence induced by ultra-violet (UV) excitation at 375 nm, and further providing a Chl/Flav ratio, which is termed the N balance index (NBI) [16]. Many studies have demonstrated a prominent relationship between Dualex-based indices and different N parameters. For example, Dualex-based Chl (R^2 = 0.49–0.90) were found to correlate with leaf chlorophyll concentrations in rice, wheat, corn, and soybean [17,18]. Zhang et al. [17] indicated a significant relationship between Flav measured by Dualex and rice LNC (R^2 = 0.52–0.83), PNC (R^2 = 0.56–0.76), and NNI (R^2 = 0.68–0.82) across different growth periods. Dualex-based NBI was also successfully used to monitor N nutrition status in corn, wheat, and other crops [19–21]. Gabriel et al. [22] compared two different leaf-clip sensors (SPAD meter and Dualex) to estimate corn LNC and indicated similar performances, with R^2 = 0.43–0.62 for the SPAD meter and R^2 = 0.42–0.68 for Dualex. Lejealle et al. [23] indicated that N balance index (NBI), the ratio of chlorophyll to flavonols that measured through by Multiplex, had an improved and more stable correlation with turfgrass LNC than Chl readings alone. Consequently, it is necessary to assess the performance of spectral indices collected from two leaf-sensors for winter wheat N status assessments.

Canopy optical sensors can collect spectral data at the canopy level compared to leaf sensor measurements at the leaf scale. Passive canopy optical sensors, including ASD Fieldspec and Cropscan, have been successfully used to monitor crop growth and to assess N status [24,25]; however, their correct function requires strict environmental conditions, such as light intensity and measurement times [26]. Active sensors, such as RapidScan CS-45 (RS sensor), possess an internal lights source that ensures effective measurements in suboptimal environmental conditions. This can be used to collect crop canopy spectrum values at 670, 730, and 780 nm wavelengths, with two default vegetation indices (normalized difference red edge (NDRE) and normalized difference vegetation index (NDVI)) synchronously. The relationship between the vegetation indices derived from the RS sensor and crop growth status has been extensively studied for various crops, including wheat and soybean [27–29].

While the SPAD meter, Dualex, and RS sensor have been widely used for estimations of crop growth and N status, the comparative assessment of these three portable sensors to real-timely diagnose winter wheat N nutrition have not been studied. The aims of this study were: (a) to evaluate the performance of the six sensor-based indices (leaf level: SPAD, Chl, Flav, and NBI; canopy level: NDRE, NDVI) for non-destructive estimates of N status of winter wheat; (b) to establish winter wheat N diagnostic models based on optimum leaf and canopy sensor-based indices, respectively; and (c) to plot N diagnosis maps temporally and spatially across all growth stages. These results can improve the non-destructive diagnosis of crop N nutrition, and could be used to guide appropriate N management strategies.

2. Materials and Methods

2.1. Experimental Design

Experiment 1 (2015–2016), 2 (2016–2017), and 3 (2017–2018) were performed at the Sihong (Figure 1; 33.37° N, 118.26° E), Rugao (Figure 1; 32.27° N, 120.75° E), and Xinghua (Figure 1; 33.08° N, 119.98° E) Experimental Stations, respectively, in the Jiangsu province of China. All experiments were performed using different wheat cultivars and N application rates with three replicates in a randomized complete block design. Plants density was 225 seedlings per square meter. The N fertilizer (granular urea with 46% N) was applied in two batches: 50% prior to sowing and 50% at the stem elongation stage. Additionally, 105 kg ha^{-1} P$_2$O$_5$ and 135 kg ha^{-1} K$_2$O were applied to all experiment plots. Detailed information is shown in Table 1.

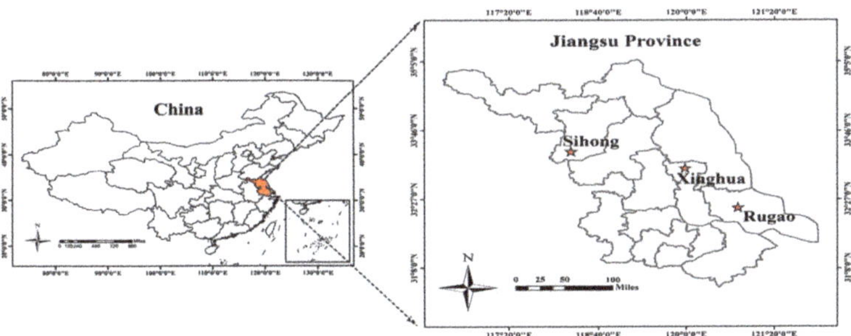

Figure 1. Three study sites in the Jiangsu province of China.

Table 1. Basic information about the three field experiments.

Experiment No. Year	Location	Plot Size (m²)	Cultivar	N Rate (kg ha^{-1})	Sampling Stage (Date)
1 2015–2016	Sihong (33.37° N,118.26° E)	42 (6 m × 7 m)	Xumai30 (XM30) Huaimai20 (HM20)	0 90 180 270 360	Jointing (5 April) Booting (15 April) Heading (22 April) Flowering (26 April) Filling (4 May)
2 2016–2017	Rugao (32.27° N, 120.75° E)	30 (5 m × 6 m)	Yangmai15 (YM15) Yangmai16 (YM16)	0 150 300	Jointing (27 March) Booting (11 April) Flowering (22 April) Filling (7 May)
3 2017–2018	Xinghua (33.08° N, 119.98° E)	63 (7 m × 9 m)	Zhenmai12 (ZM12) Yangmai23 (YM23) Ningmai13 (NM13)	0 90 180 270 360	Jointing (9 April) Booting (15 April) Flowering (24 April) Filling (9 May)

2.2. Spectral Data Collection

Three different optical sensors were used to collect wheat leaf and canopy spectral data at jointing, booting, flowering, and filling stages, respectively. The SPAD meter (Figure 2a; Minolta Camera Co., Osaka, Japan) and Dualex (Figure 2b; Dualex Scientific, Force-A Co., Orsay, France) were used to measure wheat leaf spectral parameters. The canopy sensor (Figure 2c; Holland Scientific Inc., Lincoln, NE, USA) was used to obtain wheat canopy spectral data. Detailed information of the three optical sensors is shown in Table 2.

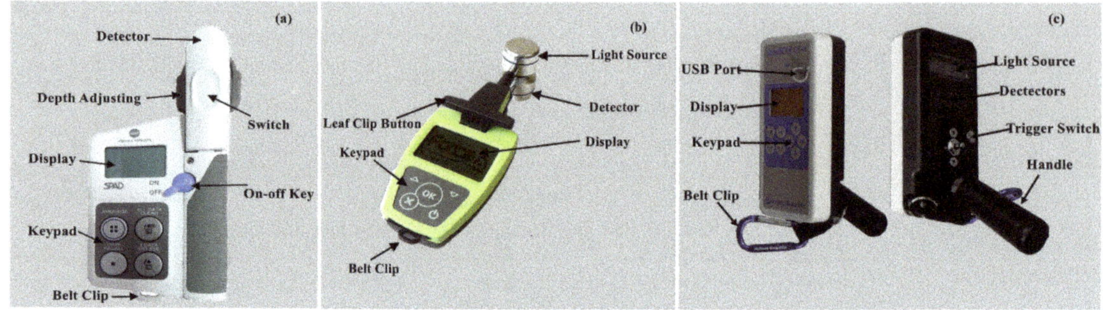

Figure 2. Images of the (**a**) SPAD-502 meter, (**b**) Dualex 4 Scientific+ sensor, and (**c**) RapidSCAN CS-45 sensor.

Table 2. Characteristics of the three optical sensors.

Sensor Information	Chlorophyll Meter	Fluorescence Sensor	Reflectance Sensor
Sensor name	SPAD-502	Dualex 4 Scientific+	RapidScan CS-45
Manufacturer	Minolta Camera Co. (Osaka, Japan)	Force-A (Orsay, France)	Holland Scientific (Lincoln, NE, USA)
Measurement scale	Leaf	Leaf	Canopy
Field of view	-	-	10°–45°
Working height	-	-	0.3–3.0 m
Measurement area	6 mm^2	20 mm^2	Dependent on measurement height
Measuring Principle	Transmittance	Fluorescence	Reflectance
Spectral band	Red (650 nm) and near infrared (940 nm)	UV (375 nm), red (655 nm), red-edge (710 nm), and near infrared (850 nm)	Red (670 nm), red-edge (730 nm), and near infrared (780 nm)
Output parameter	SPAD value	Chl, Flav, NBI	Reflectance (670, 730, 780 nm); NDRE, NDVI
Abbreviation	SPAD meter	Dualex	RS sensor

The first, second, and third fully expanded leaves (measurement location: 1/3, 1/2, and 2/3 of the distance from the leaves base) from the top of the plant were used for SPAD meter and Dualex measurements, and ten representative plants were randomly selected in each plot. All measurements were averaged to represent the leaf sensor data of each plot. The RS active sensor was held manually approximately 0.80 m above the canopy and at a constant speed in each plot (about 0.5 m s^{-1}). The RS sensor measurement path was parallel to the plant row. Three rows of wheat were randomly selected to obtain the two default vegetation indices of NDRE ((NIR − RE)/(NIR + RE)) [30] and NDVI ((NIR − R)/(NIR + R)) [31], with average vegetation indices collected to represent the spectral data of each plot.

2.3. Plant Sampling and Measurements

The plots in the field experiments are often small (42, 30 and 63 m^2 in Experiments 1, 2, and 3, respectively) and grown evenly; therefore, it is conventional to take representative samples and optical sensor measurements at different locations in each experimental plot [8,32]. Plant sampling was synchronously performed upon the completion of spectral measurements. Twenty wheat plants were randomly sampled and destructively separated into stems, leaves, and spikes. Each sub-sample was oven-dried at 105 °C for 30 min to stop all metabolic processes and samples were dried at 80 °C until reaching a constant weight. Samples were weighed and the leaf dry matter (LDM), stem dry matter (SDM), and spike dry matter (SpDM) were determined. The leaf N concentration (LNC), stem N concentration (SNC), and spike N concentration (SpNC) were measured using the standard Kjeldahl method [33].

LNA (kg ha^{-1}) was used to measure N accumulation in the leaves (Equation (1)). PNA (kg ha^{-1}) was calculated as the sum of leaf, stem, and spike N accumulation (Equation (2)). Plant N content (PNC (%); Equation (3)) was determined as the ratio of PNA (kg ha^{-1}) and plant biomass (kg ha^{-1}):

$$LNA = LDM \times LNC \tag{1}$$

$$PNA = LDM \times LNC + SDM \times SNC + SpDM \times SpNC \tag{2}$$

$$PNC = \frac{PNA}{LDM + SDM + SpDM} \tag{3}$$

The N_c curve was employed as described by Jiang et al. [27]. The NNI could be calculated based on Equation (5):

$$N_c = 4.17 \times W^{-0.39} \tag{4}$$

$$NNI = N_a / N_c \tag{5}$$

where W is the weight of the plant (Mg ha^{-1}), Na is the actual plant N concentration, and Nc is the critical plant N concentration.

2.4. Data Analysis

Data obtained from experiments 1, 2, and 3 were used for the analysis of variance between the six sensor-based indices and N status parameters using SPSS 24 software. The exponential relationship between six sensor-based indices and LNC, LNA, PNC, PNA, and NNI were calibrated and validated with 10-fold cross-validation procedure based on the data from experiments 1–3. Model performance was evaluated using the coefficients of determination (R^2), root mean square error (RMSE), and the relative error (RE (%)). The GraphPad Prism 8 software was used to plot the diagrams.

$$RMSE = \sqrt{\frac{1}{n} \times \sum_{i=1}^{n} (P_i - O_i)^2} \tag{6}$$

$$RE(\%) = 100 \times \sqrt{\frac{1}{n} \times \sum_{i=1}^{n} \left(\frac{P_i - O_i}{O_i}\right)^2} \tag{7}$$

where n is the number of samples, O_i is the measured value, and P_i is the predicted value.

N deficiency (NNI < 0.95), N optimal (0.95 $\leq$ NNI $\leq$ 1.05), and N excessive (NNI > 1.05) were used for diagnostic analysis [8,34]. The diagnostic category of predicted NNI using NBI and NDRE models was compared to those of observed NNI by areal agreement and the Kappa coefficient [35]. The areal agreement means the percentage of the two groups having same diagnostic category. The Kappa coefficient range of 0.21–0.40, 0.41–0.60, and 0.61–0.80 indicates fair, moderate, and substantial strength, respectively, of the diagnostic agreement [36]. N diagnostic maps were plotted using ArcGIS 10.3 software.

$$Kappa\ Coefficent = \frac{Observed\ Accuracy - Chance\ Agreement}{1 - Chance\ Agreement} \tag{8}$$

3. Results and Analysis

3.1. Variability of Nitrogen Status Indicators

Agronomic data from experiments 1, 2, and 3 were used for statistical analysis. The results showed that LNC, LNA, PNC, PNA, and NNI varied across growth stages, N levels, cultivars, and site-years (Table 3). LNA (coefficient of variation (CV) = 61.04%) was most variable across all growth stages, followed by PNA (CV = 50.16%), NNI (CV = 40.45%), and PNC (CV = 37.23%). LNC had a minimal CV of 26.44%. The analysis also indicated that the CV of the five N parameters were variable across growth stages. For example, the LNC was most variable at the jointing stage (CV = 30.10%), and had similar CV values

(23.35–23.94%) to the other three growth stages. In contrast, the CV of PNC (27.59–34.36%), PNA (42.66–59.98%), and NNI (33.10–46.37%) gradually decreased as the growth stage progressed. The great variability of those N indicators will help to assess the ability of the optical sensors when monitoring and diagnosing wheat N status.

Table 3. Descriptive statistics of leaf N concentration (LNC), leaf N accumulation (LNA), plant N concentration (PNC), plant N accumulation (PNA), and N nutrition index (NNI) across all growth stages.

Parameter	Growth Stage	N	Min.	Max.	SD [a]	CV [b] (%)
LNC (%)	Jointing	93	1.78	5.22	1.03	30.10
	Booting	93	2.14	5.39	0.84	23.35
	Flowering	93	2.01	5.32	0.86	23.90
	Filling	93	1.59	4.31	0.72	23.94
	All	372	1.59	5.39	0.90	26.44
LNA (kg ha^{-1})	Jointing	93	8.64	158.33	40.01	64.23
	Booting	93	11.04	156.86	36.02	57.39
	Flowering	93	11.55	123.44	26.75	50.93
	Filling	93	5.51	90.58	22.23	55.39
	All	372	5.51	144.86	33.22	61.04
PNC (%)	Jointing	93	1.06	3.50	0.71	34.36
	Booting	93	0.85	3.17	0.62	32.23
	Flowering	93	0.71	2.61	0.50	31.34
	Filling	93	0.68	2.04	0.36	27.59
	All	372	0.68	3.50	0.64	37.23
PNA (kg ha^{-1})	Jointing	93	15.39	257.46	61.75	59.98
	Booting	93	21.27	274.88	58.92	51.29
	Flowering	93	28.63	276.51	57.71	46.14
	Filling	93	33.48	268.35	57.07	42.66
	All	372	15.39	276.51	59.77	50.16
NNI	Jointing	93	0.34	1.92	0.45	46.37
	Booting	93	0.30	1.84	0.40	40.89
	Flowering	93	0.33	1.65	0.34	37.20
	Filling	93	0.34	1.41	0.28	33.10
	All	372	0.30	1.92	0.38	40.45

[a] SD indicates the standard deviation of the mean. [b] CV indicates the coefficient of variation (%).

3.2. Dynamic Changes of Six Sensor-Based Indices under Different N Treatments

The dynamic changes in the six sensor-based indices with days after sowing (DAS) across all growth periods of XM30 in Experiment 1 are shown in Figure 3. All six sensor-based indices, excluding Flav, exhibited similar trends as the wheat growth progressed. These spectral indices treated with high N application rates generally exceeded these treated with low N. The SPAD initially increased and then gradually decreased under low N treatments (0, 90 and 180 kg N ha^{-1}) as each growth stage progressed. Under conditions of high N application (270 and 360 kg N ha^{-1}), the SPAD value rapidly increased and remained high before declining, indicating the SPAD values reached saturation. For Dualex, the values of Chl and NBI increased gradually, reached peak values at 186 DAS, and then decreased during plant aging. Trends for the Flav were the opposite of that for the Chl and NBI, where the Flav initially decreased and then increased after 186 DAS. For the RS sensor, the default vegetation indices of NDRE and NDVI increased slowly and then gradually declined. Curves at 270 and 360 kg N ha^{-1} applications were close to overlapping, indicating that the plant growth achieved a non-N limited status.

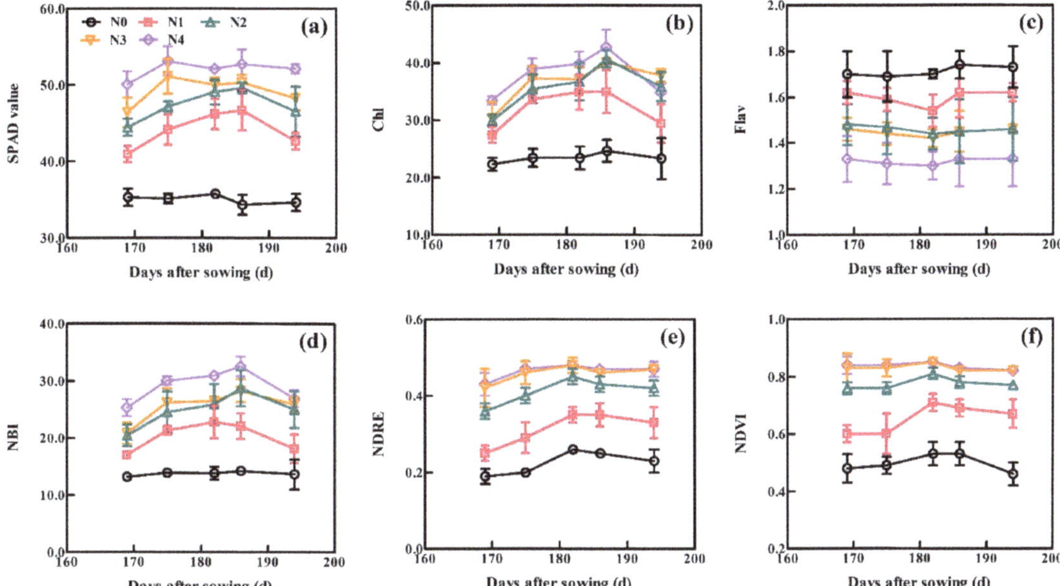

Figure 3. Dynamic variation in (**a**) SPAD, (**b**) chl, (**c**) Flav, (**d**) NBI, (**e**) NDRE, and (**f**) NDVI at the indicated days after sowing (DAS). Data were obtained from Experiment 1 using the XM30 cultivar. Vertical bars at each growth stage represent the standard error.

3.3. Relationship between the Six Sensors-Based Indices and Four N Indicators

Nitrogen indicators, such as LNC, LNA, PNC, and PNA, were collected in four spectral sensing stages: jointing, booting, flowering, and filling. Based on the data obtained from experiments 1–3, the quantitative exponential relationship between the six sensor-based indices and four N indicators were systematically analyzed, and the 10-fold cross-validation results were showed in Table 4. The results showed SPAD had R^2 values of 0.25–0.60, 0.29–0.54, 0.28–0.57, and 0.23–0.54, RMSE values of 0.46–0.78, 17.69–29.85 kg ha^{-1}, 0.24–0.55, and 41.29–52.96 kg ha^{-1}, and RE values of 16.25–25.08%, 53.04–80.50%, 18.22–33.85%, and 56.56–82.88% for LNC, LNA, PNC, and PNA estimation, respectively, at single and all growth stages. The Chl index of Dualex had a similar performance to the SPAD value, with R^2 values of 0.25–0.68, 0.27–0.52, 0.29–0.69, and 0.24–0.53, RMSE values of 0.44–0.79, 15.40–28.81 kg ha^{-1}, 0.21–0.54, and 38.86–50.07 kg ha^{-1}, and RE values of 17.20–22.44%, 54.64–81.25%, 17.93–32.28%, and 39.77–77.54% for LNC, LNA, PNC, and PNA estimation, respectively, at single and all growth stages. The NBI index performed relatively better than the other two Dualex-based indices (Chl and Flav) for estimating LNC (R^2 = 0.36–0.79, RMSE = 0.39–0.67, RE = 13.43–20.02%), LNA (R^2 = 0.49–0.70, RMSE = 14.35–23.50 kg ha^{-1}, RE = 48.33–67.38%), PNC (R^2 = 0.49–0.76, RMSE = 0.19–0.46, RE = 15.63–27.44%), and PNA (R^2 = 0.53–0.72, RMSE = 34.63–39.32 kg ha^{-1}, RE = 35.46–65.43%) at the single growth stage and across all growth stages. For the RS active canopy sensor, the default vegetation index of NDRE was more closely associated with LNC (R^2 = 0.61–0.79, RMSE = 0.39–0.53, RE = 11.84–15.79%), LNA (R^2 = 0.66–0.87, RMSE = 12.39–19.79 kg ha^{-1}, RE = 24.72–42.12%), PNC (R^2 = 0.51–0.74, RMSE = 0.18–0.45, RE = 14.81–27.34%), and PNA (R^2 = 0.64–0.87, RMSE = 24.09–36.39 kg ha^{-1}, RE = 20.56–36.21%) than NDVI at single growth stage and across all growth stages.

Table 4. Ten-fold cross-validation results for the pair-wise exponential relationship between the six sensor-based indices and leaf N concentration (LNC), leaf N accumulation (LNA), plant N concentration (PNC), and plant N accumulation (PNA) at different growth stages across experiments 1–3.

Parameter	Sensor	Index	Jointing Stage			Booting Stage			Flowering Stage			Filling Stage			All Stage		
			R^2	RMSE	RE (%)	R^2	RMSE	RE (%)	R^2	RMSE	RE (%)	R^2	RMSE	RE (%)	R^2	RMSE	RE (%)
LNC (%)	SPAD	SPAD	0.44	0.77	25.08	0.59	0.53	16.25	0.25	0.78	21.68	0.60	0.46	16.28	0.39	0.70	21.21
	Dualex	NBI	0.79	0.47	13.43	0.66	0.50	14.48	0.36	0.67	20.02	0.71	0.39	14.19	0.61	0.56	16.74
		Chl	0.68	0.58	18.66	0.57	0.55	16.27	0.25	0.79	22.44	0.65	0.44	17.20	0.42	0.68	20.00
		Flav	0.66	0.60	19.21	0.43	0.63	19.22	0.27	0.74	21.34	0.50	0.50	19.24	0.51	0.62	20.70
	RS	NDRE	0.79	0.47	15.02	0.75	0.42	11.84	0.61	0.53	15.79	0.70	0.39	13.80	0.70	0.49	15.03
		NDVI	0.72	0.55	18.61	0.56	0.55	17.05	0.56	0.57	17.86	0.62	0.44	19.45	0.64	0.54	19.03
LNA (kg ha^{-1})	SPAD	SPAD	0.36	29.85	80.50	0.54	24.86	67.24	0.29	24.32	62.51	0.46	17.69	53.04	0.29	29.11	78.15
	Dualex	NBI	0.70	23.50	67.38	0.63	22.31	52.52	0.49	18.47	55.47	0.58	14.35	48.33	0.53	23.46	59.49
		Chl	0.51	28.81	81.25	0.50	26.12	69.43	0.27	24.73	69.82	0.52	15.40	54.64	0.31	28.51	78.51
		Flav	0.50	30.23	83.25	0.49	28.32	70.21	0.49	18.95	58.39	0.50	15.55	60.67	0.53	23.43	68.19
	RS	NDRE	0.87	15.04	24.72	0.77	17.71	24.75	0.78	12.39	24.81	0.66	13.62	42.12	0.67	19.79	40.55
		NDVI	0.86	16.27	37.82	0.68	20.55	37.09	0.74	13.65	33.63	0.66	15.49	44.25	0.63	21.43	42.00
PNC (%)	SPAD	SPAD	0.36	0.53	31.75	0.50	0.44	26.44	0.30	0.42	29.56	0.57	0.24	18.22	0.28	0.55	33.85
	Dualex	NBI	0.76	0.36	17.22	0.65	0.36	21.85	0.50	0.35	24.04	0.70	0.19	15.63	0.49	0.46	27.44
		Chl	0.64	0.44	21.24	0.51	0.43	26.02	0.29	0.42	29.24	0.69	0.21	17.93	0.29	0.54	32.28
		Flav	0.61	0.45	24.84	0.47	0.45	29.61	0.32	0.41	32.47	0.44	0.27	22.63	0.45	0.47	31.85
	RS	NDRE	0.72	0.39	21.73	0.74	0.32	18.41	0.73	0.26	18.35	0.74	0.18	14.81	0.51	0.45	27.34
		NDVI	0.57	0.48	28.37	0.55	0.41	27.36	0.50	0.35	27.65	0.50	0.25	22.97	0.47	0.47	32.41
PNA (kg ha^{-1})	SPAD	SPAD	0.36	52.96	82.88	0.54	41.29	58.97	0.23	50.33	65.98	0.34	46.28	56.56	0.30	49.14	80.28
	Dualex	NBI	0.72	35.06	65.43	0.67	34.63	44.69	0.53	39.32	38.80	0.57	37.34	35.46	0.59	39.32	57.96
		Chl	0.50	46.95	77.54	0.52	41.78	59.44	0.24	50.07	56.16	0.53	38.86	39.77	0.38	47.87	74.69
		Flav	0.50	46.22	78.85	0.52	41.89	63.11	0.44	43.10	51.64	0.48	41.07	39.66	0.48	44.09	72.33
	RS	NDRE	0.87	24.09	20.56	0.75	30.26	24.00	0.76	27.96	21.18	0.64	36.39	36.21	0.68	34.72	32.53
		NDVI	0.82	28.22	33.07	0.65	35.41	34.41	0.57	37.59	33.69	0.62	37.24	37.99	0.62	37.50	36.91

3.4. Relationship between the Optimal Index of Each Sensor and N Nutrition Index

The NBI index performed consistently well for the assessment of leaf (LNC and LNA) and plant (PNC and PNA) N status across three Dualex-based indices during all growth stages, and the default vegetation index NDRE of the RS sensor also displayed a consistently high correlation with LNC, LNA, PNC, and PNA. Hence, the optimal index (SPAD, NBI, and NDRE) of the SPAD meter, Dualex, and RS sensor were selected for establishing the relationship with NNI. Based on the data collected from experiments 1–3, the quantitative exponential relationship between the SPAD, NBI, NDRE, and NNI were systematically analyzed, and the 10-fold cross-validation results were showed in Table 5. The NDRE of the RS sensor performed best for monitoring NNI across different cultivars at jointing (R^2 = 0.75–0.96, RMSE = 0.14–0.19, RE = 13.72–29.27%), booting (R^2 = 0.73–0.97, RMSE = 0.09–0.24, RE = 10.04–24.18%), flowering (R^2 = 0.76–0.86, RMSE = 0.12–0.20, RE = 12.19–24.69%), filling (R^2 = 0.74–0.96, RMSE = 0.07–0.21, RE = 10.18–28.02%), and all (R^2 = 0.67–0.87, RMSE = 0.12–0.26, RE = 18.53–24.39%) growth stages, followed by the Dualex-based index of NBI, with R^2 of 0.53–0.88, 0.59–0.87, 0.33–0.87, 0.72–0.84, and 0.56–0.75, RMSE of 0.13–0.29, 0.11–0.33, 0.15–0.33, 0.13–0.19, and 0.16–0.29, RE of 19.33–36.40, 16.53–48.32, 16.74–52.83, 12.78–23.49, and 21.25–43.45 for NNI estimation across seven cultivars at jointing, booting, flowering, filling, and all growth stage, respectively. The SPAD had a slightly worse performance for estimating NNI at jointing (R^2 = 0.29–0.81, RMSE = 0.16–0.48, RE = 19.65–77.65%), booting (R^2 = 0.38–0.84, RMSE = 0.18–0.37, RE = 18.51–63.60%), flowering (R^2 = 0.26–0.81, RMSE = 0.23–0.39, RE = 24.47–62.99%), filling (R^2 = 0.48–0.87, RMSE = 0.13–0.28, RE = 15.45–45.56%), and all (R^2 = 0.37–0.73, RMSE = 0.18–0.37, RE = 27.28–59.45%) growth stages. The seven cultivars performed differently at different growth stages, the XM30 had a consistent well validation results based on SPAD (R^2 = 0.64–0.81, RMSE = 0.20–0.24, RE = 21.74–34.38%), NBI (R^2 = 0.73–0.87, RMSE = 0.15–0.19, RE = 16.68–23.04%), and NDRE (R^2 = 0.83–0.96, RMSE = 0.13–0.16, RE = 13.89–22.62%) at single and all growth stages. The YM15 performed poorly at jointing stage (R^2 = 0.29–0.75, RMSE = 0.13–0.16, RE = 22.62–27.51%), while it achieved a relatively good performance at booting (R^2 = 0.72–0.89, RMSE = 0.10–0.18, RE = 10.75–18.54%), flowering (R^2 = 0.81–0.87, RMSE = 0.13–0.23, RE = 21.56–45.14%), and filling (R^2 = 0.72–0.77, RMSE = 0.12–0.15, RE = 15.92–23.49%) stages among three optimal sensor indices. Therefore, the exponential relationship between the SPAD, NBI, NDRE, and NNI across seven cultivars using data from experiments 1–3 was constructed and shown in Figure 4, which had an R^2 value of 0.41–0.65 for SPAD, 0.66–0.85 for NBI, and 0.76–0.87 for NDRE at single and all growth stages.

3.5. N Diagnosis of Winter Wheat Based on the SPAD, NBI, and NDRE at Different Growth Stages

To evaluate the diagnosis accuracy of the SPAD, NBI, and NDRE models, experimental plots in Experiment 1–3 were divided into three categories: N deficient (NNI < 0.95), N optimal (0.95 ≤ NNI ≤ 1.05), and N excessive (NNI > 1.05) based on the diagnosis threshold values. The results in Table 6 indicated that the diagnosis accuracy ranged from 0.54 to 0.71 for SPAD, 0.66 to 0.84 for NBI, and 0.72 to 0.86 for NDRE. The kappa coefficient ranged from 0.30 to 0.52 for SPAD, 0.42 to 0.72 for NBI, and 0.53 to 0.75 for NDRE across all growth stages. Based on the evaluation criteria, the NBI and NDRE models performed moderately well for the diagnosis of N status during each growth stage; the SPAD model performed moderately well at the jointing and booting stages, but had fair diagnosis agreement at the flowering and filling stages.

Table 5. Ten-fold cross-validation results for the pair-wise exponential relationship between the SPAD, NBI, NDRE, and NNI across seven cultivars at different growth stages in experiment 1–3.

Index	Cultivar	Jointing Stage			Booting Stage			Flowering Stage			Filling Stage			All Stage		
		R^2	RMSE	RE (%)	R^2	RMSE	RE (%)	R^2	RMSE	RE (%)	R^2	RMSE	RE (%)	R^2	RMSE	RE (%)
SPAD	XM30	0.81	0.22	34.38	0.70	0.22	21.74	0.64	0.24	24.47	0.76	0.20	26.01	0.73	0.21	31.36
	HM20	0.31	0.48	77.65	0.41	0.32	52.62	0.58	0.28	48.29	0.48	0.28	45.56	0.48	0.33	55.81
	YM15	0.29	0.16	27.51	0.72	0.18	18.54	0.81	0.23	45.14	0.77	0.13	20.78	0.63	0.18	32.28
	YM16	0.32	0.22	52.65	0.45	0.27	63.60	0.78	0.36	62.99	0.78	0.20	29.82	0.56	0.29	59.45
	ZM12	0.67	0.42	27.09	0.84	0.25	18.51	0.36	0.39	29.42	0.78	0.17	15.45	0.44	0.37	27.28
	YM23	0.79	0.42	29.93	0.38	0.37	29.04	0.26	0.27	26.18	0.87	0.20	17.45	0.42	0.37	27.91
	NM13	0.50	0.28	19.65	0.68	0.24	29.45	0.56	0.28	29.08	0.78	0.14	15.67	0.41	0.28	27.95
	All varieties	0.40	0.36	48.06	0.53	0.27	35.75	0.35	0.30	38.68	0.50	0.20	26.44	0.37	0.30	38.41
NBI	XM30	0.83	0.18	23.04	0.87	0.15	16.68	0.75	0.18	16.74	0.77	0.15	18.94	0.73	0.19	21.56
	HM20	0.85	0.13	19.35	0.64	0.18	23.03	0.78	0.15	22.58	0.78	0.13	20.76	0.74	0.15	21.90
	YM15	0.55	0.13	24.46	0.87	0.11	16.53	0.87	0.20	40.36	0.72	0.15	23.49	0.75	0.16	28.27
	YM16	0.79	0.17	36.40	0.54	0.24	48.32	0.75	0.33	52.83	0.77	0.13	18.37	0.58	0.25	43.45
	ZM12	0.66	0.29	21.47	0.76	0.23	22.69	0.49	0.28	21.96	0.78	0.13	12.78	0.64	0.24	21.25
	YM23	0.88	0.24	20.47	0.59	0.33	25.54	0.33	0.27	24.07	0.84	0.19	16.18	0.65	0.29	23.27
	NM13	0.53	0.28	19.33	0.76	0.19	20.40	0.58	0.26	24.52	0.78	0.13	14.19	0.56	0.24	22.46
	All varieties	0.78	0.22	23.12	0.71	0.22	27.03	0.54	0.23	28.44	0.72	0.15	17.73	0.65	0.23	26.43
NDRE	XM30	0.96	0.14	13.94	0.95	0.13	13.89	0.83	0.15	17.53	0.85	0.16	22.62	0.87	0.16	18.53
	HM20	0.87	0.18	19.91	0.73	0.17	19.43	0.81	0.13	15.62	0.74	0.17	28.02	0.75	0.16	21.44
	YM15	0.75	0.15	22.62	0.89	0.10	10.75	0.86	0.13	21.56	0.76	0.12	15.92	0.72	0.16	24.39
	YM16	0.83	0.15	29.27	0.97	0.09	10.04	0.83	0.14	24.69	0.96	0.07	10.18	0.85	0.12	20.12
	ZM12	0.91	0.18	15.28	0.77	0.22	20.23	0.76	0.20	16.75	0.75	0.17	15.22	0.71	0.24	20.68
	YM23	0.93	0.19	13.72	0.74	0.24	18.44	0.84	0.12	12.19	0.83	0.21	18.15	0.67	0.26	19.93
	NM13	0.81	0.17	16.47	0.79	0.21	24.18	0.79	0.15	15.26	0.86	0.10	11.66	0.70	0.20	20.47
	All varieties	0.87	0.17	18.48	0.79	0.18	18.11	0.81	0.15	17.31	0.79	0.14	16.82	0.73	0.20	20.66

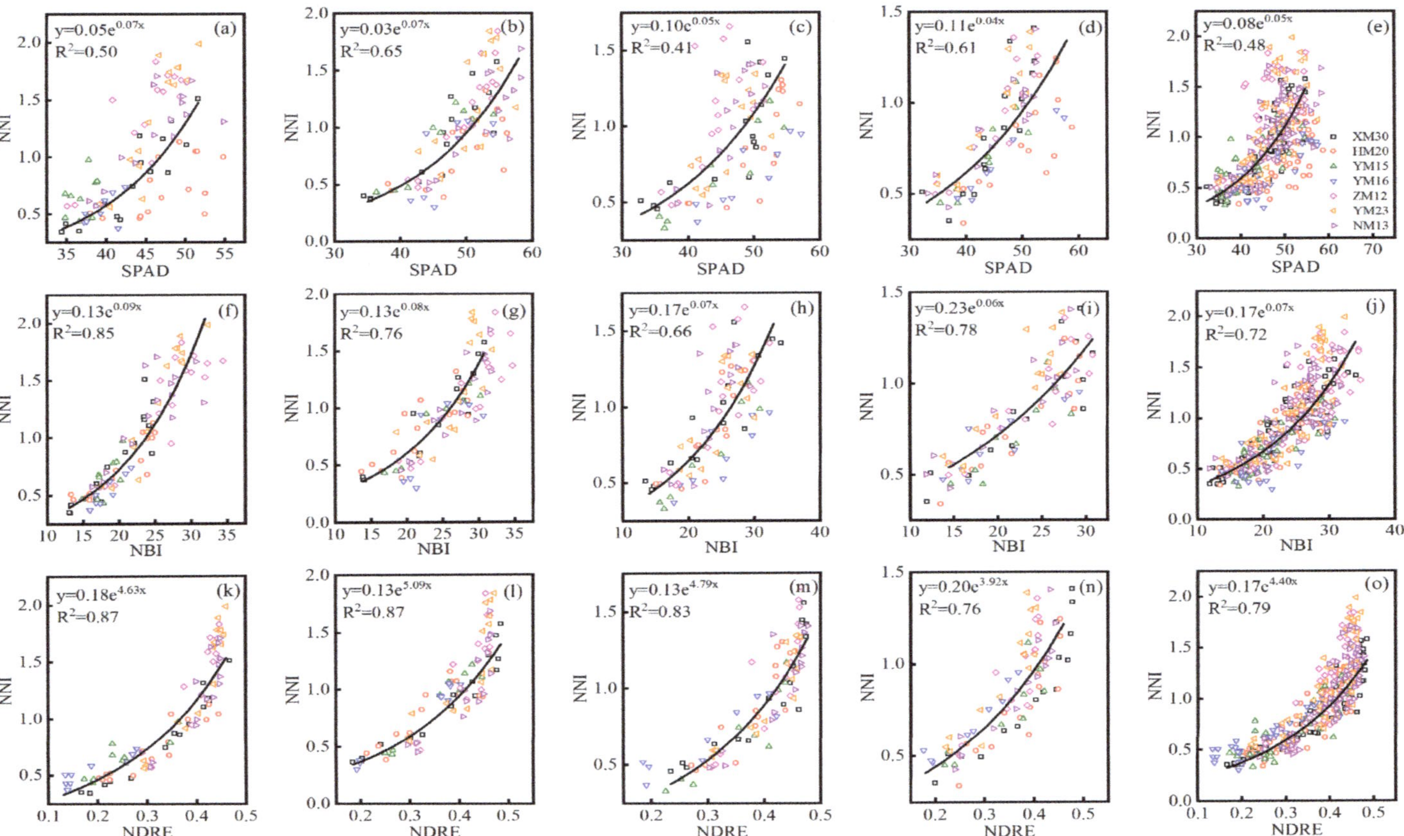

Figure 4. The exponential relationship between the SPAD and N nutrition index (NNI) across experiments 1–3 at (**a**) jointing, (**b**) booting, (**c**) flowering, (**d**) filling, and (**e**) all growth stages/the exponential relationship between the NBI and N nutrition index (NNI) across experiments 1–3 at (**f**) jointing, (**g**) booting, (**h**) flowering, (**i**) filling, and (**j**) all growth stages/the exponential relationship between the NDRE and N nutrition index (NNI) across experiments 1–3 at (**k**) jointing, (**l**) booting, (**m**) flowering, (**n**) filling, and (**o**) all growth stages. Black lines indicate regression lines.

Table 6. Areal agreement and kappa coefficient for SPAD, NBI, and NDRE at different growth stages.

Index	Areal Agreement				Kappa Coefficient			
	Jointing	Booting	Flowering	Filling	Jointing	Booting	Flowering	Filling
SPAD	0.71	0.70	0.54	0.61	0.52	0.50	0.30	0.34
NBI	0.84	0.77	0.66	0.71	0.72	0.61	0.42	0.49
NDRE	0.86	0.84	0.80	0.72	0.75	0.69	0.65	0.53

The N diagnosis maps based on the SPAD, NBI, and NDRE models at each of the growth stages during 2016 are shown in Figure 5. The N diagnosis data of each field plot were variable across the growth stages, with a large variation in winter wheat N status observed at differing values of N application. The N diagnosis map based on NBI and NDRE had similar performance, which showed that experimental plots with 270 or 360 kg N ha^{-1} were well- or over-fertilized, which fall into the N optimal or N excessive categories, respectively, across different growth stages. In contrast, experimental plots with 0 or 90 kg N ha^{-1} showed deficient fertilization, falling into the N deficient category at each growth stage. The N diagnosis map based on SPAD showed a relatively worse diagnosis results, which classified several low N treatments (90 kg N ha^{-1}) as N excessive category at the jointing and flowering stages, and classified several high N treatments (360 kg N ha^{-1}) in the N-deficient category at the filling stage. Overall, the sensor values at each of the experimental plot showed varying crop growth and N status performance due to variable N rates.

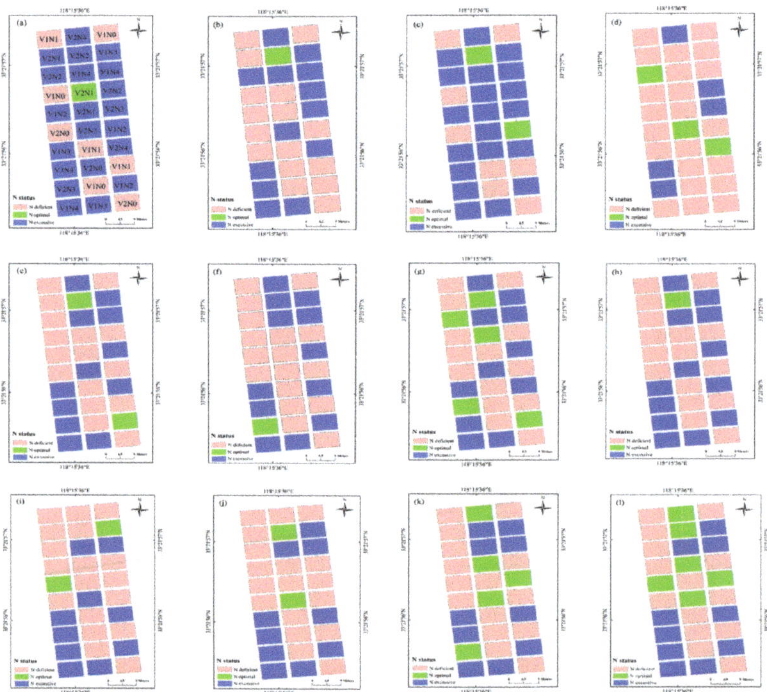

Figure 5. N diagnosis maps (experiment 1) based on the SPAD at the (**a**) jointing, (**b**) booting, (**c**) flowering, and (**d**) filling stage. N diagnosis maps based on the NBI at (**e**) jointing, (**f**) booting, (**g**) flowering, and (**h**) filling stage. N diagnosis maps based on the NDRE at (**i**) jointing, (**j**) booting, (**k**) flowering, and (**l**) filling stage. V1 and V2 in (**a**) represent XM30 and HM20 cultivars. N1, N2, N3, and N4 in (**a**) represent 0, 90, 180, 270, and 360 kg N ha^{-1} treatments, respectively, in Experiment 1.

4. Discussion

4.1. Wheat N Status Assessments Based on the Leaf and Canopy Sensors

The SPAD meter consists of a single leaf spectrometer and uses a red spectrum at 650 nm to estimate the chlorophyll concentration in the leaves, which significantly correlates with crop N status [13]. In Figure 3a, the SPAD value increased as wheat growth progressed, then remained high under high N levels (270 and 360 kg N ha^{-1}), which may have a saturated or near-saturated status in conditions of high N supply. Yue et al. [14] indicated that the response of the SPAD readings to wheat PNC showed a saturated phenomenon with an N supply that gradually increased to excessive amounts, i.e., to 420 kg N ha^{-1}, consistent with our study. The Dualex uses two near-infrared (710 and 850 nm) bands to estimate the chlorophyll content. Chlorophyll fluorescence (375 nm) of the sensor can be used to monitor leaf flavonoid content [37]. Zhang et al. [17]. showed that the Dualex (R^2 = 0.87) outperformed the SPAD meter for estimations of rice chlorophyll (R^2 = 0.77), and could mitigate the influence of saturating conditions under high N concentrations. Figure 3c showed wheat crops under N deficiency accumulate higher levels of Flav, the response of which contrasts Chl (Figure 3b). Gabriel et al. [22]. indicated that complementary polyphenol information (as Flav) can improve maize N deficiency predictions. The NBI index performed relatively better than the other two Dualex-based indices (Chl and Flav) for LNC (Table 4: R^2 = 0.36–0.79, RMSE = 0.39–0.67, RE = 13.43–20.02%), LNA (Table 4: R^2 = 0.49–0.70, RMSE = 14.35–23.50 kg ha^{-1}, RE = 48.33–67.38%), PNC (Table 4: R^2 = 0.49–0.76, RMSE = 0.19–0.46, RE = 15.63–27.44%), and PNA (Table 4: R^2 = 0.53–0.72, RMSE = 34.63–39.32 kg ha^{-1}, RE = 35.46–65.43%) estimation across single and all growth stages. The NBI can partially alleviate the influence of gradients along the plant leaf, and can accentuate the difference amongst the levels of plant N deficit due to the inverse dependence between Chl and Flav on the plant N nutritional status [38]. Previous studies have also shown that the NBI performed well for the assessment of N status and growth in muskmelon, consistent with this study [39].

The measured area of the SPAD meter was 6 mm^2 and the sampling area of each plot was 540 mm^2 based on our measurement method (90 individual measurements in each experiment plot). For Dualex, the sampling area was 1800 mm^2 (the measured area was 20 mm^2). The smaller sampling area for the SPAD meter was susceptible to factors such as blade veins and internal variation of the field. Accordingly, the SPAD values were less accurate for the estimation of crop N nutrition. The field of view of the RS sensor is 45° by 10° and the scan width of sensor perpendicular to the flight direction is 0.66 m (measurement height of 0.80 m). As a result, the scan area was a minimum of 5.94 square meters for each plot (e.g., Experiment 3). The larger sampling area in each plot was more representative of crop growth status. The NDRE derived from the RS sensor was closely associated with LNC (Table 4: R^2 = 0.61–0.79, RMSE = 0.39–0.53, RE = 11.84–15.79%), LNA (Table 4: R^2 = 0.66–0.87, RMSE = 12.39–19.79 kg ha^{-1}, RE = 24.72–42.12%), PNC (Table 4: R^2 = 0.51–0.74, RMSE = 0.18–0.45, RE = 14.81–27.34%), and PNA (Table 4: R^2 = 0.64–0.87, RMSE = 24.09–36.39 kg ha^{-1}, RE = 20.56–36.21%) at single growth stage and across all growth stages. Other studies indicated that the active canopy sensor (r = 0.73–0.86) displayed a higher accuracy in predicting the grapevine biomass and yield compared to the chlorophyll meter (r = 0.62–0.76) [40], which is consistent with this study.

4.2. Wheat N Nutrition Diagnosis Based on the Optimal Indices (SPAD, NBI, and NDRE) of Three Sensors

The estimation and diagnosis of N nutrition is a key consideration in precision wheat management [41]. NNI acts as an optimal N diagnostic indicator, with the development of remote sensing technologies making real-time estimations of NNI possible. In this study, the NNI diagnosis models based on the optimal indices (SPAD: R^2 = 0.41–0.65; NBI: R^2 = 0.66–0.85; NDRE: R^2 = 0.76–0.87) of three sensors were conducted during each growth stage, permitting the N diagnosis at all stages of wheat growth development. During the construction of the model, the R^2 of the relationship between NNI and NDRE at the flowering (Figure 4: R^2 = 0.83) and filling (Figure 4: R^2 = 0.76) stages were slightly lower than during the jointing (Figure 4: R^2 = 0.87) and booting (Figure 4: R^2 = 0.87) stages, which may be due to the emergence of a spike that influenced the estimated performance of the NDRE models at the late growth period. Similar results have also been shown in studies of rice and wheat [28,42]. The 10-fold cross-validation results of the relationship between three optimal indices and NNI across seven wheat cultivars showed that the Dualex-based NBI performed better than SPAD, and NDRE of RS canopy sensor performed better than two leaf-sensor indices at single and all growth stages. The seven wheat cultivars performed differently at different stage, which may be due to the difference of variety characteristics and climatic condition among different years [43,44]. The XM30 had a consistent good performance at each growth stage, the seedling of XM30 half creep flat on the ground with dark green and board leaves, which had a strong tilling ability and relative loose plant type. Winter wheat N status can be diagnosed based on NNI predictions using established diagnostic models and NNI threshold values. The NBI model (Table 6) had an areal agreement of 0.66 to 0.84 and a kappa coefficient of 0.42 to 0.72 across different growth stages. Similar studies have shown that the NBI measured using a Multiplex 3 sensor performs well for the diagnosis of rice N status during stem elongation (areal agreement = 75%, kappa coefficient = 0.595) and heading (areal agreement = 80%, kappa coefficient = 0.590) stages [45]. The NDRE model (Table 6: areal agreement = 0.72–0.86, kappa coefficient = 0.53–0.75) based on the RS sensor performed to a relatively higher level than the SPAD and NBI model, which also indicated that the larger sampling area by the canopy optical sensor was considerably more representative for wheat growth status [39]. Other studies have shown that the NDRE model based on the RS sensor can accurately measure rice N status at the panicle initiation and jointing stage (areal agreement = 0.55, kappa coefficient = 0.30) and heading growth stage (areal agreement = 0.76, kappa coefficient = 0.54) [46].

The N diagnosis plot maps based on the spectral indices directly reflected the N status of each plot (Figure 5), which may assist farmer for precise crop N management. As an example, the N diagnosis map and NNI diagnosis model could be coupled with the fertilization topdressing model to consider both temporal and spatial variability of crop growth and N nutrition [47,48], after which the N management decisions could be optimized to improve N use efficiency and increase economic benefits [49]. The spectral sensing method could non-destructively quantify and visualize the real time crop growth status compared to traditional leaf color-based diagnosis and chemical diagnosis, such as Kjeldahl digestion, which required laborious and time-consuming preparation [37]. The handheld spectral sensing used in this study was relatively limited when used for data collection in a large area compared to drone- and tractor-based sensing; the tractor-based Yara N-sensor was successfully used to estimate maize aboveground N uptake (R^2 = 0.57–0.84) and dry matter yields (R^2 = 0.67–0.91) at different growth stages [50]. Argento et al. [51] used the spectral index of NDRE derived from a UAV-mounted multispectral camera to guide N fertilizer application for winter wheat in a 2-hectare area, and to evaluate the sensitivity of NDRE with wheat dry matter (R^2 = 0.72), NNI (R^2 = 0.75) across different growth stages. In addition, the combination of ground and areal remote sensing data may be a promising method in crop growth monitoring, Zheng et al. [52] demonstrated the integration (R^2 = 0.72–0.75) of ground-based narrow band vegetation indices with UAV-based textural information exhibited a significant improvement for rice PNC estimation compared to individual UAV data (R^2 = 0.59–0.70). Due to the time and labor required for synchronous sensor data and agronomic indicators, this study analyzed growth stages that were initiated at the jointing stage. Monitoring frequency prior to the jointing stage can be increased in future research to improve these diagnosis models. In addition, similar studies should be performed for different wheat cultivars in other eco-sites, further enhancing the practicability of the optical sensors during field production.

5. Conclusions

Our results demonstrate that three portable optical sensors (SPAD meter, Dualex, and RS sensor) could be used to estimate and diagnose the N status of wheat. The Dualex-based NBI had a relatively well performance among four leaf-sensor indices, while NDRE of RS sensor performed best for LNC, LNA, PNC, and PNA estimation across different growth stages due to larger sampling area of canopy sensor. The areal agreement of the NNI diagnosis models ranged from 0.54 to 0.71 for SPAD, 0.66 to 0.84 for NBI, and 0.72 to 0.86 for NDRE, and kappa coefficient ranged from 0.30 to 0.52 for SPAD, 0.42 to 0.72 for NBI, and 0.53 to 0.75 for NDRE across all growth stages. We conclude that the use of sensor-based diagnostic models is appropriate for the rapid and non-destructive diagnosis of N nutrition of winter wheat.

Author Contributions: Conceptualization, J.J. and X.L.; methodology, J.J. and X.L.; validation, J.J., C.W. and X.L.; formal analysis, J.J. and C.W.; investigation, J.J., C.W. and X.L.; writing—original draft preparation, J.J., C.W., Z.F. and H.W.; writing—review and editing, Q.C., Y.T., Y.Z. and W.C.; visualization, Q.C., Y.T., Y.Z., W.C. and X.L.; supervision, X.L.; project administration, X.L.; funding acquisition, X.L. All authors have read and agreed to the published version of the manuscript.

Funding: This work was funded by the National Key Research and Development Program of China (2020YFE0202900), Jiangsu Agricultural Science and Technology Innovation (CX (20) 3072), National Natural Science Foundation of China (32071903), Jiangsu Key Research and Development Program (BE2019386), and the earmarked found for Jiangsu Agricultural Industry Technology System (JATS[2020]415, JATS[2020]135).

Institutional Review Board Statement: Not applicable.

Informed Consent Statement: Not applicable.

Acknowledgments: We would like to thank Youyan Zhou, Yan Liang, Jiayi Zhang, Fanglin Xiang, Yan Yan, and Meng Zhou for their help with the field data collection.

Conflicts of Interest: The authors declare no conflict of interest.

References

1. Cassman, K.G.; Dobermann, A.; Walters, D.T. Agroecosystems, nitrogen-use efficiency, and nitrogen management. *AMBIO* **2002**, *31*, 132–140. [CrossRef] [PubMed]
2. Miao, Y.; Stewart, B.A.; Zhang, F. Long-term experiments for sustainable nutrient management in China. A review. *Agron. Sustain. Devlop.* **2011**, *31*, 397–414. [CrossRef]
3. Miao, Y.; Mulla, D.J.; Hernandez, J.A.; Wiebers, M.; Robert, P.C. Potential impact of precision nitrogen management on corn yield, protein content, and test weight. *Soil Sci. Soc. Am. J.* **2007**, *71*, 1490–1499. [CrossRef]
4. Mulla, D.J. Twenty five years of remote sensing in precision agriculture: Key advances and remaining knowledge gaps. *Biosyst. Eng.* **2013**, *114*, 358–371. [CrossRef]
5. Dong, R.; Miao, Y.; Wang, X.; Chen, Z.; Yuan, F.; Zhang, W.; Li, H. Estimating Plant Nitrogen Concentration of Maize Using a Leaf Fluorescence Sensor across Growth Stages. *Remote Sens.* **2020**, *12*, 1139. [CrossRef]
6. Justes, E.; Mary, B.; Meynard, J.M.; Machet, J.M.; Thelier-Huche, L. Determination of a critical nitrogen dilution curve for winter wheat crops. *Ann. Bot.* **1994**, *74*, 397–407. [CrossRef]
7. Lemaire, G.; Jeuffroy, M.-H.; Gastal, F. Diagnosis tool for plant and crop N status in vegetative stage: Theory and practices for crop N management. *Eur. J. Agron.* **2008**, *28*, 614–624. [CrossRef]
8. Xia, T.; Miao, Y.; Wu, D.; Shao, H.; Khosla, R.; Mi, G. Active Optical Sensing of Spring Maize for In-Season Diagnosis of Nitrogen Status Based on Nitrogen Nutrition Index. *Remote. Sens.* **2016**, *8*, 605. [CrossRef]
9. Prost, L.; Jeuffroy, M.H. Replacing the nitrogen nutrition index by the chlorophyll meter to assess wheat N status. *Agron. Sustain. Dev.* **2007**, *27*, 321–330. [CrossRef]
10. Souza, R.; Pea-Fleitas, M.T.; Thompson, R.B.; Gallardo, M.; Padilla, F.M. Assessing Performance of Vegetation Indices to Estimate Nitrogen Nutrition Index in Pepper. *Remote Sens.* **2020**, *12*, 763. [CrossRef]
11. Watson, M.E.; Galliher, T.L. Comparison of Dumas and Kjeldahl methods with automatic analyzers on agricultural samples under routine rapid analysis conditions. *Commun. Soil Sci. Plant Anal.* **2001**, *32*, 2007–2019. [CrossRef]
12. Xue, X.; Wang, J.; Wang, Z.; Guo, W.; Zhou, Z. Determination of a critical dilution curve for nitrogen concentration in cotton. *J. Plant Nutr. Soil Sci.* **2007**, *170*, 811–817.

13. Muñoz-Huerta, R.F.; Guevara-Gonzalez, R.G.; Contreras-Medina, L.M.; Torres-Pacheco, I.; Prado-Olivarez, J.; Ocampo-Velazquez, R.V. A Review of Methods for Sensing the Nitrogen Status in Plants: Advantages, Disadvantages and Recent Advances. *Sensors* **2013**, *13*, 10823–10843. [CrossRef] [PubMed]

14. Yue, X.; Hu, Y.; Zhang, H.; Schmidhalter, U. Evaluation of Both SPAD Reading and SPAD Index on Estimating the Plant Nitrogen Status of Winter Wheat. *Int. J. Plant Prod.* **2020**, *14*, 67–75. [CrossRef]

15. Ravier, C.; Quemada, M.; Jeuffroy, M. Use of a chlorophyll meter to assess nitrogen nutrition index during the growth cycle in winter wheat. *Field Crop. Res.* **2017**, *214*, 73–82. [CrossRef]

16. Cerovic, Z.G.; Masdoumier, G.; Ben Ghozlen, N.; Latouche, G. A new optical leaf-clip meter for simultaneous non-destructive assessment of leaf chlorophyll and epidermal flavonoids. *Physiol. Plant.* **2012**, *146*, 251–260. [CrossRef] [PubMed]

17. Zhang, K.; Liu, X.; Ma, Y.; Zhang, R.; Cao, Q.; Zhu, Y.; Cao, W.; Tian, Y. A Comparative Assessment of Measures of Leaf Nitrogen in Rice Using Two Leaf-Clip Meters. *Sensors* **2020**, *20*, 175. [CrossRef] [PubMed]

18. Dong, T.; Shang, J.; Chen, J.M.; Liu, J.; Qian, B.; Ma, B.; Morrison, M.J.; Zhang, C.; Liu, Y.; Shi, Y.; et al. Assessment of Portable Chlorophyll Meters for Measuring Crop Leaf Chlorophyll Concentration. *Remote Sens.* **2019**, *11*, 2706. [CrossRef]

19. Tremblay, N.; Wang, Z.; Bélec, C. Evaluation of the Dualex for the assessment of corn nitrogen status. *J. Plant Nutr.* **2007**, *30*, 1355–1369. [CrossRef]

20. Tremblay, N.; Bélec, C.; Jenni, S.; Fortier, E.; Mellgren, R. The Dualex—A New Tool to Determine Nitrogen Sufficiency in Broccoli. *Acta Hortic.* **2009**, *824*, 121–132. [CrossRef]

21. Tremblay, N.; Wang, Z.; Bélec, C. Performance of Dualex in spring wheat for crop nitrogen status assessment, Yield predication and estimation of soil nitrate content. *J. Plant Nutr.* **2010**, *33*, 57–70. [CrossRef]

22. Gabriel, J.L.; Quemada, M.; Alonso-Ayuso, M.; Lizaso, J.I.; Martín-Lammerding, D. Predicting N Status in Maize with Clip Sensors: Choosing Sensor, Leaf Sampling Point, and Timing. *Sensors* **2019**, *19*, 3881. [CrossRef] [PubMed]

23. Lejealle, S.; Evain, S.; Cerovic, Z.G. Multiplex: A new diagnostic tool for management of nitrogen fertilization of turfgrass. In Proceedings of the 10th International Conference on Precision Agriculture, Denver, CO, USA, 18–21 July 2010.

24. Chen, P.; Haboudane, D.; Tremblay, N.; Wang, J.; Vigneault, P.; Li, B. New spectral indicator assessing the efficiency of crop nitrogen treatment in corn and wheat. *Remote Sens. Environ.* **2010**, *114*, 1987–1997. [CrossRef]

25. Feng, W.; Yao, X.; Zhu, Y.; Tian, Y.C.; Cao, W. Monitoring leaf nitrogen status with hyperspectral reflectance in wheat. *Eur. J. Agron.* **2008**, *28*, 394–404. [CrossRef]

26. Wang, Z.J.; Wang, J.H.; Liu, L.Y.; Huang, W.J.; Zhao, C.J.; Wang, C.Z. Prediction of grain protein content in winter wheat (*Triticum aestivum* L.) using plant pigment ratio (PPR). *Field Crop. Res.* **2004**, *90*, 311–321. [CrossRef]

27. Jiang, J.; Wang, C.; Wang, Y.; Cao, Q.; Tian, Y.; Zhu, Y.; Cao, W.; Liu, X. Using an Active Sensor to Develop New Critical Nitrogen Dilution Curve for Winter Wheat. *Sensors* **2020**, *20*, 1577. [CrossRef] [PubMed]

28. Aranguren, M.; Castellón, A.; Aizpurua, A. Topdressing nitrogen recommendation in wheat after applying organic manures: The use of field diagnostic tools. *Nutr. Cycl. Agroecosyst.* **2018**, *110*, 89–103. [CrossRef]

29. Miller, J.J.; Schepers, J.S.; Shapiro, C.A.; Arneson, N.J.; Eskridge, K.M.; Oliveira, M.C.; Giesler, L.J. Characterizing soybean vigor and productivity using multiple crop canopy sensor readings. *Field Crop. Res.* **2018**, *216*, 22–31. [CrossRef]

30. Barnes, E.M.; Clarke, T.R.; Richards, S.E.; Colaizzi, P.D.; Thompson, T. Coincident detection of crop water stress, nitrogen status, and canopy density using ground based multispectral data. In Proceedings of the Fifth International Conference on Precision Agriculture, Bloomington, MN, USA, 16–19 July 2000; pp. 16–19.

31. Tucker, C. Red and photographic infrared linear combination for monitoring vegetation. *Remote Sens. Environ.* **1979**, *8*, 127–150. [CrossRef]

32. Zhang, J.; Liu, X.; Liang, Y.; Cao, Q.; Tian, Y.; Zhu, Y.; Cao, W.; Liu, X. Using a Portable Active Sensor to Monitor Growth Parameters and Predict Grain Yield of Winter Wheat. *Sensors* **2019**, *19*, 1108. [CrossRef]

33. Bremner, J.M.; Mulvaney, C.S. Nitrogen -total. In Methods of Soil Analysis. In *Chemical and Microbial Properties*; Page, A.L., Miller, R.H., Keeney, D.R., Eds.; American Society of Agronomy, and Soil Science Society: Madison, WI, USA, 1982; pp. 595–624.

34. Cao, Q.; Miao, Y.; Shen, J.; Yuan, F.; Cheng, S.; Cui, Z. Evaluating Two Crop Circle Active Canopy Sensors for In-Season Diagnosis of Winter Wheat Nitrogen Status. *Agronomy* **2018**, *8*, 201. [CrossRef]

35. Lillesand, T.; Kiefer, R.; Chipman, J. *Remote Sensing and Image Interpretation*, 7th ed.; Wiley: Hoboken, NJ, USA, 2015; pp. 575–581.

36. Landis, J.R.; Koch, G.G. The measurement of observer agreement for categorical data. *Biometrics* **1977**, *33*, 159–174. [CrossRef]

37. Padilla, F.M.; Gallardo, M.; Teresa Peña-Fleitas, M.; de Souza, R.; Thompson, R.B. Proximal Optical Sensors for Nitrogen Management of Vegetable Crops: A Review. *Sensors* **2018**, *18*, 2083. [CrossRef]

38. Cartelat, A.; Cerovic, Z.G.; Goulas, Y.; Meyer, S.; Lelarge, C.; Prioul, J.L.; Barbottin, A.; Jeuffroy, M.H.; Gate, P.; Agati, G.; et al. Optically assessed contents of leaf polyphenolics and chlorophyll as indicators of nitrogen deficiency in wheat (*Triticum aestivum* L.). *Field Crop. Res.* **2005**, *91*, 35–49. [CrossRef]

39. Padilla, F.M.; Teresa Pena-Fleitas, M.; Gallardo, M.; Thompson, R.B. Evaluation of optical sensor measurements of canopy reflectance and of leaf flavonols and chlorophyll contents to assess crop nitrogen status of muskmelon. *Eur. J. Agron.* **2014**, *58*, 39–52. [CrossRef]

40. Taskos, D.G.; Koundouras, S.; Stamatiadis, S.; Zioziou, E.; Nikolaou, N.; Karakioulakis, K.; Theodorou, N. Using active canopy sensors and chlorophyll meters to estimate grapevine nitrogen status and productivity. *Precis. Agric.* **2015**, *16*, 77–98. [CrossRef]

41. Aula, L.; Omara, P.; Nambi, E.; Oyebiyi, F.B.; Raun, W.R. Review of Active Optical Sensors for Improving Winter Wheat Nitrogen Use Efficiency. *Agronomy* **2020**, *10*, 1157. [CrossRef]
42. Wang, Y.; Zhang, K.; Tang, C.; Cao, Q.; Tian, Y.; Zhu, Y.; Cao, W.; Liu, X. Estimation of Rice Growth Parameters Based on Linear Mixed-Effect Model Using Multispectral Images from Fixed-Wing Unmanned Aerial Vehicles. *Remote Sens.* **2019**, *11*, 1371. [CrossRef]
43. Ata-Ul-Karim, S.T.; Cao, Q.; Zhu, Y.; Tang, L.; Rehmani, A.; Cao, W. Non-destructive Assessment of Plant Nitrogen Parameters Using Leaf Chlorophyll Measurements in Rice. *Front. Plant Sci.* **2016**, *7*, 1829. [CrossRef] [PubMed]
44. Kaur, P.; Kaur, A.; Nigan, R.; Gill, A.; Singh, J.; Sandhu, S. Spectral indices of wheat cultivars at different growth stages under Punjab conditions. *J. Agrometeorol.* **2018**, *19*, 160–165.
45. Huang, S.; Miao, Y.; Yuan, F.; Cao, Q.; Ye, H.; Lenz-Wiedemann, V.I.S.; Bareth, G. In-Season Diagnosis of Rice Nitrogen Status Using Proximal Fluorescence Canopy Sensor at Different Growth Stages. *Remote Sens.* **2019**, *11*, 1184. [CrossRef]
46. Lu, J.; Miao, Y.; Shi, W.; Li, J.; Yuan, F. Evaluating different approaches to non-destructive nitrogen status diagnosis of rice using portable RapidSCAN active canopy sensor. *Sci. Rep.* **2017**, *7*, 14073. [CrossRef] [PubMed]
47. Ata-Ul-Karim, S.; Liu, X.; Lu, Z.; Zheng, H.; Cao, W.; Zhu, Y. Estimation of nitrogen fertilizer requirement for rice crop using critical nitrogen dilution curve. *Field Crop. Res.* **2017**, *201*, 32–40. [CrossRef]
48. Zhang, K.; Yuan, Z.; Yang, T.; Lu, Z.; Cao, Q.; Tian, Y.; Zhu, Y.; Cao, W.; Liu, X. Chlorophyll meter-based nitrogen fertilizer optimization algorithm and nitrogen nutrition index for in-season fertilization of paddy rice. *Agron. J.* **2020**, *112*, 288–300. [CrossRef]
49. Bean, G.M.; Kitchen, N.R.; Camberato, J.J.; Ferguson, R.B.; Fernandez, F.G.; Franzen, D.W.; Laboski, C.A.M.; Nafziger, E.D.; Sawyer, J.E.; Scharf, P.C.; et al. Active-Optical Reflectance Sensing Corn Algorithms Evaluated over the United States Midwest Corn Belt. *Agron. J.* **2018**, *110*, 2552–2565. [CrossRef]
50. Mistele, B.; Schmidhalter, U. Spectral measurements of the total aerial N and biomass dry weight in maize using a quadrilateral-view optic. *Field Crop. Res.* **2008**, *106*, 94–103. [CrossRef]
51. Argento, F.; Anken, T.; Abt, F.; Vogelsanger, E.; Walter, A.; Liebisch, F. Site-specific nitrogen management in winter wheat supported by low-altitude remote sensing and soil data. *Precis. Agric.* **2021**, *22*, 364–386. [CrossRef]
52. Zheng, H.; Cheng, T.; Li, D.; Yao, X.; Tian, Y.; Cao, W.; Zhu, Y. Combining Unmanned Aerial Vehicle (UAV)-Based Multispectral Imagery and Ground-Based Hyperspectral Data for Plant Nitrogen Concentration Estimation in Rice. *Front. Plant Sci.* **2018**, *9*, 936. [CrossRef]

Article

Comparison of Selected Dimensionality Reduction Methods for Detection of Root-Knot Nematode Infestations in Potato Tubers Using Hyperspectral Imaging

Janez Lapajne, Matej Knapič and Uroš Žibrat *

Plant Protection Department, Agricultural Institute of Slovenia, 1000 Ljubljana, Slovenia;
janez.lapajne@kis.si (J.L.); matej.knapic@kis.si (M.K.)
* Correspondence: uros.zibrat@kis.si; Tel.: +386-1-280-5131

Abstract: Hyperspectral imaging is a popular tool used for non-invasive plant disease detection. Data acquired with it usually consist of many correlated features; hence most of the acquired information is redundant. Dimensionality reduction methods are used to transform the data sets from high-dimensional, to low-dimensional (in this study to one or a few features). We have chosen six dimensionality reduction methods (partial least squares, linear discriminant analysis, principal component analysis, RandomForest, ReliefF, and Extreme gradient boosting) and tested their efficacy on a hyperspectral data set of potato tubers. The extracted or selected features were pipelined to support vector machine classifier and evaluated. Tubers were divided into two groups, healthy and infested with *Meloidogyne luci*. The results show that all dimensionality reduction methods enabled successful identification of inoculated tubers. The best and most consistent results were obtained using linear discriminant analysis, with 100% accuracy in both potato tuber inside and outside images. Classification success was generally higher in the outside data set, than in the inside. Nevertheless, accuracy was in all cases above 0.6.

Keywords: hyperspectral imaging; dimensionality reduction; LDA; PLS; PCA; RandomForest; ReliefF; XGB; *Meloidogyne*; *Solanum tuberosum*

Citation: Lapajne, J.; Knapič, M.; Žibrat, U. Comparison of Selected Dimensionality Reduction Methods for Detection of Root-Knot Nematode Infestations in Potato Tubers Using Hyperspectral Imaging. *Sensors* **2022**, *22*, 367. https://doi.org/10.3390/s22010367

Academic Editors: Jiyul Chang and Sigfredo Fuentes

Received: 30 November 2021
Accepted: 27 December 2021
Published: 4 January 2022

Publisher's Note: MDPI stays neutral with regard to jurisdictional claims in published maps and institutional affiliations.

1. Introduction

Quarantine pests are of major importance for agriculture and the food industry, and are being officially monitored and controlled [1]. Among these, root-knot nematodes (RKN) of the genus *Meloidogyne* present the most destructive group of plant-parasitic nematodes. They can infest a broad range of host plants, and are alone responsible for approximately 5% of global crop losses. These are soil-borne parasites, where they infest the host plants' root system and cause non-specific symptoms on above-ground parts of plants. Furthermore, they can cause latent (asymptomatic) infestations in potato tubers [2], which pose an additional threat in seeding material, as they could be spread over larger areas quite quickly. The parasite *M. luci* has been originally described by Carneiro et al. from samples from Brazil, Chile, and Iran [3], and has since been found several times in Europe as well [4]. Even though *M. luci* belongs to the group of tropical RKNs, it can survive winter in fields under temperate and Mediterranean climates [5]. It is therefore considered an emerging pest in Europe and was included in the alert list of harmful organisms in 2017 [6].

Since RKNs cause non-specific symptoms, laboratory diagnoses are required for accurate identification. Traditionally RKN species are identified morphometrically, and by analysing dehydrogenase and esterase isozyme phenotypes [7]. These methods require the isolation of mature females from plant tissue, making them unsuitable for a large number of samples. First visible symptoms are presented as reduced plant growth. Infections start in small areas of the crop, but can over the years become full field infestations if not appropriately handled. Particularly latent infestations of potato seed tubers have the potential to facilitate this process and lead to infestations over larger areas in just one or two seasons. These characteristic of RKN infestations show a clear need for detection of infestations in early stages [8].

Precision agriculture helps reduce the spread of diseases, and includes well-established practices to mitigate losses [9]. The plant immune system reacts to stressors by changing their biophysical and biochemical makeup, which in turn affects their spectral properties [10]. Currently the most common remote sensing method, used for plant diseases detection, is hyperspectral imaging (HSI) [1]. Nowadays, HSI is used in various applications, such as biotechnology, agriculture, environmental monitoring, and chemistry [11]. In HSI, reflected light is captured and data stored in several spectral bands, with a high spectral resolution (bandwidths are typically around 4–5 nm). Consequently, a large number of spectral bands are acquired at each capture, for the entire spectrum the sensors record (typically between 400 and 2500 nm).

Hyperspectral imaging has been used extensively for assessing plant root and tuber quality. The published research can generally be divided into three interlinked groups, according to the investigated properties: (1) physical properties (e.g., colour and texture), (2) chemical constituents (e.g., proteins and polysaccharides), and (3) pest and disease detection. Research into the latter is mostly focused on early detection of infections and infestations in above-ground parts of plants [12]. Biotic and abiotic stressors can cause changes in spectral signatures, by triggering various defense mechanisms, such as production of specific metabolites, induction of hypersensitive reactions, and changes in plant tissues [7]. The most extensive use of remote sensing of potato tubers has been for quality assessment, e.g., detection of defects [13] and bruises [14], and chemometric analyses, such as sugar [15], cellulose and starch content [16]. On the other hand, only a handful of studies deal with pest and disease detection in potato tubers. For example, Dacal-Nieto et al. and Huang et al. used hyperspectral imaging and support vector machines to detect hollow heart disease [17,18], and Zhou et al. used partial least squares as a pre-processing step in linear discriminant analysis to detect blackheart [19]. A partial least squares discriminant analysis approach was used by Garhwal et al. to detect zebra chip disease [20], and Al Riza et al. used a combination of genetic algorithms for feature selection and partial least squares to detect common scab [21]. But results aren't always so clear-cut, as Zhao et al. found that infrared and thermal imaging did not distinguish between healthy and *Liberibacter solanacearum* infected tubers in storage [22].

Hyperspectral data shows a high level of collinearity between spectral bands, leading to high redundancy and decreases the signal-to-noise ratio. Furthermore, the data sets are of high dimensionality, which increases the difficulty of knowledge discovery and pattern recognition. Dimensionality reduction is therefore an obligatory and crucial step in HSI data pre-processing [10].

With dimensionality reduction (DR) methods we retain the descriptive power of the data, but reduce the number of dimensions. This process removes some patterns in the data, but the features of interest remain. The large number of spectral bands can cause reduced discriminating ability of the HSI features. This problem is especially severe when the available training set consists of a small amount of samples (referred as the curse of dimensionality). Dimensionality reduction is therefore a crucial step that transforms the data to lower dimensional space, while preserving relevant information [23]. Generally, DR methods can be grouped into two sets: (1) feature extraction, and (2) feature selection. Feature extraction methods transform the whole feature space to a lower dimensional one, while feature selection approach picks out the most significant features from the whole feature space. Furthermore, contrary to feature extraction, feature selection preserves physical characteristics of the original feature space. But, feature selection information is lost, as features are removed from further analysis. Furthermore, feature extraction is less prone to overfitting and often results in better classification accuracy [23,24]. However, there is no standard approach which would yield the best possible result for any specific dataset [25].

In recent years researchers have shown big interest in developing new methods and tools for processing of hyperspectral data. Trends show that many authors decide to choose feature selection over feature extraction DR algorithms. However conventional feature extraction algorithms are still present, due to a better performance in some cases. Moghimi et al. tested the performance of NaCl treated wheat with different feature selection methods [11]. They aggregated all with ensemble method to increase robustness and accuracy. The transformed data were classified using quadratic discriminants analysis (QDA) and validated using 5-fold cross-validation. Similarly, AlSuwaidi et al. also used feature selection method for crop disease detection [26]. As classification method they used support vector machines (SVM). Moghadam et al. showed that SVM is one of the best classification algorithms used for plant disease detection due to its generalization ability [1]. Their feature extraction method is used based on probabilistic topic modelling. Collected features with reduced dimensionality were pipelined to Latent Dirichlet Allocation model for plant leaves disease detection. On the other hand, Jin et al. directly used convolutional neural networks (CNN) for classification, without any dimensionality reduction [27]. Feature extraction problems for HSI are oftentimes solved by using convolutional neural networks [28,29]. However, CNNs usually need more training data in comparison with conventional methods [28,30], limiting their applicability.

This study was motivated by the need to find a dimensionality reduction method for detection of *M. luci*-infested potato tubers using hyperspectral imaging. The DR method ses were twofold: (1) DR methods can achieve good classification accuracy even with should enable accurate identification in combination with support vector machines (SVM), with only one extracted feature, or a very limited number thereof. Our hypothesis only one feature, and (2) data from the outside of tubers will achieve better classification success. Since only a limited amount of information is available about spectral and chemical effects of nematode infestations on potato tubers, we tested two tuber processing methods. We selected six dimensionality reduction methods, three from each group: Partial least squares (PLS), Linear discriminant analysis (LDA), Principal component analysis (PCA), RandomForest (RF), ReliefF (RFF), and Extreme gradient boosting (XGB). Of these, LDA provided the best results, as it achieved the highest classification accuracy in both external and internal images of potato tubers. All DR methods achieved better success with data from outside of tubers, except for LDA, where the results were equal.

2. Materials and Methods

2.1. Tuber Cultivation and Preparation

The tubers were obtained from an experiment on potato (*Solanum tuberosum* cv., variety Desiree) infestation with *M. luci*, which was established from June to September 2018 in a glasshouse at the Agricultural Institute of Slovenia. A total of 20 day-old plants were transplanted to 13 cm-diameter pots (V = 1 L) and supported with 1 m plastic-coated stakes. The substrate of 10 randomly selected plants was inoculated with *M. luci* at the beginning of the experiment [31]. Roots of tomato plants, infested by *M. luci* (i.e., egg-masses were visible on the root surface; the parasites were from the collection at the Agricultural institute of Slovenia) were cut into pieces and mixed. A subset of infested roots was then weighed and nematode eggs were collected in suspension, in accordance with Hussey & Barker [32]. The number of eggs was determined visually under a stereomicroscope (Nikon SMZ800). Infested roots were introduced to the substrate to a final concentration of 250×10^3 eggs/plant. The presence of *M. luci* was confirmed with isoenzyme analysis [4]. The microplot experiment was completed at the end of the growing season at 97 days after inoculation.

Potato tubers were harvested at the end of the growing season in 2018 and stored in boxes in a dark storage room with ventilation and a temperature of $18 \pm 2\ ^\circ$C for the incubation period, until further processing. Tubers from infested pots were visually checked for signs of infection (surface galls) (Figure 1). Visibly decaying tubers were excluded from further analysis. The tubers were divided into two groups, inoculated and healthy, of 5 tubers each.

Figure 1. Inoculated and healthy potato tuber. (**a**) inoculated tubers, and (**b**) healthy tubers. Note the galls on the surface of the infested tubers. The difference in size is not necessarily symptomatic.

2.2. Hyperspectral Image Acquisition

For hyperspectral imaging tubers were sliced in half and placed on a black background. This way both the outside and inside of tubers could be imaged simultaneously. Reflectance spectra in 448 bands in the VNIR (visible to near infrared) and SWIR (short-wave infrared) regions were acquired using two Norsk Elektro Optikk AS (Oslo, Norway) pushbroom hyspex cameras, VNIR-1600 (400–988 nm, 160 bands, bandwidth 3.6 nm, spatial resolution at 1 m distance 0.1 mm) and SWIR-384 (950–2500 nm, 288 bands, bandwidth 5.4 nm, spatial resolution 0.25 mm), mounted vertically above the samples at a distance of 1 m. The samples were illuminated with two calibrated halogen lamps with homogeneous light intensity between 400 and 2500 nm, placed above the samples next to the cameras. The lamps were switched on 15 min before image acquisition to stabilize the light source's temperature drift and establish spatial uniformity of illumination [33]. A calibrated diffuse grey reference plate with 20% reflectance (SphereOptics, Herrsching, Germany) was included in each image and used to calculate reflectance. The signal-to-noise ratio was increased by scanning each line three times and calculating the average. Hyperspectral images were radiometrically calibrated to radiance units (W sr^{-1}m^{-2}).

2.3. Pre-Processing and Analysis

The data analysis workflow consisted of five stages (Figure 2). First, radiometrically corrected images were loaded into working memory. Second, images were segmented and these segments were then used to calculate reflectance values and mean spectra of each sample. The segmented image of each tuber was then divided into six equal parts. Reflectance values and mean spectra for each of these sub-segments were extracted. Then, data was split into training (4 potatoes) and test sets (1 potato). This process was repeated 5 times. In the next stage, we applied dimensionality reduction algorithms to extract the most relevant, features. In addition to DR methods, we also included a data set without any dimensionality reduction (labelled "None"). In the last stage, the chosen features were tested using support vector machine classification.

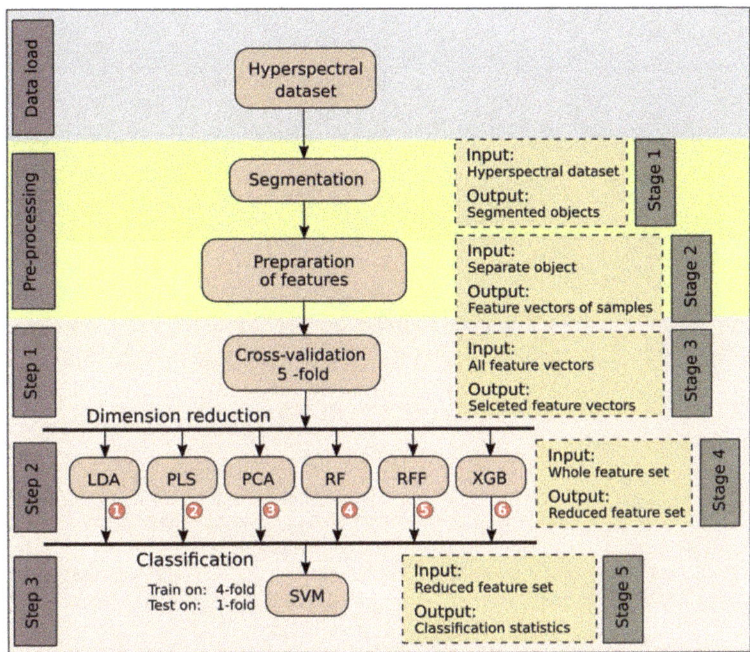

Figure 2. Analysis flowchart for proposed procedure. Data load: Image acquisition. Pre-processing: Image segmentation and feature preparation techniques. Step 1: Separation of training and validation dataset with cross-validation. Step 2: Dimensionality reduction with LDA, PCA, RF, RFF, XGB or PLS. Step 3: Classification with the use of SVM and 5-fold cross-validation.

2.3.1. Segmentation

Image segmentation was performed using spectral information divergence (SID). It uses a divergence measure to match HSI image pixels to reference pixels [34]. In HSI each pixel consists of multiple values which form a discrete signal. For j-th pixel we can write:

$$\mathbf{x}^{(j)} = (x_1, x_2, \ldots, x_D)^{\mathrm{T}} \tag{1}$$

where x_i represents the value of spectral band B_i acquired at wavelength λ_i. Index D represents a number of spectral bands. Probabilistic measure p_i can be calculated for each element x_i. Probabilities for all elements are then written into vector of probabilities $\mathbf{p}$:

$$\begin{aligned} p_i &= p(x_i) = \frac{x_i}{\sum_{i=1}^{D} x_i} \\ \mathbf{p}\left(\mathbf{x}^{(j)}\right) &= (p_1, p_2, \ldots, p_D)^{\mathrm{T}} \end{aligned} \tag{2}$$

Relative entropy can be calculated between $\mathbf{p}$ and $\mathbf{q}$ probability vectors with Kullback–Leibler information measure:

$$KL\left(\mathbf{p}\left(\mathbf{x}^{(j)}\right) \,||\, \mathbf{q}\left(\mathbf{r}^{(C_k)}\right)\right) = \sum_{i=1}^{D} \mathrm{p}_i \cdot log\frac{p_i}{q_i} \tag{3}$$

where $\mathbf{q}\left(\mathbf{r}^{(C_k)}\right)$ represents probability measure for reference vector $\mathbf{r}$ of k-th segmentation class. In our case, possible segmentation classes are included in a set: $C_k \in$ {potato tuber, background, reference panel}. Reference vectors for each segmentation class separately are constructed from manually selected area of pixels. An area for segmentation class C_k is defined as: $S_k = \left(\mathbf{x}^{(1)}, \mathbf{x}^{(2)}, \ldots, \mathbf{x}^{(N)} \right)$ where N represents number of pixel vectors included in area corresponding to segmentation class k. Reference values can be calculated from pixels for each spectral band:

$$r_i = \frac{1}{N} \sum_{j=1}^{N} x_{ij} \quad i = (1, 2, \ldots, D) \tag{4}$$

Reference vector is then defined as:

$$\mathbf{r}^{(C_k)} = (r_1, r_2, \ldots, r_D)^{\mathrm{T}} \tag{5}$$

Probability vector $\mathbf{q}\left(\mathbf{r}^{(C_k)} \right) = (q_1, q_2, \ldots, q_D)$ can then be calculated by (3) for all segmentation classes. When reference vectors are known, SID values can be calculated [35]:

$$SID\left(\mathbf{p}\left(\mathbf{x}^{(j)} \right), \mathbf{q}\left(\mathbf{r}^{(C_k)} \right) \right) = KL\left(\mathbf{p}\left(\mathbf{x}^{(j)} \right) \mid\mid \mathbf{q}\left(\mathbf{r}^{(C_k)} \right) \right) + KL\left(\mathbf{q}\left(\mathbf{r}^{(C_k)} \right) \mid\mid \mathbf{p}\left(\mathbf{x}^{(j)} \right) \right) \tag{6}$$

Equation (6) assigns divergence value to each pixel constructing the HSI image. Pixels from an HSI image are classified to segmentation class C_k with smallest divergence value. The greater the similarity of pixel to reference signal, the smaller the value of divergence. With the use of additional thresholding it is possible to fully separate predefined segmentation classes. Thresholding values were chosen with trial-and-error approach. Segmentation masks for each class can then be built based on calculated divergence values (Figure 3).

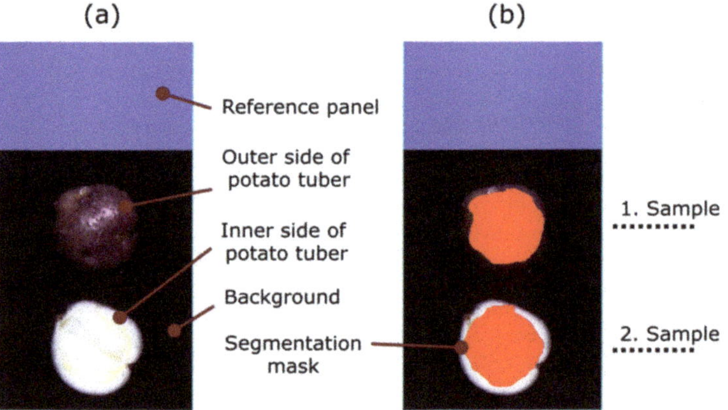

Figure 3. Hyperspectral image of tubers prior to segmentation. (**a**) With labelled segmentation class which construct the image. (**b**) Same image with applied segmentation masks of potato tubers and labelled consecutive sample number.

2.3.2. Preparation of Features

Feature vectors can be created from segmentation masks. Pixels that correspond to the same segmentation class C_k define observing object $\mathbf{o}^{(z)} = \left(\mathbf{x}^{(1)}, \mathbf{x}^{(2)}, \mathbf{x}^{(j)}, \ldots, \mathbf{x}^{(N)} \right)^{T}$, where N represents number of connected pixels located within the segmentation mask. Feature vector $\mathbf{O}$ is calculated from defined objects with arithmetic mean for D spectral bands. For z-th object it can be calculated using following equations:

$$O_i^{(z)} = \frac{1}{N} \sum_{j=1}^{N} x_{ij}^{(z)} \quad i = (1, 2, \ldots, D)$$
$$\mathbf{O}^{(z)} = (O_1, O_2, \ldots, O_D)^{\mathrm{T}} \tag{7}$$

Spectral reflectance was calculated for tuber data, using a 50% grey reference panel. For each HSI image we calculate feature vector $\mathbf{O}^{(0)}$, which represents the reference panel. Reflectance values of reference panel were provided by the manufacturer (SphereOptics, Germany). We assigned those values to vector $\mathbf{R}^{(0)}$. For arbitrary object z, in the same image where $\mathbf{O}^{(0)}$ was calculated, converted feature vector $\mathbf{R}^{(z)}$ can be calculated as:

$$
\begin{aligned}
R_i^{(z)} &= R_i^{(0)}/O_i^{(0)} \cdot O_i^{(z)} \quad i = (1,2,\ldots,D) \\
\mathbf{R}^{(z)} &= (R_1, R_2, \ldots, R_D)^{\mathrm{T}}
\end{aligned}
\tag{8}
$$

An important note to add is that object of reference panel $\mathbf{o}^{(0)}$ is filtered before usage in Equation (8). The reason is to remove outliers for more accurate calculation of feature vector $\mathbf{O}^{(0)}$. Outliers are filtered with median absolute deviation (MAD). Values outside $\pm 2\text{MAD}$ were removed before further calculation. MAD can be calculated by Equation (9), for each spectral band: $i = (1,2,\ldots,D)$. In equation $\mathbf{x}^{(i)}$ represents vector of all pixels at i-th spectral band. Value labeled as $\tilde{x}^{(i)}$ represents median value of this vector. With $\mathbf{I}$ raw vector of ones is labelled.

$$
\begin{aligned}
\mathbf{x}^{(i)} &= \left\{ o_{ij}^{(0)} \,\middle|\, i \in D, 1 \leq j \leq N \right\}_{1 \times N} \\
\tilde{x}^{(i)} &= \mathrm{median}\left(\mathbf{x}^{(i)} \right) \\
\mathrm{MAD}^{(i)} &= \mathrm{median}\left(\left| \mathbf{x}^{(i)} - \tilde{x}^{(i)} \mathbf{I} \right| \right)
\end{aligned}
\tag{9}
$$

The feature vector $\mathbf{R}^{(z)}$ was then smoothed using a Savitzky–Golay filter to emphasize small spectral variations the same way as in Schafer [36]. Savitzky–Golay filter is based on local least squares polynomial approximation. It was shown that it reduces noise while maintaining shape and important information in a feature vector [26]. In this study the filter window length was 15, polynomial order was 2 and second order derivatives were used. Parameters were chosen with regard to the highest exhibition of performance evaluated with classification accuracy.

Dimensionality Reduction

In this paper three feature extraction methods (Principal component analysis, Linear discriminant analysis, and Partial least squares), and three feature selection methods (RandomForest, Extreme Gradient Boosting, and ReliefF) are taken into consideration. Main reason for pre-processing the data with the use of stated algorithms is to reduce the number of dimensions in initial space. From all algorithms we extracted only the most prominent features.

Principal Component Analysis

Principal component analysis is an unsupervised linear transformation technique used in machine learning applications and multivariate statistics. It is widely used across different fields, most prominently for feature extraction, dimensionality reduction and visualization. It helps identify patterns in data based on the correlation between features. PCA aims to find the directions of maximum variance in high-dimensional data and projects it onto a new subspace with equal or fewer dimensions than the original one [37]. This is achieved based on a covariance matrix formulation of centered and normalized data. Axes of original coordinate system are transformed so that newly created axes describe maximal covariance of the data. Each axis is described by an eigenvector, whose variance corresponds to its eigenvalue. Eigenvectors and eigenvalues of covariance matrix Σ can be calculated using Singular value decomposition (SVD), which can be written as:

$$
\Sigma = \frac{1}{n-1} X^T X = P \Lambda P^T = \sum_{j=1}^{m} \lambda_j p_j p_j^{\mathrm{T}}
\tag{10}
$$

In Equation (10) data instances are included in matrix $X \in \mathbb{R}^{n \times m}$, where n represents number of data instances and m number of attributes in each instance. Matrix $P \in \mathbb{R}^{m \times m}$ represents m orthogonal basis vectors pj, j = 1, ... , m and Λ diagonal matrics composed of eigenvalues λ_j, j = 1, ... , m. To each eigenvalue λ_j belongs particular basis vector p_j. Matrix of eigenvectors P is organized so that column vectors are sorted by decreasing magnitude of eigenvalues $\lambda_1 < \lambda_2 < \cdots < \lambda_m$. In other words, eigenvectors are sorted by decreasing amount of information they provide.

Since HIS data contains many correlated features (i.e., spectral bands), the data set can be fully described by using only a subset of eigenvectors of covariance matrix Σ. The general assumption is that part of the information can be explained with k eigenvectors, which we call principal components. Various criteria can be used to determine the number of principal components, e.g., percentage of explained variance in the data. Mathematically we can write:

$$\frac{\sum_{j=1}^{s} \lambda_j}{\sum_{j=1}^{m} \lambda_j} \geq \epsilon \tag{11}$$

where ϵ represents predefined threshold. Usually it is set to 0.95 to keep 95% of initial variance of the data. Another option is to directly choose desired number of principal component. The covariance matrix can then be approximated by neglecting eigenvectors with small corresponding eigenvalues. In other words, we neglect p_j, where $j = s+1, \ldots, m$. Approximated covariance matrix can then be calculated as:

$$\Sigma_s = \sum_{j=1}^{s} \lambda_j p_j p_j^{\mathrm{T}} \tag{12}$$

where vectors p_j, $j = 1, \ldots, s$ define principal directions in which the data extends and is weighted by corresponding eigenvalues.

Linear Discriminant Analysis

Linear discriminant analysis is a robust classification method, but can also be used for dimension reduction and data visualization. Unlike PCA, which tries to maximize variance, it is a supervised machine learning method that computes decision boundaries which enhance the separation between multiple classes used.

It tries to separate different classes by maximizing distances between projected means and minimizing projected variance. Both optimization problems are incorporated in one single criterion function which can be, for binary classification, written as:

$$\max_{w} \; J(w) = \frac{(m_1 - m_2)^2}{s_1^2 + s_2^2} \tag{13}$$

where $(m_1 - m_2)^2$ represents the difference in means between the two classes and $s_1^2 + s_2^2$ the total scatter (standard deviations) of the two classes. The goal of LDA is to find the vector w that maximizes criterion function $J(w)$.

In LDA it is assumed that all K classes have equal covariance. Following this assumption we can obtain the following discriminant function for k-th class:

$$\delta_k(x) = x^T \Sigma^{-1} \mu_k - \frac{1}{2} \mu_k^T \Sigma^{-1} \mu_k + \log \pi_k \tag{14}$$

Which predicts the class with the highest value of $\delta_k(x)$ given an input $x \in \mathbb{R}^{p \times 1}$. In Equation (14) $\Sigma \in \mathbb{R}^{p \times p}$ represents common covariance matrix, $\mu_k \in \mathbb{R}^{p \times 1}$ the mean of inputs for class k and π_k prior distribution of class k. Symbol p represents number of attributes in each data instance.

Features are transformed so that classes are as separate as possible from each other and that features within a class are as close as possible. Transformed dimensions are ranked based on the separation ability. Maximal number of components must be at least one fewer than the number of classes used for classification. Therefore, since we performed binary classification in this study, only the first and only linear discriminant was used [38].

Partial Least Squares

Partial least squares is a technique that transforms the initial dataset to a reduced set of uncorrelated features using a technique similar to principal component analysis. It extracts features that describe maximum correlation with target variables; i.e., they provide the greatest predictive ability. This method is especially useful when features in initial dataset are highly collinear [39].

The underlying core equations of PLS could be written as:

$$X = TP^T + E \tag{15}$$

$$Y = UQ^T + F \tag{16}$$

where $X \in \mathbb{R}^{n \times m}$ is the matrix of independent variables (with hyperspectral data spectral bands) and $Y \in \mathbb{R}^{n \times p}$ is the matrix of dependant variables (these can be measured variables or dummy coded nominal variables). Symbol n represents number of data instances, m number of attributes in each independent variable and p number of attributes in each dependent variable. Matrices T, $U \in \mathbb{R}^{n \times l}$ respectively represent projections (scores) of X and Y. Matrices $P \in \mathbb{R}^{m \times l}$ and $Q \in \mathbb{R}^{p \times l}$ respectively represent orthogonal loadings matrices of X and Y. Symbol l represents a user-defined number of latent factors used in for regression. Model is optimized in such a way that the first score in X has maximum covariance with the first score in Y. Therefore, we can predict the first score in Y from the first score in X.

Partial least squares has some advantages over basic ordinary least square (OLS) solution. It is able to dispose correlated variables and model their shared and underlying information. In contrast to many machine learning methods, it can directly model multiple dependent variables at the same time. Several variants of PLS exist; we used Partial least squares discriminant analysis, which is an extension of PLS regression, the foundation for other variants.

ReliefF

ReliefF is an extension of the basic Relief algorithm, and is a generally well-performing attribute selector. It can provide a combined view of relevance and conditional dependencies between attributes. The algorithm prescribes a separate weight (w_j) to each attribute, where higher values correspond to more important attributes. The basic idea of the algorithm is that is penalizes attributes which provide different result of the same class in comparison with its nearest neighbours [40].

At the beginning, ReliefF sets all attribute weights w_j^i to zero, these are then iteratively adapted. Then, it selects a random observation x_r and k-nearest observations for each class. All the weights are updated for each nearest neighbours x_q by equations:

$$w_j^i = w_j^{i-1} - \frac{\Delta_j\left(x_r, x_q\right)}{m} \cdot d_{rq} \tag{17}$$

$$w_j^i = w_j^{i-1} + \frac{p_{yq}}{1 - p_{yr}} \cdot \frac{\Delta_j\left(x_r, x_q\right)}{m} \cdot d_{rq} \tag{18}$$

where w_j^i represents the weight of the j-th attribute at iteration i, m is the total number of iterations, p_{yq} and p_{yr} are prior probabilities of classes where x_q and x_r respectively belong, and d_{rq} is the distance function, which is subject to scaling. Symbol $\Delta_j\left(x_r, x_q\right)$ represents the difference between prediction values of observations x_r and x_q for j-th attribute F_j. For continuous attributes it is calculated as:

$$\Delta_j(x_r, x_q) = \frac{\left|x_{rj} - x_{qj}\right|}{\max(F_j) - \min(F_j)} \tag{19}$$

RandomForest

Random forest is an ensemble technique that combines multiple de-correlated decision trees. Decision trees are fitted on various randomly chosen subsets of a given dataset. Overall performance of the model is increased by aggregating predictions from all trees and performing a majority vote for each class in classification problems.

In the training phase of Random forest a technique called bootstrap aggregation or bagging is used.

Given training data instances in a matrix $X \in \mathbb{R}^{n \times m}$ and corresponding labels $y \in \mathbb{R}^{n \times 1}$ (where n represents number of data instances and m number of attributes in each instance), bagging repeatedly selects random data instances with replacement and fits B decision trees f to these instances. Unseen data instances x' are predicted by averaging all predictions made by individual decision trees:

$$\hat{y} = \frac{1}{B} \sum_{b=1}^{B} f_b(x') \tag{20}$$

where $\hat{y}$ represents approximated predicted output. Bagging decreases variance without increase of bias. This leads to more accurate performance even if each individual decision tree is highly sensitive to noise. Furthermore, Random forest also includes feature bagging, i.e. selection of a random subset of the attributes in the training set. A small number of attributes may have a very strong prediction power for the response variable. Consequently, these attributes would be selected many times causing decision trees to become correlated. We used the Gini index as split criterion and for assessing variable importance. For each Random forest 100 trees were constructed [41].

Extreme Gradient Boosting

Gradient boosting is one of the most powerful and flexible machine learning methods, which can be applied to various machine learning problems. It refers to a class of ensemble methods used for predictive modelling problems. Similarly to Random forests, it is constructed from decision tree models (weak learners). Unlike Random forest, weak learners are added one at a time to correct errors produced by prior decision trees. This type of error correction is called boosting, where models are iteratively trained with gradient descent optimization of any differentiable loss function. For instance, a squared error may be used for regression problems and logarithmic loss for classification problems. New decision trees are trained on error residuals produced by initial learner. Intuitively, newly trained models are influenced more by misclassified observations or by areas where they are performing poorly. The contribution from all decision trees are aggregated to make the final prediction [42].

Simplified optimization could be mathematically written as follows. First model is initialized with a constant value with minimization of loss function $L(x_i, y_i, \theta)$:

$$\hat{f}_{(0)}(X) = \operatorname*{argmin}_{\theta} \sum_{i=1}^{n} L(x_i, y_i, \theta) \tag{21}$$

where $X \in \mathbb{R}^{n \times m}$ is a matrix of input data instances with corresponding labels $y \in \mathbb{R}^{n \times 1}$. Symbols n and m represent number of data instances and number of attributes in each instance, respectively. Based on the weak learner from the previous iteration, gradients and hessians are calculated and then a new weak learner is fitted using optimization problems. At the end of iteration, the model transfer function is updated as:

$$\hat{f}_{(m)}(X) = \hat{f}_{(m-1)}(X) + \hat{f}_{(m)}(X) \tag{22}$$

where $m = (1, 2, \ldots, M)$, where M is the total number of weak learners. Unseen data instances x' are then predicted by summation of all predictions made by individual decision trees:

$$\hat{y} = \hat{f}(x') = \hat{f}_{(M)}(x') = \sum_{m=0}^{M} \hat{f}_{(m)}(x') \tag{23}$$

2.4. Support Vector Machines

Extracted or selected features from dimensionality reduction were pipelined to support vector machine classificator. Classification models were therefore built on reduced data sets, consisting of only the most prominent feature. In this study performance is tested on radial-basis kernel function for data transformation [43]. Hyperparameter tuning (gamma and C) was performed using a grid search, whereupon combinations yielding the best accuracy were retained.

Trained SVM classifier was evaluated using mean accuracy. It was iteratively trained and tested 5 times for each DR algorithm. Accuracies from all iterations were then averaged. For this reason an objective criteria is devised for comparison between all DR methods:

$$\bar{c}^{(m)} = \frac{1}{F}\frac{1}{P} \sum_{f=1}^{F} \sum_{p=1}^{P} \Gamma\left(y_{\mathrm{p}}, \hat{y}_{\mathrm{p}}^{(m)}\right)$$

$$\Gamma\left(y_{\mathrm{p}}, \hat{y}_{\mathrm{p}}\right) = \begin{cases} 1; & \text{if } y_{\mathrm{p}} = \hat{y}_{\mathrm{p}}^{(m)} \\ 0; & \text{else} \end{cases} \tag{24}$$

$$y_{\mathrm{p}}, \hat{y}_{\mathrm{p}}^{(m)} \in \{\text{Healthy}, \text{Inoculated}\}$$

In Equation (24) mean accuracy is labelled as $\bar{c}^{(m)}$. Superscript represents m-th DR method belonging to a set $M_{\mathrm{m}} \in \{\text{PCA, LDA, PLS, RF, RFF, XGB}\}$. Mean accuracy is calculated from comparison between predicted $\hat{y}_{\mathrm{p}}$ and known y_{p} labels of potato tuber, which can be either healthy or inoculated. It is calculated on test feature vectors for P predictions and F folds. In our case $F = 5$ and $P = 12$. Precision, recall and F1-score were calculated using equations in [44]. All analyses were performed in Python, using libraries scikit-learn [45] and XGBoost [46].

3. Results and Discussion

The first extracted features, or limited set of selected features, proved to be sufficient for accurate detection of infested potato tubers. Spectral differences between inoculated and healthy tubers were more pronounced in images of their outside (Figure 4). Spectral signatures of the outside of tubers show a larger variability in infested tubers, than in healthy ones. These differences are more pronounces in the SWIR region, where inoculated tubers uniformly exhibited higher reflectance than healthy tubers. In contrast, data from inside tubers shows comparatively little variability, regardless of inoculation status. The high variability in outside images could be a consequence of tuber surface characteristics. Healthy tubers are comparatively smooth, while galls cover the surface of inoculated tubers. This leads to a more varied viewing geometry, which was accounted for in pre-processing of the images.

We used the first two principal components for data visualization in a generated feature space. Feature vectors were separated into 5 cross-validation folds, and PCA performed on each fold of the training data, and applied to both train and test sets. The generated features from all folds were pooled to generate scatter plots of the first two PCA components (Figure 5). The first two components explain more than 80% of the variance in the data (93% for outside, and 84% for inside tubers). These scatter plots show a better distinction between healthy and inoculated tubers for data from the outside of tubers. Yet any linear separability does not appear to be present, at least not in the first two PCA components.

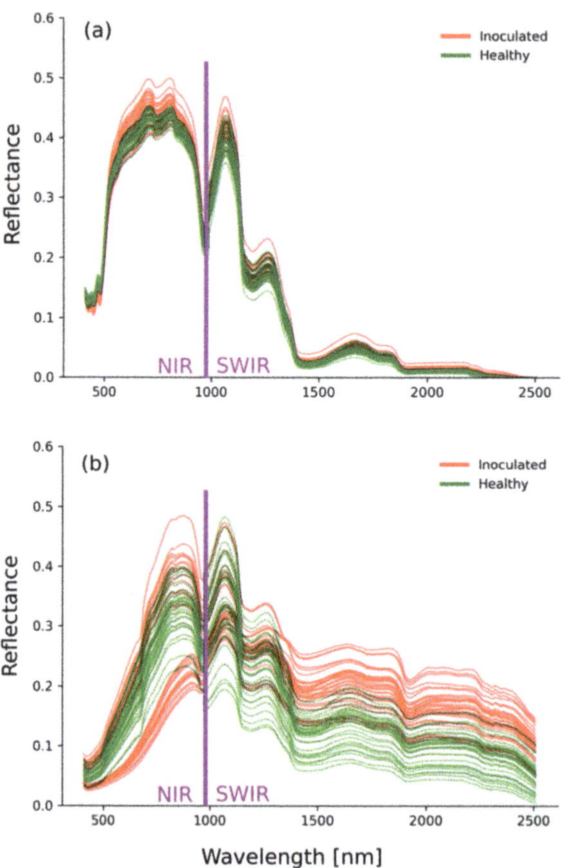

Figure 4. Reflectance feature vectors for (**a**) inside and (**b**) outside of potato tuber. Green colour refers to healthy and red to inoculated specimens. Separation between VNIR and SWIR cameras is marked with violet colour.

Females of root-knot nematodes reside within a few millimeters below tuber skin, in the vascular ring, where they form comparatively large egg-sacs [47]. When these grow enough, they form galls on tuber surface. Even though some evidence exists that RKNs change the chemistry of the entire tuber, i.e., also the starchy insides [2], we expected classification success to be higher in outside data. This hypothesis was confirmed for all DR methods (Figure 6). All methods achieved a mean accuracy of at least 0.6 (Tables 1 and 2). LDA showed the most consistent results, as it achieved a mean accuracy of 1.0 in both inside and outside tuber data. Overall RF came second, with XGB yielding very similar accuracies. PLS came next, followed by PCA, and lastly ReliefF achieved the worst results. Data was also analysed without any DR. Even though this method was capable of achieving good accuracy, a mean of 0.9 from both data sets, it is computationally much more burdensome. With SVMs solving the quadratic problem involves inverting the kernel matrix, with a complexity of up to n^3, where n is the feature space [43]. RandomForest, XGB and ReliefF suffer from the same problem, as they only select features, i.e. they generate a subset of the original feature space. Furthermore, even though all three feature selection methods are robust, they can suffer from overfitting and should be optimized accordingly [48]. In this regard feature extraction methods are beneficial, since they generate a new feature space, with lower dimensionality.

Increasing the number of features has an expected effect, of increasing classification accuracies (Figure 7). The most profound effect is observable in PCA and RFF, while RF shows the smallest change. Unlike the other five methods, only one feature gets extracted by LDA in binary classification. Interestingly, PCA on inside data decreases accuracy with the first three components. Accuracy then increases with more features, but still remains bellow PLS and LDA accuracies. With more features extracted, only PLS and XGB achieve a 100% accuracy in both inside and outside tuber data. Our results indicate that even with an extreme reduction, to just one feature, identification accuracies are still acceptable to excellent (mean accuracy between 0.8 and 1.0).

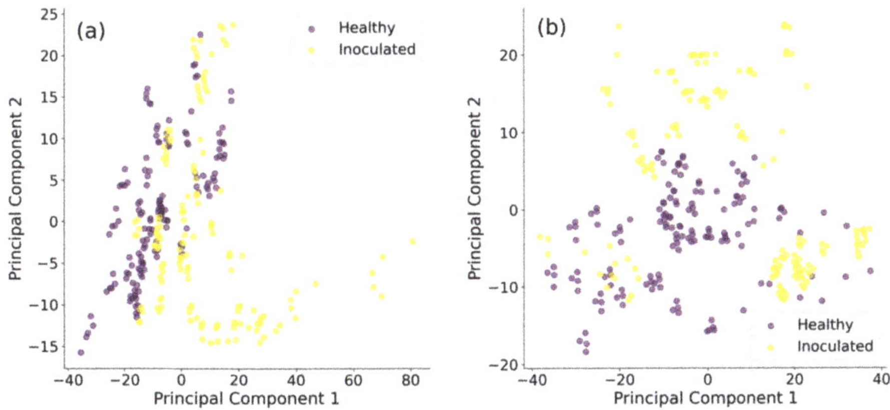

Figure 5. Scatter plot of first two principal components for (**a**) inside and (**b**) outside of potato tubers. Yellow colour refers to healthy and purple colour to inoculated specimens.

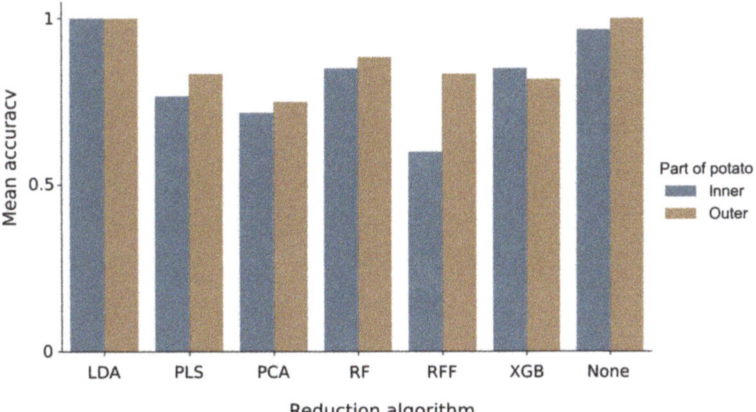

Figure 6. Mean accuracy for several dimensional reduction algorithms. Accuracy of classifier trained on data from: outer side of potato (orange columns), inner side of potato (blue columns).

Table 1. Comparison of classification results with only one feature for selected DR methods for outside tuber data.

	Outer Side of Potato Tuber				
Method	**Class**	**Precision**	**Recall**	**F1-Score**	**Accuracy**
LDA	Healthy	1.00	1.00	1.00	1.00
	Inoculated	1.00	1.00	1.00	
PLS	Healthy	0.88	0.77	0.82	0.83
	Inoculated	0.79	0.90	0.84	
PCA	Healthy	0.86	0.60	0.71	0.75
	Inoculated	0.69	0.90	0.78	
RF	Healthy	0.93	0.83	0.88	0.88
	Inoculated	0.85	0.93	0.89	
RFF	Healthy	0.88	0.77	0.82	0.83
	Inoculated	0.79	0.90	0.84	
XGB	Healthy	0.81	0.83	0.82	0.82
	Inoculated	0.83	0.80	0.81	
None	Healthy	1.00	1.00	1.00	1.00
	Inoculated	1.00	1.00	1.00	

Table 2. Comparison of classification results with only one feature for selected DR methods for inside tuber data.

	Outer Side of Potato Tuber				
Method	**Class**	**Precision**	**Recall**	**F1-Score**	**Accuracy**
LDA	Healthy	1.00	1.00	1.00	1.00
	Inoculated	1.00	1.00	1.00	
PLS	Healthy	0.75	0.80	0.77	0.77
	Inoculated	0.79	0.73	0.76	
PCA	Healthy	0.70	0.77	0.73	0.72
	Inoculated	0.74	0.67	0.70	
RF	Healthy	0.89	0.80	0.84	0.85
	Inoculated	0.82	0.90	0.86	
RFF	Healthy	0.64	0.47	0.54	0.60
	Inoculated	0.58	0.73	0.65	
XGB	Healthy	0.89	0.80	0.84	0.85
	Inoculated	0.82	0.90	0.86	
None	Healthy	0.94	1.00	0.97	0.97
	Inoculated	1.00	0.93	0.97	

Compared to no DR, dimensionality reduction using PLS and PCA on outside data reduced detection accuracy (Table 2). On the other hand, in a data set with less pronounced patterns, such as tuber insides, classification accuracy was increased by using PLS. One of the purposes of dimensionality reduction is to generate or retain only those features, which are informative for the problem under study. This way the signal-to-noise ratio can be improved, leading to better model performance.

Dimensionality reduction algorithms use different metrics to asses feature importance. For example, principal components analysis uses a correlation matrix between generated components and original features. Correlations above or below 0.7 or −0.7, respectively, are considered as relevant. In PLS correlations can also be considered, but a more accurate assessment of feature importance is possible using variable importance in projection analysis (VIP) [49]. VIP coefficients reflect the relative importance of each variable for each variate in the prediction model. Variable importance in LDA was calculated as LDA scalings, i.e., the eigenvectors of the components. Important to note here is that LDA is a discriminant analysis method and as such it maximizes the between-group variance. The eigenvectors of the comparison matrix of between and within group's sum of squares and cross-products describe how much the original variables contribute to the new component(s). Gini importance, used with RandomForests and Extreme gradient boosting, provides a relative ranking of the original features, and is a by-product of the training of the classifier [50]. Lastly, ReliefF assigns feature relevance depending on the difference between this feature and two neighbours of the same and opposite classes [51]. Each of these methods provides their own metric of variable importance. In order to directly compare all methods, we normalized their values to a range of 0 to 1.

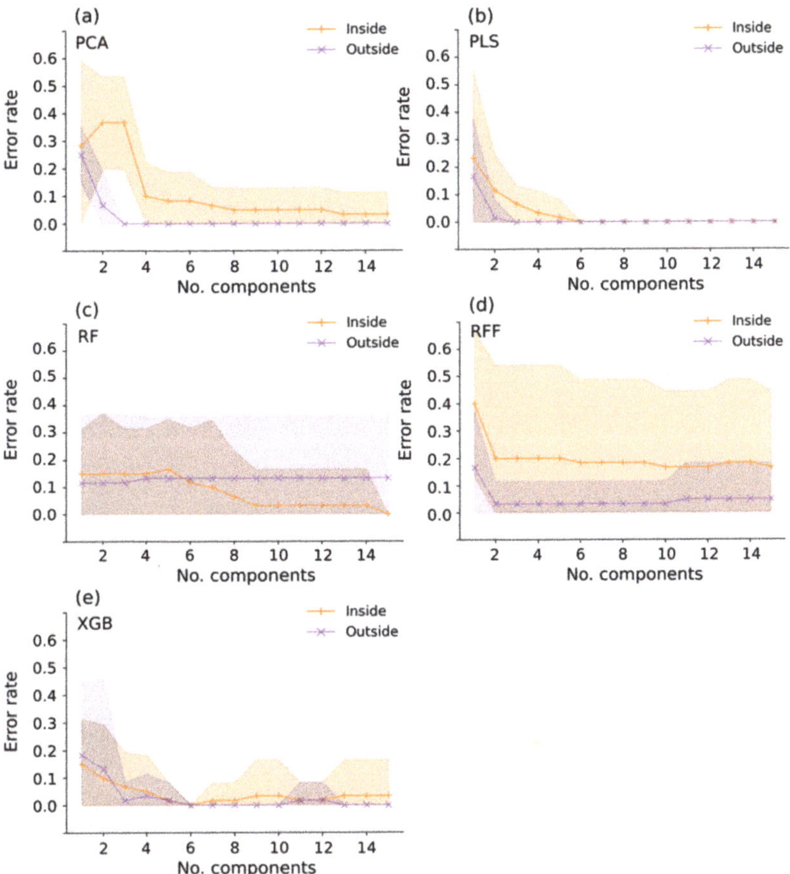

Figure 7. Influence of increasing number of features on classification error rates for both inside and outside tuber data. Since LDA generates only one feature in binary classification it wasn't included in this figure. (**a**) Principal component analysis, (**b**) Partial least squares, (**c**) Random forest, (**d**) ReliefF, and (**e**) Extreme gradient boosting.

RandomForest and XGB identified several relevant wavelengths, distributed comparatively evenly throughout the spectrum, in both data sets. Similarly, LDA also identified a large number of relevant wavelengths, but unlike RF, these were not evenly distributed in the outside data set. In this set the importance of variable shifted towards the SWIR region. Only ReliefF showed a different pattern, compared to the other five methods. Interestingly, while PLS, LDA, RF, and XGB show a similar grouping of relevant wavelengths in the same SWIR regions (1500–1600 nm, 1850–2000 nm, and 2300–2450 nm), PCA found relevant regions between these groups (1600–1800 nm, and 2100–2200 nm). On the other hand, ReliefF found relevant wavelengths in the range 1000–1400 nm, i.e., in shorter wavelengths than the other methods. With inside data, only PCA found relevant regions in the SWIR part of the spectrum, above 1500 nm (1600–1850 nm), while the remaining methods found a strong grouping of relevant wavelengths closer to the VNIR region, between 1000 and 1200 nm (Figure 8). This spectral region is linked to various hydrocarbons, both aliphatic and aromatic [52]. The region identified by PCA is also linked to different hydrocarbons (aliphatic, aromatic, and methyl), but also alcohols, amines and proteins. In the outside data set, regions linked to water, polysaccharides, aromatic amines (1850–2000 nm), and lipids and glucose (2300–2450 nm) were identified as relevant. In order to fully test the accuracy of the different variable importance measures, employed by the dimensionality reduction methods, more detailed chemometric analyses of potato tubers are needed.

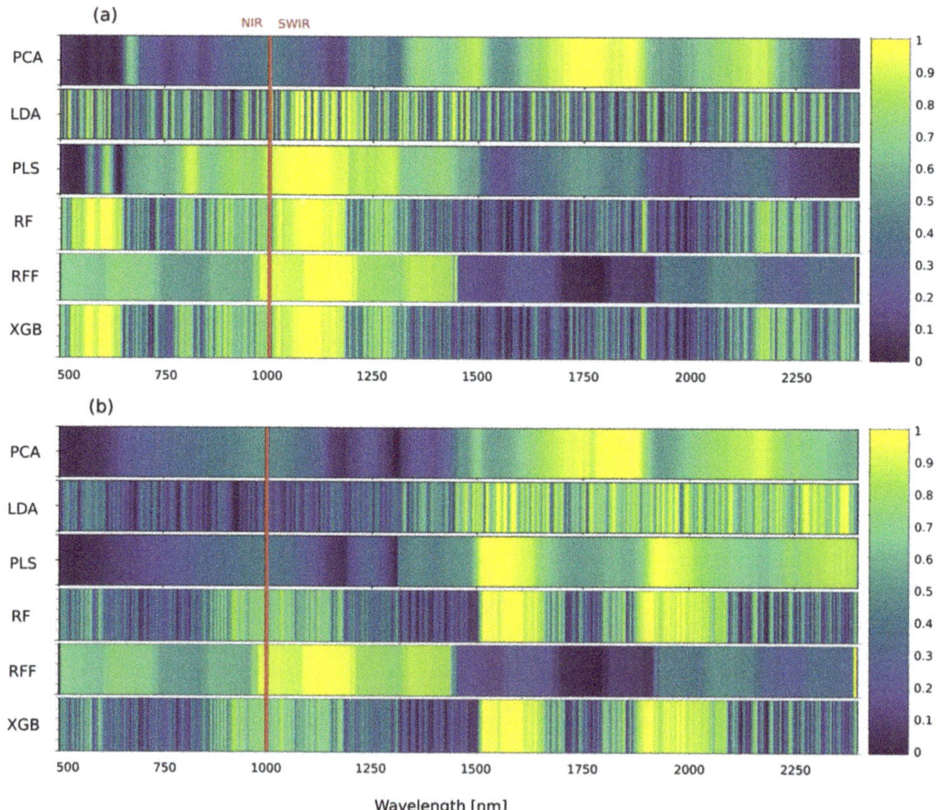

Figure 8. Importance of individual wavelengths, as determined by the different DR methods. (**a**) inside, and (**b**) outside of potato tuber. Brighter colours represent higher importance. Separation between VNIR and SWIR cameras is marked with red color.

Differences on the inside of tubers are most likely of a chemical nature, while the outside is influenced by both differences in chemistry as well as surface texture. While surface texture effects can be reduced using normalization and Savitzky–Golay derivatives, they can still affect the classifications. The infested tubers used in this study had visible symptoms (bulges on the outside, approximately 3–5 mm in diameter), covering at most 50% of the surface. In our case, they added another distinguishing dimension, leading to better classification accuracy with data from tuber surfaces. Nevertheless, the feasibility of hyperspectral imaging for detection of latent (i.e., without visible symptoms) infestations with nematodes in potatoes has been shown by Žibrat et al. [2]. In that case surface texture didn't influence spectral signatures; therefore the observed differences are exclusively due to differences in chemistry between healthy and infested tubers. From a practical application standpoint, even if tubers in production spot checks would have to be halved and imaged, the throughput of such a method would be much higher than with molecular analyses (e.g., real–time PCR). So the benefit of using hyperspectral imaging for detecting infestations, even with processed tubers, is evident.

Molecular spectra are the result of motions of atomic nuclei. They can rotate, vibrate, wag and move together or apart along a straight line (this type of movement is called stretching). Vibrations follow a functional description, i.e. the type of vibration determines the frequency at which it absorbs energy. The amplitude of absorption is determined by absorptivity and the number of molecules in the beam path of a particular sensor. Changes in spectral responses follow Beer's Law, which states that the absorbance is equal to product of absorptivity of a molecule and the concentration of molecules. The above means that the light absorbed by plant tissue depends on the chemical composition of that tissue, the concentration of individual molecule species, and their interactions [52]. We've identified several groups of molecules, which account for the differences between healthy and infested potato tubers (such as aliphatic and aromatic hydrocarbons). The two spectral regions, VNIR and SWIR, are generally linked to different characteristics of plants, but there is some overlap. In the VNIR region we mostly get information about pigments and structure, e.g., morphological structure of plant leaves. In addition, in wavelengths above 700 nm there is also information related to hydrocarbons (mostly aliphatic) and alcohols, these are generally the third and fourth overtones of the C-H stretch. The SWIR region carries information on plant biophysical properties (e.g., hydrocarbons and proteins). Changes in chemistry can therefore be measured in both. Since we do not know how exactly nematode infestations change the chemistry of potato tubers, we decided to use both systems. The wider spectral range of the combined system enables us to better search for the effects of nematode infestations. With the currently available information we can only speculate which exact compounds account for the observed differences between inoculated and healthy potato tubers.

We identified several relevant wavelengths from each DR method. The latter use different methods for deciding, which wavelength is relevant, so we would recommend to look for overlaps between the methods. Those wavelengths which were identified as relevant by different methods are good candidates for a multispectral sensor. Different bandpass filters are commercially available, with different spectral ranges and bandwidths. So it might not be necessarily needed to develop new filters. Using these filters we would effectively get a multispectral data set. New classification models would then have to be developed using this reduced data. Using this procedure we could assess the importance of each new wide band and determine which ones significantly increase classification success. Potentially this would mean that by reducing a hyperspectral dataset to a few-band (~5 bands) multispectral one we would still get acceptable classification accuracy. Production of such a dedicated multispectral system would be much cheaper, than a hyperspectral one, with similar classification success.

4. Conclusions

In this study we have shown that discrimination between healthy and inoculated potatoes with quarantine pests is possible based on hyperspectral image analysis. We successfully reduced the initial hyper-dimensional feature space to one–dimensional (or few–dimensional) with the use of dimensionality reduction algorithms, and still obtained high classification accuracies. These results suggest that a comparatively low–cost imaging system utilizing band–pass filters could be designed for the specific purpose of identifying tubers infested by root–knot nematodes. But the study was performed on a very small data set and further analyses are needed to fully test this idea.

Author Contributions: Conceptualization, J.L., U.Ž., M.K.; methodology, J.L. and U.Ž.; formal analysis, J.L. and U.Ž.; writing—original draft preparation, J.L. and U.Ž.; writing—review and editing, U.Ž. and M.K.; project administration, U.Ž.; funding acquisition, U.Ž and M.K. All authors have read and agreed to the published version of the manuscript.

Funding: This research was supported by funds from the Slovenian Research Agency (ARRS) (MR 38128, MR 54720, P4-0072). Some of the research was conducted using equipment financed by EU-FP7 project CropSustaIn, grant agreement FP7-REGPOT-CT2012-316205.

Data Availability Statement: The data presented in this study are available on request from the corresponding author. The data are not publicly available due to their use in ongoing research projects.

Acknowledgments: We would like to thank Nik Susič, Saša Širca, and Barbara Gerič Stare for providing the plant material and performing nematode species identification.

Conflicts of Interest: The authors declare no conflict of interest.

References

1. Moghadam, P.; Ward, D.; Goan, E.; Jayawardena, S.; Sikka, P.; Hernandez, E. Plant Disease Detection Using Hyperspectral Imaging. In Proceedings of the 2017 International Conference on Digital Image Computing: Techniques and Applications, Sydney, Australia, 29 November–1 December 2017; pp. 1–8.
2. Žibrat, U.; Gerič Stare, B.; Knapič, M.; Susič, N.; Lapajne, J.; Širca, S. Detection of root-knot nematode *Meloidogyne luci* infestation of potato tubers using hyperspectral remote sensing and real-time PCR molecular methods. *Remote Sens.* **2021**, *13*, 1996. [CrossRef]
3. Carneiro, R.M.D.G.; Correa, V.R.; Almeida, M.R.A.; Gomes, A.C.M.M.; Deimi, A.M.; Castagnone-Sereno, P.; Karssen, G. *Meloidogyne luci* n. sp. (Nematoda: Meloidogynidae), a root-knot nematode parasitising different crops in Brazil, Chile and Iran. *Nematology* **2014**, *16*, 289–301. [CrossRef]
4. Gerič Stare, B.; Strajnar, P.; Susič, N.; Urek, G.; Širca, S. Reported populations of *Meloidogyne ethiopica* in Europe identified as *Meloidogyne luci*. *Plant Dis.* **2017**, *101*, 1627–1632. [CrossRef]
5. Strajnar, P.; Širca, S.; Knapič, M.; Urek, G. Effect of Slovenian climatic conditions on the development and survival of the root-knot nematode *Meloidogyne ethiopica*. *Eur. J. Plant Pathol.* **2011**, *129*, 81–88. [CrossRef]
6. EPPO Alert List: Addition of *Meloidogyne luci* together with *M. ethiopica*; EPPO Reporting Service No. 11-2017, Num. Article: 2017/218. Available online: https://gd.eppo.int/reporting/article-6186 (accessed on 10 September 2021).
7. Cunha, G.T.; Visôtto, L.I.; Lopes, E.A.; Oliveira, C.M.G.; God, P.I.V.G. Diagnostic methods for identification of root-knot nematodes species from Brazil. *Crop. Prot.* **2018**, *48*, 2. [CrossRef]
8. Žibrat, U.; Širca, S.; Susič, N.; Knapič, M.; Gerič Stare, B.; Urek, G. Noninvasive Detection of Plant Parasitic Nematodes Using Hyperspectral and Other Remote Sensing Systems. In *Hyperspectral Remote Sensing*; Pandey, P.C., Srivastava, P.K., Balzter, H., Bhattacharya, B., Petropoulos, G.P., Eds.; Elsevier: Amsterdam, The Netherlands, 2020; pp. 357–375.
9. Zhang, W.; Qibing, Z.; Min, H.; Ya, G.; Jianwei, Q. Detection and Classification of Potato Defects Using Multispectral Imaging System Based on Single Shot Method. *Food Anal. Methods* **2019**, *12*, 2920–2929. [CrossRef]
10. Thomas, S.; Kuska, M.T.; Bohnenkamp, D.; Brugger, A.; Alisaac, E.; Wahabzada, M.; Behmann, J.; Mahlein, A.-K. Benefits of Hyperspectral Imaging for Plant Disease Detection and Plant Protection: A Technical Perspective. *J. Plant Dis Prot.* **2018**, *125*, 5–20. [CrossRef]
11. Moghimi, A.; Yang, C.; Marchetto, P.M. Ensemble Feature Selection for Plant Phenotyping: A Journey from Hyperspectral to Multispectral Imaging. *IEEE Access.* **2018**, *6*, 56870–56884. [CrossRef]
12. Federico, M.; Riccardo, S.; Salvatore, D.; Stefano, P.; Giuseppe, S. Advanced methods of plant disease detection: A review. *Agron. Sustain. Dev.* **2015**, *35*, 1–25. [CrossRef]
13. Ji, Y.; Sun, L.; Li, Y.; Ye, D. Detection of bruised potatoes using hyperspectral imaging technique based on discrete wavelet transform. *Infrared Phys. Technol.* **2019**, *103*, 103054. [CrossRef]
14. Ji, Y.; Sun, L.; Li, Y.; Li, J.; Liu, S.; Xie, X.X.Y. Non-destructive classification of defective potatoes based on hyperspectral imaging and support vector machine. *Infrared Phys. Technol.* **2019**, *99*, 71–79. [CrossRef]
15. Ayvaz, H.; Santos, A.M.; Moyseenko, J.; Kleinhenz, M.; Rodriguez-Saona, L.E. Application of a portable infrared instrument for simultaneous analysis of sugars, asparagine and glutamine levels in raw potato tubers. *Plant Foods Hum. Nutr.* **2015**, *70*, 215–220. [CrossRef]
16. Su, W.H.; Sun, D.W. Chemical imaging for measuring the time series variations of tuber dry matter and starch concentration. *Comput. Electron. Agric.* **2017**, *140*, 361–373. [CrossRef]
17. Dacal-Nieto, A.; Formella, A.; Carrión, P.; Vazquez-Fernandez, E.; Fernández-Delgado, M. Non–Destructive Detection of Hollow Heart in Potatoes Using Hyperspectral Imaging. In *Computer Analysis of Images and Patterns, Proceedings of the 14th International Conference, CCAIP 2011, Seville, Spain, 29–31 August 2011*; Real, P., Diaz-Pernil, D., Molina-Abril, H., Berciano, A., Kropatsch, W., Eds.; Springer: Berlin/Heidelberg, Germany, 2011; pp. 180–187. [CrossRef]
18. Huang, T.; Li, X.Y.; Xu, M.L.; Jin, R.; Ku, J.; Xu, S.M. Non-destructive detection research for hollow heart of potato based on semi-transmission hyperspectral imaging and SVM. *Spectrosc. Spect. Anal.* **2015**, *35*, 198–202. [CrossRef]

19. Zhou, Z.; Zeng, S.; Li, X.; Zheng, J. Nondestructive detection of blackheart in potato by visible/near infrared transmittance spectroscopy. *J. Spectrosc.* **2015**, *2015*, 786709. [CrossRef]
20. Garhwal, A.S.; Pullanagari, R.R.; Li, M.; Reis, M.M.; Archer, R. Hyperspectral imaging for identification of Zebra Chip disease in potatoes. *Biosyst. Eng.* **2020**, *197*, 306–317. [CrossRef]
21. Riza, D.F.A.; Suzuki, T.; Ogawa, Y.; Kondo, N. Diffuse reflectance characteristic of potato surface for external defects discrimination. *Postharvest Biol. Technol.* **2017**, *133*, 12–19. [CrossRef]
22. Zhao, Z.; Prager, S.M.; Cruzado, R.K.; Liang, X.; Cooper, W.R.; Hu, G. Characterizing Zebra Chip symptom severity and identifying spectral signatures associated with 'Candidatus Liberibacter solanacearum'-infected potato tubers. *Am. J. Potato Res.* **2018**, *95*, 584–596. [CrossRef]
23. Sawant, S.S.; Manoharan, P. A Survey of Band Selection Techniques for Hyperspectral Image Classification. *J. Spectr. Imaging* **2020**, *9*, a5. [CrossRef]
24. Ballabio, D.; Consonni, V. Classification tools in chemistry. Part 1: Linear models. *Anal. Methods* **2013**, *5*, 3790–3798. [CrossRef]
25. Chong, I.G.; Jun, C.H. Performance of some variable selection methods when multicollinearity is present. *Chemom. Intell. Lab. Syst.* **2005**, *78*, 103–112. [CrossRef]
26. AlSuwaidi, A.; Grieve, B.; Yin, H. Feature-Ensemble-Based Novelty Detection for Analyzing Plant Hyperspectral Datasets. *IEEE J. Sel. Top. Appl. Earth Obs. Remote Sens.* **2018**, *11*, 1041–1055. [CrossRef]
27. Jin, X.; Jie, L.; Wang, S.; Qi, H.J.; Li, S.W. Classifying Wheat Hyperspectral Pixels of Healthy Heads and Fusarium Head Blight Disease Using a Deep Neural Network in the Wild Field. *Remote Sens.* **2018**, *10*, 395. [CrossRef]
28. Liu, B.; Yu, X.; Zhang, P.; Yu, A.; Fu, Q.; Wei, X. Supervised Deep Feature Extraction for Hyperspectral Image Classification. *IEEE Trans. Geosci. Remote Sens.* **2017**, *56*, 1909–1921. [CrossRef]
29. Chen, Y.; Zhu, L.; Ghamisi, P.; Jia, X.; Li, G.; Tang, L. Hyperspectral Images Classification with Gabor Filtering and Convolutional Neural Network. *IEEE Geosci. Remote Sens. Lett.* **2017**, *14*, 2355–2359. [CrossRef]
30. Golhani, K.; Balasundram, S.K.; Vadamalai, G.; Pradhan, B. A Review of Neural Networks in Plant Disease Detection Using Hyperspectral Data. *Inf. Process. Agric.* **2018**, *5*, 354–371. [CrossRef]
31. Susič, N.; Žibrat, U.; Sinkovič, L.; Vončina, A.; Razinger, J.; Knapič, M.; Sedlar, A.; Širca, S.; Gerič, S.B. From genome to field-Observation of the multimodal nematicidal and plant growth-promoting effects of *Bacillus firmus* I-1582 on tomatoes using hyperspectral remote sensing. *Plants* **2020**, *9*, 592. [CrossRef]
32. Hussey, R.S.; Barker, K.R. Comparison of methods of collecting inocula for *Meloidogyne* spp., including a new technique. *Plant Dis. Rep.* **1973**, *57*, 1025–1028.
33. Piqueras, S.; Burger, J.; Tauler, R.; Juan, A. Relevant aspects of quantification and sample heterogeneity in hyperspectral image resolution. *Chemometr. Intell. Lab. Syst.* **2012**, *117*, 169–182. [CrossRef]
34. Chang, C.-I. An Information-Theoretic Approach to Spectral Variability, Similarity, and Discrimination for Hyperspectral Image Analysis. *IEEE Trans. Inf. Theory* **2000**, *46*, 1927–1932. [CrossRef]
35. Zhang, E.; Zhang, X.; Yang, S.; Wang, S. Improving Hyperspectral Image Classification Using Spectral Information Divergence. *IEEE Geosci. Remote. Sens. Lett.* **2014**, *11*, 249–253. [CrossRef]
36. Schafer, R.W. What Is a Savitzky-Golay Filter?[lecture Notes]. *IEEE Signal Process. Mag.* **2011**, *28*, 111–117. Available online: https://inst.eecs.berkeley.edu/~{}ee123/sp14/docs/SGFilter.pdf (accessed on 10 September 2021). [CrossRef]
37. Wold, S.; Esbensen, K.; Geladi, P. Principal Component Analysis. *Chemometr. Intell. Lab. Syst.* **1987**, *2*, 37–52. [CrossRef]
38. Tharwat, A.; Gaber, T.; Ibrahim, A.; Hassanien, A.E. Linear Discriminant Analysis: A Detailed Tutorial. *AI Commun.* **2017**, *30*, 169–190. [CrossRef]
39. Garthwaite, P.H. An Interpretation of Partial Least Squares. *J. Am. Stat. Assoc.* **1994**, *89*, 122–127. [CrossRef]
40. Kononenko, I.; Simec, E.; Robnik-Sikonja, M. Overcoming the myopia of inductive learning algorithms with ReliefF. *Appl. Intell.* **1997**, *7*, 39–55. Available online: https://citeseerx.ist.psu.edu/viewdoc/summary?doi.10.1.1.56.4740 (accessed on 10 September 2021). [CrossRef]
41. Breiman, L. Random forests. *J. Mach. Learn.* **2001**, *45*, 5–32. [CrossRef]
42. Mason, L.; Baxter, J.; Bartlett, P.L.; Frean, M. Boosting algorithms as gradient descent. In Proceedings of the NIPS'99: 12th International Conference on Neural Information Processing Systems, Cambridge, MA, USA, 29 November–4 December 1999; pp. 512–518. [CrossRef]
43. Noble, W.S. What Is a Support Vector Machine? *Nat. Biotechnol.* **2006**, *24*, 1565–1567. [CrossRef]
44. Fawcett, T. An Introduction to Roc Analysis. *Pattern Recognit. Lett.* **2005**, *35*, 299–309. [CrossRef]
45. Pedregosa, F.; Varoquaux, G.; Gramfort, A.; Michel, V.; Thirion, B.; Grisel, O.; Blondel, M.; Prettenhofer, P.; Weiss, R.; Dubourg, V.; et al. Scikit-learn: Machine learning in Python. *J. Mach. Learn. Res.* **2011**, *12*, 2825–2830.
46. Chen, T.; Guestrin, C. XGBoost: A scalable tree boosting system. In Proceedings of the 22nd ACM SIGKDD International Conference on Knowledge Discovery and Data Mining, New York, NY, USA, 13–17 August 2016; pp. 785–794. [CrossRef]
47. Viaene, N.; Mahieu, T.; Peña, E. Distribution of *Meloidogyne chitwoodi* in potato tubers and comparison of extraction methods. *Nematology* **2007**, *9*, 143–150. [CrossRef]
48. Bordes, A.; Ertekin, S.; Weston, J.; Bottou, L. Fast kernel classifiers with online and active learning. *J Mach Learn.* **2005**, *6*, 1579–1619. [CrossRef]

49. Hastie, T.; Tibshirani, R.; Friedman, J. *The Elements of Statistical Learning*, 2nd ed.; Springer: New York, NY, USA, 2008; pp. 43–99. ISBN 0-387-95284-5.
50. Menze, B.H.; Kelm, B.M.; Masuch, R.; Himmelreich, U.; Bachert, P.; Petrich, W.; Hamprecht, F.R. A comparison of random forest and its Gini importance with standard chemometric methods for the feature selection and classification of spectral data. *BMC Bioinform.* **2009**, *10*, 213. [CrossRef]
51. Kononenko, I. Estimating attributes: Analysis and extensions of Relief. In *Machine Learning: ECML-94 1994*; Bergadano, F., De Raedt, L., Eds.; Lecture Notes in Computer Science (Lecture Notes in Artificial Intelligence); Springer: Berlin/Heidelberg, Germany, 1994; Volume 784, pp. 171–182. [CrossRef]
52. Workman, J., Jr.; Weyer, L. *Practical Guide and Spectral Atlas for Interpretative Near-Infrared Spectroscopy*, 2nd ed.; CRC Press: New York, NY, USA, 2012; pp. 231–249, ISBN 9781439875254.

sensors

Article

Monitoring Wheat Powdery Mildew Based on Hyperspectral, Thermal Infrared, and RGB Image Data Fusion

Ziheng Feng [1,2], Li Song [1], Jianzhao Duan [1], Li He [1], Yanyan Zhang [1], Yongkang Wei [1] and Wei Feng [1,*]

[1] State Key Laboratory of Wheat and Maize Crop Science, Agronomy College, Henan Agriculture University, Zhengzhou 450046, China; fengziheng@stu.henau.edu.cn (Z.F.); songli2021@stu.henau.edu.cn (L.S.); djz2020@stu.henau.edu.cn (J.D.); xmzxhl@henau.edu.cn (L.H.); Yanyan2021@stu.henau.edu.cn (Y.Z.); Weiyongkang@stu.henau.edu.cn (Y.W.)

[2] Information and Management Science College, Henan Agricultural University, Zhengzhou 450046, China

* Correspondence: fengwei@henau.edu.cn

Abstract: Powdery mildew severely affects wheat growth and yield; therefore, its effective monitoring is essential for the prevention and control of the disease and global food security. In the present study, a spectroradiometer and thermal infrared cameras were used to obtain hyperspectral signature and thermal infrared images data, and thermal infrared temperature parameters (TP) and texture features (TF) were extracted from the thermal infrared images and RGB images of wheat with powdery mildew, during the wheat flowering and filling periods. Based on the ten vegetation indices from the hyperspectral data (VI), TF and TP were integrated, and partial least square regression, random forest regression (RFR), and support vector machine regression (SVR) algorithms were used to construct a prediction model for a wheat powdery mildew disease index. According to the results, the prediction accuracy of RFR was higher than in other models, under both single data source modeling and multi-source data modeling; among the three data sources, VI was the most suitable for powdery mildew monitoring, followed by TP, and finally TF. The RFR model had stable performance in multi-source data fusion modeling (VI&TP&TF), and had the optimal estimation performance with 0.872 and 0.862 of R^2 for calibration and validation, respectively. The application of multi-source data collaborative modeling could improve the accuracy of remote sensing monitoring of wheat powdery mildew, and facilitate the achievement of high-precision remote sensing monitoring of crop disease status.

Keywords: wheat powdery mildew; machine learning; information fusion; remote sensing monitoring

Citation: Feng, Z.; Song, L.; Duan, J.; He, L.; Zhang, Y.; Wei, Y.; Feng, W. Monitoring Wheat Powdery Mildew Based on Hyperspectral, Thermal Infrared, and RGB Image Data Fusion. *Sensors* **2022**, 22, 31. https://doi.org/10.3390/s22010031

Academic Editors: Jiyul Chang and Sigfredo Fuentes

Received: 11 November 2021
Accepted: 19 December 2021
Published: 22 December 2021

Publisher's Note: MDPI stays neutral with regard to jurisdictional claims in published maps and institutional affiliations.

1. Introduction

In recent years, multiple crop diseases and insect pests have emerged, with considerable impacts on yield and productivity following local outbreaks. According to the United Nations Food and Agriculture Organization (FAO), 20%–40% of crops globally are damaged by disease and insect pests annually [1]. Powdery mildew is the major wheat disease; it causes considerable yield reductions or even no harvest, posing a major threat to wheat production and global food security. Conventional methods of monitoring wheat disease are time-consuming and laborious, and are associated with mechanical damage to crops. Therefore, it is essential to identify and develop approaches of carrying out rapid and damage-free wheat disease monitoring.

Plant disease and insect pest infestations lead to biomass reductions, leaf structure destruction, and chlorophyll and water content reductions. Shifts in chlorophyll, water, and other biochemical components in plant tissues would inevitably yield diverse absorption and reflectance characteristics on the plant reflectance spectrum curve, which provides a theoretical basis and facilitates the real-time monitoring of wheat diseases using remote sensing technologies [2]. In recent years, with continuous advancements in remote sensing technologies, numerous scholars have applied technologies to monitor wheat diseases. Generally, different crops, varieties, and diseases exhibit diverse spectral characteristics, which leads to varying reflectance sensitivities at different bands following disease infestation [3]. Consequently, the identification of crop diseases and crop disease incidence estimation can be achieved based on changes in spectral responses and reflectance characteristics [4–6]. Researchers have previously developed disease monitoring indices following the extraction of disease-sensitive bands for monitoring the infestation of crops by bacterial diseases, such as powdery mildew index (PMI) [7], double green vegetation index [8], and red edge vegetation stress index (RVSI) [9].

The modeling algorithms applied in remote sensing influence the accuracy of remote sensing technologies. Today, the algorithms applied in disease and pest monitoring with remote sensing technologies are mainly empirical models and machine learning algorithms [10,11]. Among them, the empirical models are relatively simple; however, the data are easily influenced by external conditions and have poor universality. In recent years, machine learning methods have emerged, with rapid development. Crop disease monitoring models established based on machine learning methods consider training error and generalization ability, and address the challenges associated with slight changes in reflection coefficient during crop disease detection [12,13]. Gu et al. [14] used hyperspectral imaging technologies to monitor tobacco infected by tomato spotted wilt virus and reported that the combination of a successive projections algorithm (SPA) and boosted regression tree was the optimal modeling approach. In addition, Liu et al. [15] established a wheat wilt monitoring model using an improved backward propagation neural network. Wheat is a crop planted in dense rows; if only a single spectral data type is applied in disease monitoring activities, the model is often insensitive to changes in canopy spectrum reflectance, and the reflectance data can be saturated, leading to significant model prediction errors [16,17].

Texture information obtained using imaging spectroscopy tools can reflect disease spot sizes and infestation levels of bacterial diseases [18], in addition to integrating plant morphology and canopy structure information in the spectral data, which enhances the accuracy of remote sensing tools in crop disease monitoring activities [19]. Many researchers have exploited the complementary advantages of spectrum and texture information, which has significantly improved crop growth parameters and the inversion effect of disease severity [20,21]. For example, Guo et al. [22] used vegetation index (VI) and texture features (TF) data obtained using an unmanned aerial vehicle (UAV) platform to establish a wheat stripe rust monitoring model based on partial least squares regression (PLSR). TF can provide plant morphology data that could be applied in the monitoring of crop growth based on remote sensing technologies, which can address the saturation and low accuracy shortcomings associated with single spectral information source-based monitoring, and in turn enhance the robustness of a model and model inversion performance.

Infrared thermal imaging technologies have high sensitivity and early warning capacity. Wheat plants are infected by powdery mildew fungus, and the early symptoms are mostly manifested by changes in internal physiological reactions. Thermal infrared images can reveal temperature changes infected regions that cannot be discerned by visible light images [23]. Mahlein et al. [24] used an infrared thermal instrument (IRT) to measure wheat canopy temperature (CT) and found that the temperature of diseased spikelets was significantly higher than that of healthy spikelets. Therefore, infrared thermal imaging technologies can be used to monitor crop stress during growth and physiological conditions. Many researchers have begun to combine IRT with other remote sensing data sources in plant disease monitoring activities. For example, Zarco-Tejada et al. [25] confirmed that the combination of VI, sun induced fluorescence (SIF), and crop water stress index (CWSI) could be used to effectively monitor diseased trees, and the identification accuracy rate exceeds 80%. In addition, Poblete et al. [26] combined the spectrum, SIF and CWSI, which could effectively distinguish diseased and non-diseased olive trees, whereas Zhang et al. [27] used UAV multi-spectral VI in combination with CT information to estimate disease severity in disease-stressed chickpea, with significantly enhanced detection accuracy. The results of the above studies indicate that thermal infrared data can reliably reflect abnormal conditions in the CT of stressed crops, and can facilitate disease identification and disease classification when combined with other remote sensing data sources.

In the wake of rapid advancements in modern electronic information science, numerous sensors are available for application in the detection of crop morphology and canopy structure, such as reflectance spectrometers, chlorophyll fluorescence meters, and IRT and RGB cameras, which detect crop morphology and growth status based on different factors and principles [28]. However, crop information associated with a single information source is often potentially biased and has certain limitations. Data from different sensor types can be deployed synergistically to enhance target detection and recognition capabilities [29]. Compared to a single sensor data source, multi-sensor data sources can enhance the reliability and robustness of real-time detection [30]. At present, few studies have reported on the monitoring of wheat powdery mildew disease based on a synergy of spectral data and thermal infrared temperature data; in particular, there is a dearth of studies on disease monitoring using approaches that synergize VI, TF and temperature parameters (TP).

To further explore the synergistic effects of multimodal data obtained from different sensors in disease monitoring, in the present study, multimodal data on the incidence of wheat powdery mildew was obtained using hyperspectral surface spectrometer and thermal infrared camera, and compared with ground disease investigations. Multimodal data were obtained using modern modeling and inversion algorithms, such as PLSR, support vector machine regression (SVR), and random forest regression (RFR). The results of the present study could provide a technical basis for the rapid and large-scale monitoring of wheat powdery mildew, and facilitate the prevention and precise control of wheat powdery mildew, in addition to the improvement of pesticide efficiency and food safety.

2. Materials and Methods

2.1. Experimental Design

Experiment 1 (EXP.1): The experiment was conducted in the 2020–2021 wheat growing season at the Science and Education Demonstration Park (34°51′ N, 113°35′ E) of Henan Agricultural University, Zhengzhou, China. The tested varieties were varieties susceptible to wheat powdery mildew: Aikang 58 and Yumai 49–198. The first crop was corn, and the stalks were crushed and returned to the field. The soil was loam, the 0~30-cm soil contained 0.99–1.18 g kg^{-1} of total nitrogen (N), 0.023–0.034 g kg^{-1} of available phosphorus, 0.114–0.116 g kg^{-1} of available potassium, and 11.4–15.3 g kg^{-1} of organic matter. In the experiments, relatively high water and N fertilizer amounts were used to create favorable conditions for powdery mildew. The amount of N applied was 270 kg·hm^{-2}, and the irrigation amount during the wintering period-jointing stage was 900 m^3·hm^{-2}. Powdery mildew fungus was inoculated at the jointing stage, and the wheat was infected from the flowering stage, and the canopy spectrum data were obtained at the flowering and filling stages. Other field management approaches were similar to those applied locally.

Experiment 2 (EXP.2): Carried out simultaneously with experiment 1, experiment 2 was a variety comparison experiment in the field, involving Yanzhan 4110, Nongmai 18, Zhoumai 27, Jinfeng 205, Zhengmai 1342, Xumai 318, Bainong 207, and Xinmai 26. The amount of N applied was 225 kg·hm^{-2}, and the irrigation amount during the wintering period-jointing stage was 675 m^3·hm^{-2}. The experimental area was close to fences and pig farms, and the terrain was low-lying. Due to the terrain, air humidity, rainfall, and diseases in previous years, the wheat growth environment was suitable for the occurrence and spread of wheat powdery mildew, without field inoculation. Disease emergence was natural and more severe. Other field management approaches were similar to those applied in EXP.1.

2.2. Ground Data Collection

2.2.1. Investigation of Powdery Mildew

During the wheat flowering and filling periods, wheat powdery mildew incidence was investigated manually, and 77 and 37 samples were collected in EXP.1 and EXP.2, respectively. About 0.2 m^2 of the experimental area was investigated at each point, and 20 representative wheat plants were selected to test for powdery mildew infection. The survey was conducted in strict accordance with the technical specifications for crop disease monitoring (Chinese Standard: NY/T 2738.2-2015) [31]. The ratio of the leaf area covered by the diseased mycelium layer on the diseased leaf to the total leaf area was expressed based on a grading method, with eight levels representing 1%, 5%, 10%, 20%, 40%, 60%, 80%, and 100% coverage. The grid method was used to calculate the ratio of the diseased spot area to the leaf area. The operation involved using grids to cover the leaves, recording the total number of grids with disease spots, to facilitate the calculation of the ratio of the diseased spot area to leaf area. The closest value between grades was selected as the actual level. For example, at onset with a severity of less than 1%, the coverage was considered 1%. The average severity of diseased leaves was calculated as follows (1):

$$D = \frac{\sum (Di \times Li)}{L} \times 100 \tag{1}$$

where, D is the average disease severity in leaves, and the unit is percentage (%); Di is each severity value; Li is the number of diseased leaves corresponding to each severity value, and the unit is slice; and L is the total number of leaves under investigation, and the unit is slice.

On the basis of the severity of disease in leaves, the disease index (DI) is calculated to represent the average level of disease occurrence (Equation (2)).

$$DI = F \times D \times 100 \tag{2}$$

where, DI is the disease index; F is the diseased leaf rate; D is the average severity of disease in leaves.

2.2.2. Canopy Spectrum Data Measurement

From 10:00 to 14:00 (Beijing local time) with little wind and clear weather, a FieldSpec handheld spectrometer (FieldSpec Handheld 2, Analytical Spectral Devices, Boulder, CO, USA) was used to obtain wheat canopy spectrum data, and the probe was 1.0 m from the top of the wheat crop. The field of view of the spectrometer was 25°, in the 325–1075 nm band, the spectral sampling interval was 1.4 nm, and the spectral resolution was 3.0 nm. A 0.4-m × 0.4-m BaSO$_4$ calibration plate was used to calculate black and baseline reflectance. Ten spectral reflectance values were recorded at each sampling point as samples, and the average value was considered the spectral reflectance of the sampling area.

2.2.3. Thermal Infrared Image and RGB Image Acquisition

An FLIR T650sc thermal infrared camera (FLIR Systems, Inc., Wilsonville, OR, USA) was used to obtain the wheat canopy temperature (CT) and RGB images. The device has dual thermal infrared and visible light sensors, and the image resolution is 640 × 480 pixels. Synchronous with the spectral reflectance measurement, the lens was 1.0 m from the top of the wheat crop, and the thermal infrared and RGB images were obtained vertically (Figure 1).

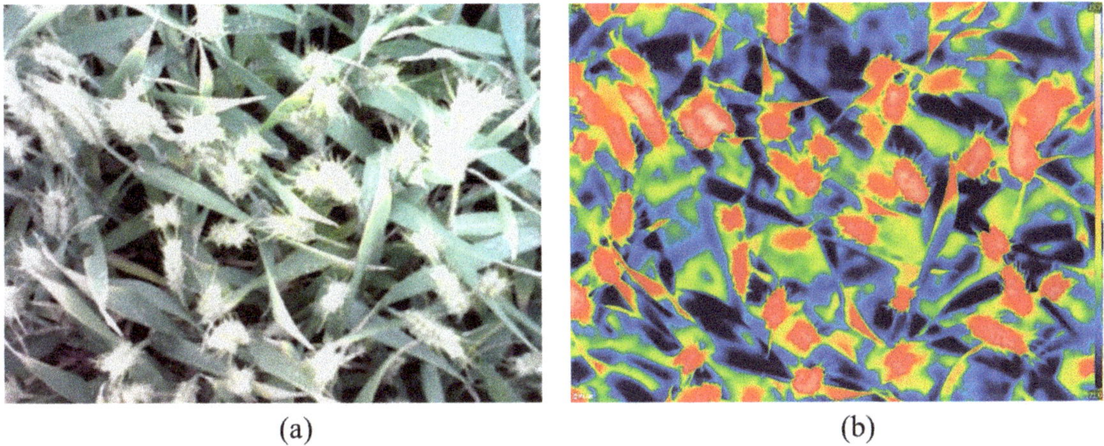

(a)   (b)

Figure 1. Wheat canopy RGB image (**a**) and thermal infrared image (**b**) obtained using thermal infrared camera.

2.3. Data Processing Methods

2.3.1. Spectral Vegetation Index (VI)

Before extracting the VIs, the bands with high noise before 400 nm and after 1000 nm were removed, and then the Savitzky-Golay function was used to smoothen the spectra in MATLAB 7.0 (The MathWorks Inc., Natick, MA, USA). VIs associated with the disease were pre-selected by consulting relevant literatures (Table 1). Considering the potential existence of the collinearity problem among VIs, SPA was used to optimize VIs and reduce their multicollinearity. SPA is a forward variable selection method that selects characteristic variables by calculating the sizes of the projection vector of the remaining variables and the selected variables, which can ensure that the linear relationship between the selected variables is minimized, so as to eliminate redundant information between variables and reduce multicollinearity, to achieve the purpose of selecting sensitive variables [32].

Table 1. Spectral vegetation indices.

Vegetation Index	Formula	References
Modified simple ration (MSR)	$MSR = (R_{800}/R_{670} - 1)/(R_{800}/R_{670} + 1)^{0.5}$	[33]
Photochemical reflectance index (PRI)	$PRI = (R_{531} - R_{570})/(R_{531} + R_{570})$	[34]
Physiological reflectance index (PHRI)	$PhRI = (R_{550} - R_{531})/(R_{550} + R_{531})$	[35]
Transformed chlorophyll absorption in reflectance index (TCARI)	$TCARI = 3(R_{700} - R_{650}) - 0.2(R_{700}/R_{500})/(R_{700}/R_{670})$	[36]
Red-edge vegetation stress index (RVSI)	$RVSI = ((R_{712} + R_{752})/2) - R_{732}$	[37]
Structural independent pigment index (SIPI)	$SIPI = (R_{800} - R_{445})/(R_{800} - R_{680})$	[38]
Visible atmospherically resistant index (VARI)	$VARI = (R_{550} - R_{670})/(R_{550} + R_{670} - R_{480})$	[39]
Renormalized difference vegetation index (RDVI)	$RDVI = (R_{800} - R_{670})/(R_{800} + R_{670})^{0.5}$	[40]
Anthocyanin reflectance index (ARI)	$ARI = (R_{550})^{-1}/(R_{700})^{-1}$	[41]
Damage sensitive spectral index 2 (DSSI2)	$DSSI2 = (R_{747} - R_{901} - R_{537} - R_{572})/(R_{747} - R_{901} + R_{537} - R_{572})$	[42]
Greenness index (GI)	$GI = R_{554}/R_{677}$	[43]
Plant senescence reflectance index (PSRI)	$PSRI = (R_{680} - R_{500})/R_{750}$	[44]
Normalized pigment chlorophyll Index (NPCI)	$NPCI = (R_{680} - R_{430})/(R_{680} - R_{430})$	[43]
Nitrogen reflectance index (NRI)	$NRI = (R_{570} - R_{670})/(R_{570} - R_{670})$	[45]
Healthy index (HI)	$HI = (R_{534} - R_{698})/(R_{534} + R_{698}) - 0.5R_{704}$	[7]
Powdery mildew index (PMI)	$PMI = (R_{520} - R_{584})/(R_{520} + R_{584}) - R_{724}$	[7]
Triangular vegetation index (TVI)	$TVI = 0.5[120(R_{750} - R_{550}) - 200(R_{670} - R_{550})]$	[46]
Green normalized difference vegetation index (GNDVI)	$GNDVI = (R_{800} - R_{550})/(R_{800} + R_{550})$	[47]
Nir shoulder region index (NSRI)	$NSRI = R_{890}/R_{780}$	[48]
Soil-adjusted vegetation index (SAVI)	$SAVI = 1.5(R_{800} - R_{670})/(R_{800} + R_{670} + 0.5)$	[49]

2.3.2. RGB Image Texture Features (TF)

The gray level co-occurrence matrix (GLCM) method, proposed by Haralick in 1973 [50] is one of the most widely used texture extraction methods. The method has the advantages of rotation invariance, multi-scale characteristics, and low computational complexity, and is widely used in image processing, pattern recognition, and remote sensing monitoring [51,52]. In ENVI (Harris, Bloomfield, CO, USA), the gray level image of the RGB image was subjected to 3×3 sliding filtering using GLCM. Eight texture feature maps in the directions of $0°$, $45°$, $90°$ and $135°$ were extracted (Figure 2, Table 2), and the average of four directions was taken as the final texture feature map. To ensure that the extracted texture features are all based on canopy vegetation, a K-means clustering algorithm is used for bare soil rejection. The soil and vegetation mask is shown in Figure 3.

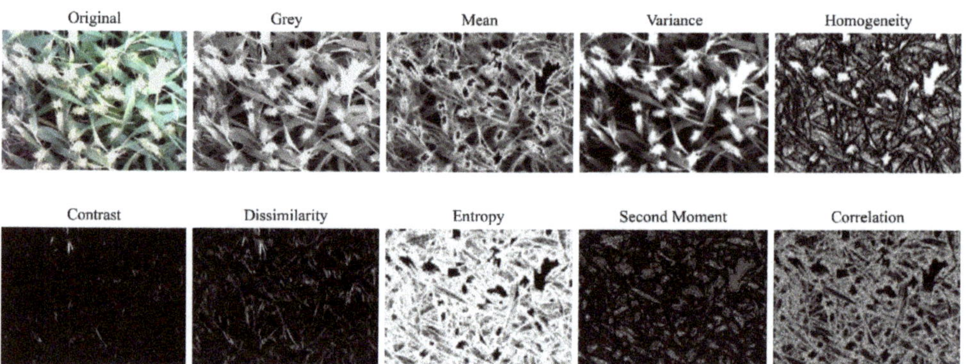

Figure 2. Eight texture feature maps of gray-level co-occurrence matrix from RGB image of wheat canopy in the $0°$ direction.

Table 2. Texture feature calculation formula.

Texture	Equation	Description
Mean, MEA	$MEA = \sum\limits_{i,j=1}^{G} (iP(i,j))$	Reflects the average of the greyscale
Variance, VAR	$VAR = \sum\limits_{i=1}^{G} \sum\limits_{J=1}^{G} (i-u)^2 P(i,j)$	Reflects the magnitude of grey scale variation
Homogeneity, HOM	$HOM = \sum\limits_{i=1}^{G} \sum\limits_{J=1}^{G} \frac{P(i,j)}{1+(i-j)^2}$	Reflects the roughness of image texture
Contrast, CON	$CON = \sum\limits_{i=1}^{G} \sum\limits_{J=1}^{G} (i-j)^2 P(i,j)$	Reflects the local variations in the gray-level co-occurrence matrix
Dissimilarity, DIS	$DIS = \sum\limits_{i=1}^{G} \sum\limits_{J=1}^{G} P(i,j)\|i-j\|$	Same as contrast, used to detect similarity
Entropy, ENT	$ENT = -\sum\limits_{i=1}^{G} \sum\limits_{J=1}^{G} P(i,j)logP(i,j)$	Reflects the degree of the gray distribution and the thickness of the texture
Second moment, SEM	$SEC = \sum\limits_{i=1}^{G} \sum\limits_{J=1}^{G} P^2(i,j)$	Reflects the homogeneity of an image's distribution of greyscale
Correlation, COR	$COR = \sum\limits_{i=1}^{G} \sum\limits_{J=1}^{G} \frac{(i-MEA_j)(j-MEA_j)P(i,j)}{\sqrt{VAR_i}\sqrt{VAR_j}}$	Reflects the length of the extension of a certain grey value in a certain direction

Note: i and j indicate the row and column number of the images, respectively; $P(i,j)$ is the relative frequency of two neighboring pixels.

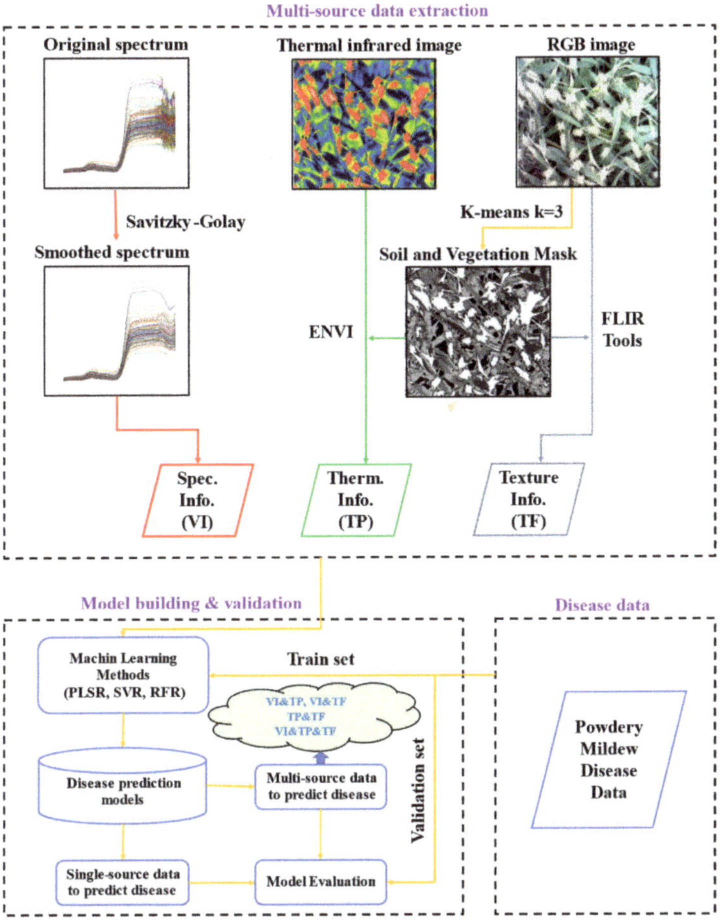

Figure 3. A workflow diagram of feature extraction and modeling.

2.3.3. Thermal Infrared Temperature Parameters (TP)

The thermal infrared image was annotated and combined with K-means clustering segmentation results (Figure 3) using FLIR Tools (FLIR Systems Inc., Wilsonville, OR, US), and the temperature parameters were extracted. Considering that CT changes with the daily change in atmospheric temperature, the canopy temperature difference (CTD), canopy temperature ratio (CTR), and normalized relative canopy temperature (NRCT) were extracted to eliminate the influence of atmospheric temperature on CT. The temperature parameter formula was as follows:

$$CTD = CT_i - AT \tag{3}$$

$$CTR = \frac{CT_i}{AT} \tag{4}$$

$$NRCT = \frac{CT_i - CT_{min}}{CT_{max} - CT_{min}} \tag{5}$$

where, AT is the atmospheric temperature, CT_i is the CT of the i-th pixel in the image, CT_{max} is the highest temperature measured in the entire experimental field, and CT_{min} is the lowest temperature measured in the entire experimental field.

2.3.4. Estimation Model

(1) PLSR

PLSR is a classic modeling method, which includes the characteristics of principal component analysis (PCA), canonical correlation analysis, and multiple linear regression analysis, and is often used for quantitative analysis in remote sensing [53]. PLSR transforms the original variables with high data redundancy into a few variables by selecting the optimal latent variables, to describe the linear model of the relationship between the predicted value and the true value.

(2) SVR

The basic idea of SVR is to use training samples to establish a regression hyperplane, and to approximate the samples to the hyperplane to minimize the total deviation from the sample point to the plane [54]. The commonly used kernel functions of the SVR algorithm include the linear kernel function, radial basis function (RBF) kernel function, polynomial kernel function, and Sigmoid kernel function. Among them, the RBF kernel function can handle the complex nonlinear problem between the independent variable and the dependent variable.

(3) RFR

RFR is a machine learning algorithm based on a classification regression tree [55]. RFR uses the bootstrap resampling method to extract multiple samples from the original sample, models each bootstrap sample into a decision tree, combines them into multiple decision trees for prediction, and then applies the majority voting method to determine the final classification result of the joint prediction model. The advantage of the method is that the training speed is relatively fast and it does not require cross-validation. In addition, the randomness of sampling and feature selection make the random forest averts overfitting [56]. It is widely used in classification and prediction in remote sensing-based monitoring activities.

2.3.5. Model Validation

With VI, TP and TF as independent variables, and DI as the dependent variable, a monitoring model for wheat powdery mildew disease index was established based on the three algorithms above. The workflow from feature extraction to model building and evaluation was demonstrated in Figure 3. To make the model evaluation results more objective, EXP.1 test data were used as the modeling set, and EXP.2 test data were used as the verification set. The accuracy of the wheat powdery mildew disease index monitoring model was evaluated based on three indicators: coefficient of determination (R^2), root mean square error (RMSE), and relative error (RE). The closer R^2 is to 1, the lower the **RMSE**, and the lower the **RE**, the higher the accuracy of the monitoring model. The formula was as follows:

$$R^2 = \frac{\sum_{i=1}^{n}(x_i - \bar{x})^2 \times (y_i - y)^2}{\sum_{i=1}^{n}(x_i - \bar{x})^2 \times \sum_{i=1}^{n}(y_i - y)^2} \tag{6}$$

$$RMSE = \sqrt{\frac{\sum_{i=1}^{n}(y_i - x_i)^2}{n}} \tag{7}$$

$$RE\ (\%) = \sqrt{\frac{1}{n} \times \sum_{i=1}^{n}\left(\frac{y_i - x_i}{x_i}\right)^2} \times 100 \tag{8}$$

where, x_i, $\bar{x}$, y_i, and y are the measured **DI**, average **DI**, predicted **DI**, and average **DI**, respectively; n is the number of samples.

3. Results

3.1. Changes in Wheat Canopy Spectra under Different Powdery Mildew Severity Levels

With an increase in DI, the spectral reflectance of the visible light band from 400 to 780 nm increased gradually, and the discrimination of DI was better (Figure 4a). The spectral reflectance of the near-infrared bands region across 780 nm–1000 nm was less distinguishable when the disease was mild; when the disease was more than moderate, the spectral reflectance increased gradually; and when the disease was severe (such as DI = 80), the spectral reflectance rose sharply due to severe damage to the canopy structure, even higher than the spectral reflectance of healthy wheat. From the perspective of the correlation between disease severity and reflectance (Figure 4b), there was a positive correlation in the visible light band from 400 to 730 nm, and a negative correlation in the near-infrared region from 730 to 1000 nm, especially at 600–700 nm (r = 0.373–0.431, probability value, $p < 0.01$) and 780–960 nm (r=−0.355−−0.294, $p < 0.01$), which can be considered as disease-sensitive bands for the real-time monitoring of disease progression.

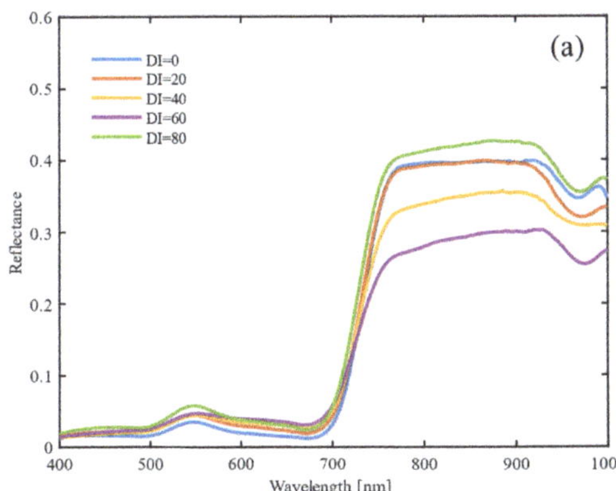

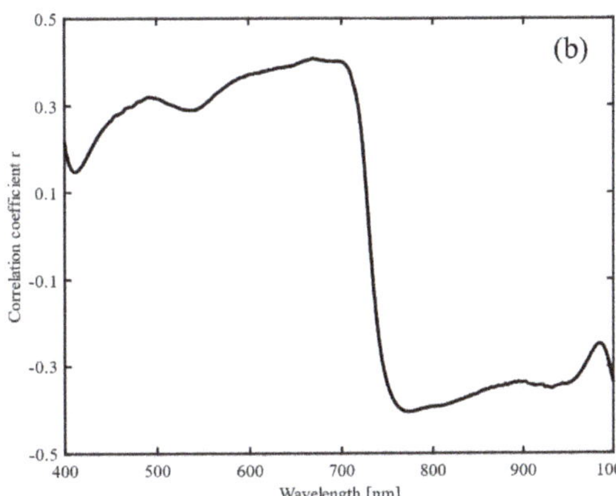

Figure 4. Spectral reflectance changes (**a**) of wheat canopy and its correlation (**b**) with disease index.

3.2. Selection of Vegetation Index

Based on the reported VIs related to plant disease, the correlations between 20 VIs and DI were analyzed. The VI with the highest correlation was NSRI (r = 0.743) (Figure 5a), followed by GI and NPCI. Because VI is a combination of bands, and there is a certain degree of information duplication between bands, there is considerable multicollinearity among the spectral parameters. Therefore, the SPA algorithm is used to optimize the VI. The minimum number of sensitive variables extracted was two, and the maximum number of sensitive variables was twenty. RMSE decreased with an increase in the number of variables. RMSE was the minimum (RMSE = 15.575) when the number of variables was 10; however, with an increase in the number of variables, the RMSE increased gradually (Figure 5b). After SPA screening, there were 10 VIs, namely NSRI, NPCI, PSRI, PRI, ARI, SIPI, PMI, MSR, RVSI, and GNDVI. According to the results, when a single VI was used to estimate the DI, the linear R^2 was low ($R^2 < 0.56$), and the error in monitoring powdery mildew disease was relatively large (Figure 6).

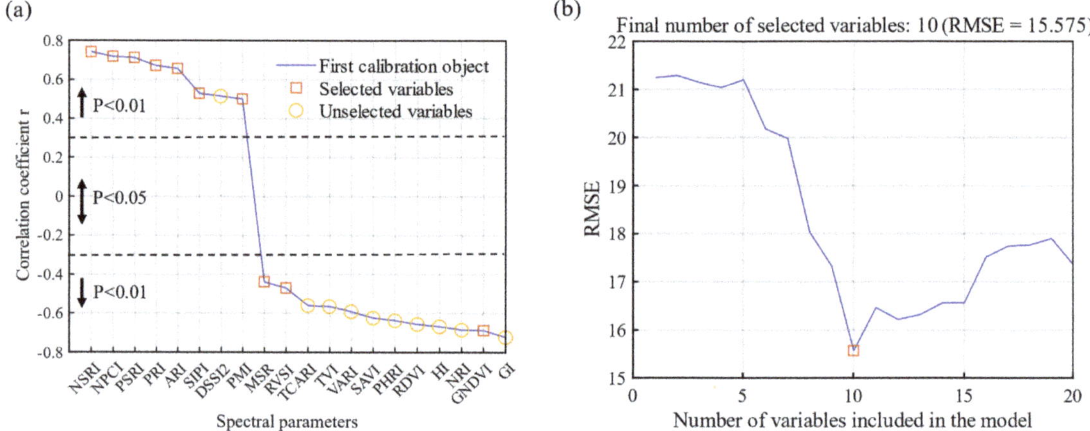

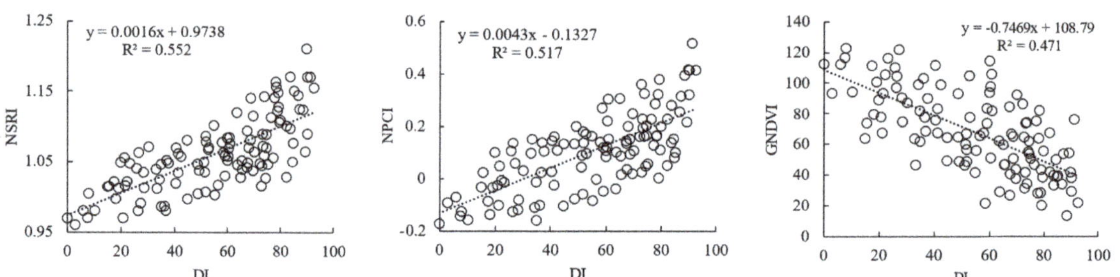

Figure 5. Root mean square error (**a**) in the optimal variables selected using successive projections algorithm (SPA) and correlation (**b**) between vegetation index (VI) and disease index (DI).

Figure 6. Linear relationship of the optimal vegetation indices with wheat disease index (DI).

3.3. Selection of Texture Feature Parameters

Analysis of the correlation between TF parameters of canopy RGB images and DI showed that the correlation coefficients were all positive. Excluding the correlation coefficient between the correlation and DI, which was not significant, all the others were significant, among which entropy was the highest ($r = 0.486$, $p < 0.01$) (Figure 7a). The eight extracted TFs were all calculated from grayscale images, and considering the potential multicollinearity among TFs, the SPA algorithm was used to optimize the variables. The RMSE was the lowest (RMSE = 18.043) when the number of TF variables was five. Five TF parameters, mean, variance, homogeneity, entropy, and second moment, were selected as the input variables in the estimation model (Figure 7b).

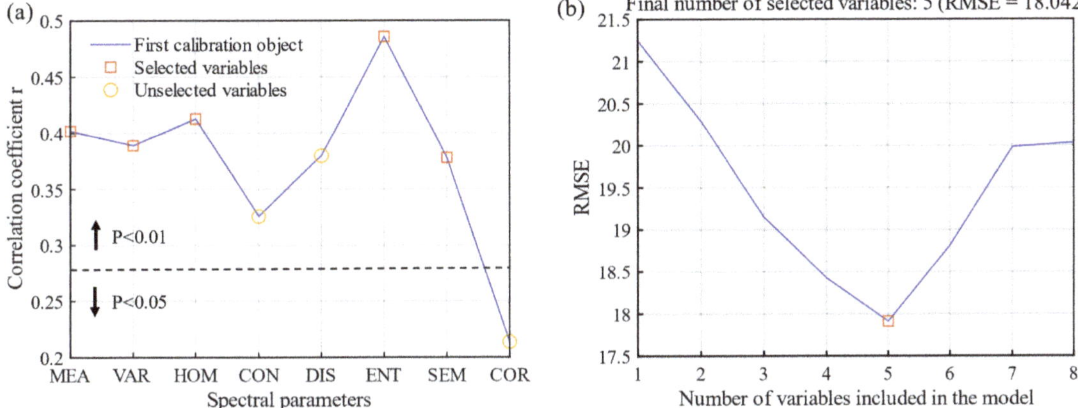

Figure 7. Correlation (**a**) between texture feature parameter and disease index (DI) and the root mean square error (**b**) for the optimal variables selected using SPA.

3.4. Selection of Thermal Infrared Temperature Parameters

The TP parameters from thermal infrared images were extracted to analyze their correlation with DI, and the results showed that the correlation coefficient between CT and DI was 0.382, and was significant ($p < 0.01$) (Figure 8). Considering CT changes with daily atmospheric temperature changes, the CTD, CTR and NRCT were extracted to eliminate the influence of atmospheric temperature on CT. After eliminating the influence of atmospheric temperature, the significant level of correlation of TP was further improved, with a very significant level ($p < 0.01$) observed. Combining the principle of strong correlation and eliminating duplicate information, two TPs, CTD, and NRCT, were selected as input variables for the subsequent steps of modeling and analysis.

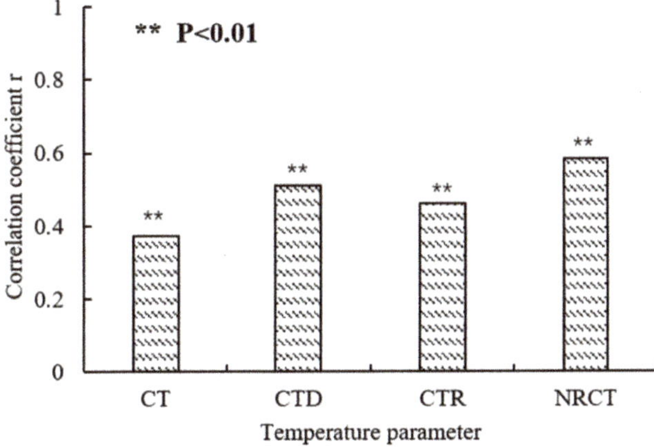

Figure 8. Correlation coefficient between temperature parameters and DI.

3.5. Comparison of Different Model Algorithms Based on Single Data Sources

With a single data source as an independent variable, three methods, including, PLSR, SVR and RFR, were used to invert the DI of wheat powdery mildew (Table 3, Figure 9). Comprehensive comparison revealed that the RFR model performed optimally, followed by the SVR and PLSR models. Based on the performance results of the three data sources, regardless of which method was used to estimate wheat DI, the performance of the VI was the best, followed by TP and TF. Based on combinations modeling methods and independent variable data type, the RFR method with VI as the independent variable was the best combination, with R^2, RMSE, and RE values of 0.690, 14.488, and 18.42%, respectively, in the calibration set, and R^2, RMSE, and RE values of 0.680, 14.298, and 18.16%, respectively, in the validation set. The SVR method with VI as an independent variable was the second best combination, with R^2, RMSE and RE values of 0.670, 14.757, and 18.69%, respectively, in the calibration set, and R^2, RMSE, and RE values of 0.666, 15.578, and 18.16%, respectively, in the validation set.

Table 3. Estimation performance of single data source model based on different algorithms.

Independent Variable Type	Number of Variables	Modeling Method	Calibration Set			Validation Set		
			R^2	RMSE	RE	R^2	RMSE	RE
VI	10	PLSR	0.666	15.014	19.24%	0.650	16.425	19.86%
		SVR	0.670	14.757	18.69%	0.666	15.578	19.28%
		RFR	0.690	14.488	18.42%	0.680	14.298	18.16%
TF	5	PLSR	0.509	17.852	32.03%	0.486	18.367	32.05%
		SVR	0.517	18.616	30.70%	0.514	17.489	30.23%
		RFR	0.537	17.621	29.18%	0.537	17.799	27.95%
TP	2	PLSR	0.553	17.420	29.27%	0.546	17.673	29.66%
		SVR	0.567	17.347	28.96%	0.571	17.094	27.13%
		RFR	0.575	16.470	27.83%	0.577	16.791	27.79%

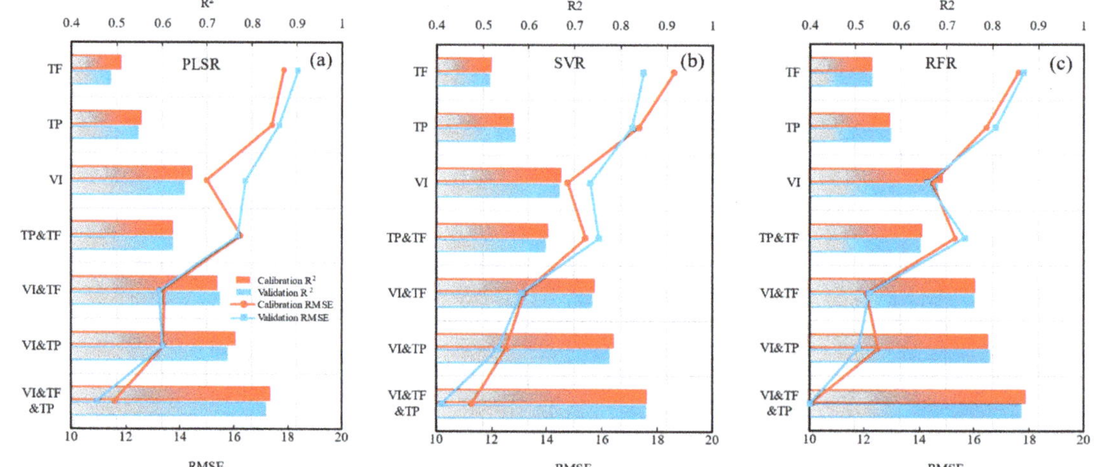

Figure 9. Prediction performance of different models with various input feature types.

3.6. Comparison of Different Model Algorithms Based on Multi-Source Data Combination

To fully exploit the information obtained from different data sources, TP, TF, and VI were combined to carry out a comparative analysis of three modeling methods (Table 4, Figure 9). After fusing TF on the basis of VI data, the average R^2, RMSE, and RE values in the calibration set were 0.743, 12.91, and 17.57%, respectively, with the R^2 value representing an average increase of 10% when compared with the R^2 of single VI data source. The mean R^2, RMSE, and RE values in the validation set were 0.742, 12.849, and 17.71%, respectively, and the R^2 value represented a 11.6% increase when compared to the R^2 of the single VI data source. The addition of TP based on VI data enhanced the accuracy of the combined model. The average R^2 values in the validation and calibration sets were 0.779 and 0.772, respectively, the RMSE values were 12.823 and 12.467, respectively, and the RE values were 15.56% and 15.62%, respectively. The R^2 values were 15.4% and 16% higher, respectively, than the R^2 values based on the single VI data source. In addition, in the combined TP and TF modelling, the estimation performance was superior to those of the single data sources, both TP or TF; however, the model was inferior to VI based on both the calibration and validation sets. The results indicate that when using single data source-based VI as a benchmark, TF has a minor positive effect on model accuracy when performing multi-data collaboration modeling, whereas TP has a relatively high positive effect on model improvement.

Table 4. Estimation performance of multi-source collaboration models based on different algorithms.

Independent Variable Type	Number of Variables	Modeling Method	Calibration Set			Validation Set		
			R^2	RMSE	RE	R^2	RMSE	RE
TP&TF	7	PLSR	0.624	16.265	21.67%	0.623	16.151	22.67%
		SVR	0.641	15.413	20.39%	0.637	15.900	20.91%
		RFR	0.646	15.328	20.29%	0.641	15.687	20.77%
VI&TF	15	PLSR	0.723	13.417	18.37%	0.728	13.236	18.15%
		SVR	0.744	13.211	17.33%	0.738	13.107	17.27%
		RFR	0.762	12.102	17.02%	0.761	12.203	17.71%
VI&TP	12	PLSR	0.763	13.385	16.08%	0.746	13.375	16.20%
		SVR	0.784	12.554	15.43%	0.776	12.221	15.59%
		RFR	0.791	12.531	15.51%	0.794	11.804	15.07%
VI&TP&TF	17	PLSR	0.840	11.606	14.07%	0.831	10.947	14.09%
		SVR	0.857	11.277	13.66%	0.854	10.200	13.07%
		RFR	0.872	10.108	12.54%	0.862	10.049	12.31%

Three modeling methods were used to synergize three data sources, VI, TF, and TP. The R^2 values of the calibration and validation sets were further improved compared with the R^2 value following combination of the two data sources. The average R^2 values of the calibration and validation sets in the model with the three data sources combined were 0.856 and 0.849, respectively, the RMSE values were 10.997 and 10.399, respectively, and the RE values were 13.42% and 13.16%, respectively. The R^2 values were 26.8% and 27.6% higher, respectively, than the R^2 values of the single VI data source model. Comparison of the modeling algorithm results showed that the RFR model had the highest R^2, the lowest RMSE and RE, and the greatest DI predictive capacity, followed by the SVR model and PLSR models (Figure 10). The R^2 values of the RFR fusion model of the three data sources were 0.872 and 0.862 in the calibration and validation sets, respectively, the RMSE values were 10.108 and 10.049, respectively, and the RE values were 12.54% and 12.31%, respectively. The above results indicate that collaborative modeling with multiple data sources is superior to single data source-based modeling, with the combined model exhibiting better fit, accuracy, and predictive ability (Figure 9).

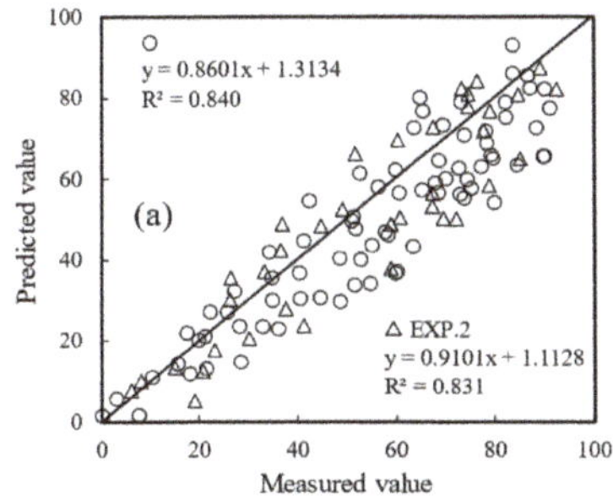

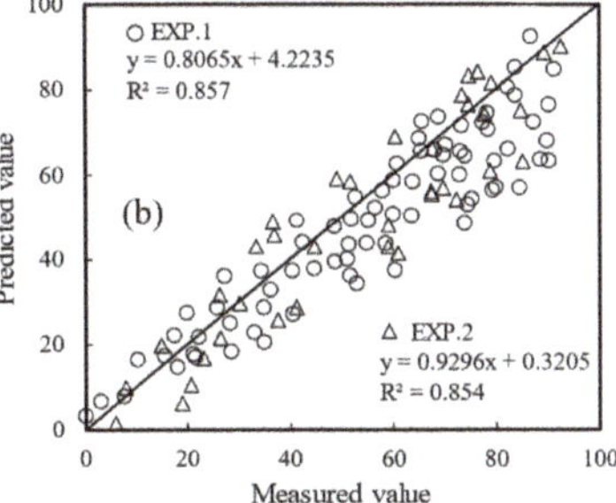

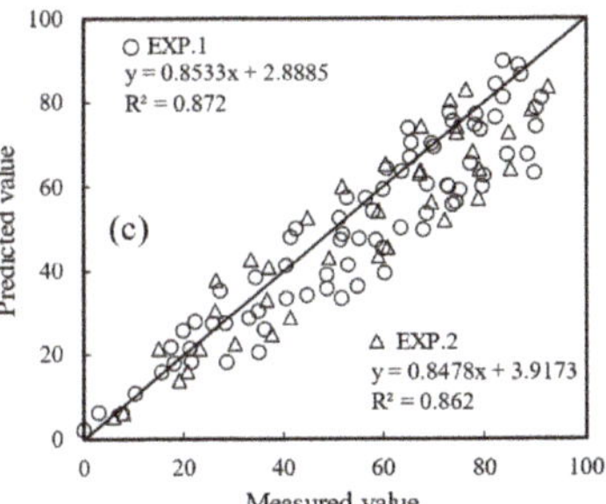

Figure 10. Comparison of three data source fusion models based on (**a**) PLSR, (**b**) SVR and (**c**) RFR algorithm.

4. Discussion

4.1. Combining VI and TF to Monitor Crop Diseases

Previous literature has confirmed the importance of reflectance spectrum data in crop disease monitoring and its application prospects. The visible light and near-infrared regions are the sensitive bands for spectral identification of different crop diseases and insect pests; furthermore, the spectral sensitivity bands of different crops and different diseases vary. The sensitive bands of wheat powdery mildew are located at 490–780 nm [57], and wheat powdery mildew monitoring is mainly based on the sensitive band [58,59], and different forms of disease VI can be established according to the reflection characteristics of the disease [7–9]. Disease emergence involves gradual development that alters internal tissue physiology and biochemistry, and, in turn, the external morphological structure, and then manifests externally as disease that can be detected by remote sensing. Due to the combined effects of internal and external factors, such as mesophyll cells, water, chlorophyll, and leaf yellowing and dryness, the ability to extract disease information from a single band is often limited. The VIs with good performance in the present study were NSRI, NPCI, and CVI. Among them, NSRI performed optimally, with a linear R^2 of only 0.552, which hardly meets the information requirements for accurate crop protection.

The onset of powdery mildew disease has a significant bottom-up characteristic. In the early and middle stages of the disease, the disease is mainly concentrated in the middle and lower levels of the plant. However, the canopy reflectance spectra data mainly originate from the upper level, which leads to lack of consistency between the collected canopy spectra data and disease characteristics, and increases the challenge of monitoring powdery mildew using canopy spectrum data only. Therefore, the use of multivariate analysis methods to identify and monitor disease has become a hotspot in quantitative remote sensing research.

In the present study, multiple VIs were used as independent variables, and three algorithms PLSR, SVR, and RFR were used to predict DI. The results showed that the RFR model had the highest monitoring accuracy; however, R^2 was still lower than 0.7. From the perspective of precise crop protection and disease prevention and control, the spectral data could not be used to monitor wheat powdery mildew reliably. Some scholars have attempted to incorporate fluorescence data in modeling when using hyperspectral data for disease monitoring, and achieved good monitoring results [60]. When a pathogen infects plants, the canopy structure changes following physiological and biochemical responses, and the TF can reflect the change in canopy structure caused by pathogen infestation to a certain degree [61,62]. Researchers have used hyperspectral VI in combination with TF to monitor wheat stripe rust, and reported that the estimation results of the two combined data sources were significantly better than that of the single data source [22]. In the present study, the VI and TF were modeled together, and model accuracy improved when compared with when the VI from a single data source was used. However, the highest accuracy of the combined model in the validation set was only 0.761, the optimization effect was limited, and it did not satisfy the requirements of accurate monitoring, which could be due to the gradual senescence of wheat leaves after the flowering period, and the increased background complexity of withered plants. Furthermore, multiple factors in some plots, such as disease, drought, senescence, and atmospheric temperature, which make it impossible to accurately distinguish whether the withered leaves and structural changes are attributed to disease stress, could have adversely influenced the modelling findings.

4.2. Combining VI and TP to Monitor Crop Disease

Thermal infrared imaging technologies have great application potential in remote sensing monitoring activities [63–65]. After crops are infected by fungi and pathogens, cell membrane permeability increases, water is lost, and plants exhibit dehydration and wilting. In addition, stomata are closed and heat loss on the leaf surface changes, which leads to leaf surface temperature response. At the onset of crop disease, changes in heat radiation energy caused by plant water loss, stomata closure, and increased respiration can be intuitively reflected in infrared heat maps; however, most studies have focused on disease classification and disease identification. Calderón et al. [66] demonstrated the capacity of using canopy temperature information and hyperspectral VI to identify olive trees with yellow dwarf disease; however, they did not estimate disease severity. In the previous monitoring research on wheat powdery mildew, no studies have reported the combination of VI and TP. The correlation analyses carried out in the present study showed that the thermal infrared temperature was sensitive to disease, and it was more effective to convert CT into CTD, CTR, and NRCT. Compared to canopy TF, temperature information had a greater role in disease monitoring. The RFR model performed best ($R^2 = 0.577$) and was slightly more accurate than the TF-RFR model; however, it was significantly less accurate than the VI-RFR model. VI as a single data source was more suitable for monitoring wheat powdery mildew, followed by canopy TP and canopy TF. To further improve the information limitations of single data sources, VI and canopy TP were modeled together (VI&TP). Model accuracies of different algorithms were higher than that of VI&TF on the whole, indicating that canopy temperature information has great application potential in disease monitoring.

Previous studies have also demonstrated that spectral information, texture information, and thermal infrared information have the ability to monitor crop diseases [22,24,26]; however, no study has reported their joint application in wheat powdery mildew monitoring. To that end, the present study conducted fusion modeling based on VI, TF, and TP (VI&TF&TP). According to the results, the combination of the three data sources had obvious advantages over single data sources or two combined data sources. Among them, the R^2 values of the three data source models based on the RFR algorithm was 0.862, which provides technical support and a reference method for the prevention and precise control of wheat powdery mildew.

4.3. Machine Learning Algorithms in Disease Monitoring

In the wake of rapid advancements in computer modeling science, machine learning technology has been applied extensively in crop disease monitoring, with the achievement of remarkable results [67,68]. Jiang et al. [69] demonstrated the high estimation capacity of the RFR model in the monitoring of mangrove disease and insect pests. In addition, Zhang et al. [70] demonstrated the good classification performance of the RFR model in the identification of wheat grains infected with *Fusarium*. In the present study, three modeling methods (PLSR, SVR, RFR) were used to monitor wheat powdery mildew DI. The RFR model performed best, regardless of whether it was based on single data source modeling or multi-source data modeling. This is mainly because the RFR algorithm has good anti-noise ability and does not easily exhibit over-fitting [71]. In the present study, SVR was used to integrate information from three data sources, and the average accuracy that the model achieved was 0.77. Considering the operation efficiency and prediction accuracy of the model, the method is effective for monitoring the disease. In contrast, the performance of the PLSR model was slightly worse, which might be because PLSR was better at addressing multicollinearity between parameters [72], and the parameters used in the present study were optimized by the SPA algorithm, which eliminated the influence of multicollinearity, resulting in an inability to maximize the performance of the PLSR model.

Although the overall performance of the RFR model in the present study was the best, the estimated value was lower than the actual value under more severe disease conditions, which was also observed in the other model algorithms. Generally, the greater the population density of wheat powdery mildew, the worse the air permeability, and the more severe the disease, which decreases the sensitivity of spectral and thermal imaging data to disease severity. Under the condition of multi-source data fusion, the saturation of the model was alleviated, which also demonstrates the effectiveness of multi-data source fusion.

When applying different model algorithms to monitor wheat powdery mildew in collaboration with VI, TF, and TP, the present study did not consider the contribution rates of different data source parameters to the model. How to use different algorithms to determine the weights of different data source parameters to further improve the accuracy of the model remains to be further studied. The occurrence and characteristics of powdery mildew are certainly associated with crop variety, growth period, and other diverse factors. Using targeted information extraction algorithms to clarify the effects and contribution levels of each influencing factor could facilitate the integration multiple effect factors to accurately monitor disease occurrence, and provide a theoretical basis for crop protection and precise operations.

5. Conclusions

Based on multi-source data fusion and machine learning, the present study explores the application potential of canopy spectral vegetation index, thermal infrared information, and texture feature information obtained using different sensors in wheat powdery mildew monitoring. In the case of wheat disease index prediction based on single data source, spectral information is better than thermal infrared information and texture features. Regardless of the modeling method, the results obtained following the fusion of data from multiple sources are more reliable than the data obtained from a single data source. When using the combination of vegetation index, thermal infrared information and texture features, higher prediction precision can be achieved. Regardless of whether single data source or multi-source data is used, the monitoring accuracy of the RFR model is higher than that of other algorithm models. Therefore, the combination of multi-source data fusion and the RFR model have broad application prospects in wheat powdery mildew monitoring, which could not only promote disease prevention and control but also reduce pesticide use and enhance the efficiency of disease prevention and control activities. However, the models identified should be tested under different crop types, growth stages, and environmental conditions, to further evaluate the robustness of the models.

Author Contributions: Conceptualization, W.F.; methodology, W.F.; software, Z.F.; validation, Z.F. and L.S.; formal analysis, J.D. and L.H.; investigation, Z.F. and L.S.; resources, Y.Z.; data curation, Y.W.; writing—original draft preparation, Z.F. and L.S.; writing—review and editing, Z.F. and L.S.; visualization, Y.Z.; supervision, Y.Z and Y.W.; project administration, J.D. and L.H.; funding acquisition, W.F. and L.H. All authors have read and agreed to the published version of the manuscript.

Funding: This research was funded by National Natural Science Foundation of China under Grant Numbers 31971791.

Institutional Review Board Statement: Not applicable.

Informed Consent Statement: Not applicable.

Data Availability Statement: Not applicable.

Acknowledgments: We gratefully acknowledge the College of Agronomy, and Information and Management Science College, Henan Agricultural University, for providing the laboratory of this study. We sincerely thank the reviewers and the editors for their helpful comments and supplementary proposal. We also thank the native English speakers who provided assistance with language correction.

Conflicts of Interest: The authors declare no conflict of interest.

References

1. Zhang, N.; Yang, G.; Pan, Y.; Yang, X.; Chen, L.; Zhao, C. A Review of advanced technologies and development for hyperspectral-based plant disease detection in the past three decades. *Remote Sens.* **2020**, *12*, 3188. [CrossRef]
2. Zhang, J.; Huang, Y.; Pu, R.; Gonzalez-Moreno, P.; Yuan, L.; Wu, K.; Huang, W. Monitoring plant diseases and pests through remote sensing technology: A review. *Comput. Electron. Agric.* **2019**, *165*, 104943. [CrossRef]
3. Feng, W.; Qi, S.; Heng, Y.; Zhou, Y.; Wu, Y.; Liu, W.; He, L.; Li, X. Canopy vegetation indices from in situ hyperspectral data to assess plant water status of winter wheat under powdery mildew stress. *Front. Plant Sci.* **2017**, *8*, 1219. [CrossRef]

4. Shi, Y.; Huang, W.; González-Moreno, P.; Luke, B.; Dong, Y.; Zheng, Q.; Ma, H.; Liu, L. Wavelet-based rust spectral feature set (wrsfs): A novel spectral feature set based on continuous wavelet transformation for tracking progressive host–pathogen interaction of yellow rust on wheat. *Remote Sens.* **2018**, *10*, 525. [CrossRef]
5. Chen, T.; Zhang, J.; Chen, Y.; Wan, S.; Zhang, L. Detection of peanut leaf spots disease using canopy hyperspectral reflectance. *Comput. Electron. Agric.* **2019**, *156*, 677–683. [CrossRef]
6. Franceschini, M.H.D.; Bartholomeus, H.; van Apeldoorn, D.F.; Suomalainen, J.; Kooistra, L. Feasibility of unmanned aerial vehicle optical imagery for early detection and severity assessment of late blight in potato. *Remote Sens.* **2019**, *11*, 224. [CrossRef]
7. Mahlein, A.-K.; Rumpf, T.; Welke, P.; Dehne, H.W.; Plümer, L.; Steiner, U.; Oerke, E.-C. Development of spectral indices for detecting and identifying plant diseases. *Remote Sens. Environ.* **2013**, *128*, 21–30. [CrossRef]
8. Feng, W.; Shen, W.; He, L.; Duan, J.; Guo, B.; LI, Y.; Wang, C.; Guo, T. Improved remote sensing detection of wheat powdery mildew using dual-green vegetation indices. *Precis. Agric.* **2016**, *17*, 608–627. [CrossRef]
9. Naidu, P.A.; Perry, E.M.; Pierce, F.J.; Mekuria, T. The potential of spectral reflectance technique for the detection of Grapevine leafroll associated virus-3 in two red-berried wine grape cultivars. *Comput. Electron. Agric.* **2009**, *66*, 38–45. [CrossRef]
10. Zhang, J.; Pu, R.; Wang, J.; Huang, W.; Yuan, L.; Luo, J. Detecting powdery mildew of winter wheat using leaf level hyperspectral measurements. *Comput. Electron. Agric.* **2012**, *85*, 13–23. [CrossRef]
11. Tian, L.; Xue, B.; Wang, Z.; Li, D.; Yao, X.; Cao, Q.; Zhu, Y.; Cao, W.; Cheng, T. Spectroscopic detection of rice leaf blast infection from asymptomatic to mild stages with integrated machine learning and feature selection. *Remote Sens. Environ.* **2021**, *257*, 112350. [CrossRef]
12. Badnakhe, M.R.; Durbha, S.S.; Jagarlapudi, A.; Gade, R.M. Evaluation of citrus gummosis disease dynamics and predictions with weather and inversion based leaf optical model. *Comput. Electron. Agric.* **2018**, *155*, 130–141. [CrossRef]
13. Ashourloo, D.; Aghighi, H.; Matkan, A.A.; Mobasheri, M.R.; Rad, A.M. An investigation into machine learning regression techniques for the leaf rust disease detection using hyperspectral measurement. *IEEE J. Sel. Top. Appl. Earth Obs. Remote Sens.* **2016**, *9*, 4344–4351. [CrossRef]
14. Gu, Q.; Sheng, L.; Zhang, T.; Lu, Y.; Zhang, Z.; Zheng, K.; Hu, H.; Zhou, H. Early detection of tomato spotted wilt virus infection in tobacco using the hyperspectral imaging technique and machine learning algorithms. *Comput. Electron. Agric.* **2019**, *167*, 105066. [CrossRef]
15. Liu, L.; Dong, Y.; Huang, W.; Du, X.; Ma, H. Monitoring wheat fusarium head blight using unmanned aerial vehicle hyperspectral imagery. *Remote Sens.* **2020**, *12*, 3811. [CrossRef]
16. Chen, P.; Nicolas, T.; Wang, J.; Philippe, V.; Huang, W.; Li, B. New index for crop canopy fresh biomass estimation. *Spectrosc. Spectr. Anal.* **2010**, *30*, 512–517. [CrossRef]
17. Zhao, Y.; Jing, X.; Huang, W.; Dong, Y.; Li, C. Comparison of sun-induced chlorophyll fluorescence and reflectance data on estimating severity of wheat stripe rust. *Spectrosc. Spectr. Anal.* **2019**, *39*, 2739–2745. [CrossRef]
18. Zhang, D.; Chen, G.; Yin, Y.; Hu, R.; Gu, C.; Pan, Z.; Zhou, X.; Chen, Y. Integrating spectral and image data to detect Fusarium head blight of wheat. *Comput. Electron. Agric.* **2020**, *175*, 105588. [CrossRef]
19. Khan, I.H.; Liu, H.; Li, W.; Cao, A.; Wang, X.; Liu, H.; Cheng, T.; Tian, Y.; Zhu, Y.; Cao, W.; et al. Early Detection of Powdery Mildew Disease and Accurate Quantification of Its Severity Using Hyperspectral Images in Wheat. *Remote Sens.* **2021**, *13*, 3612. [CrossRef]
20. Zheng, H.; Cheng, T.; Zhou, M.; Li, D.; Yao, X.; Tian, Y.; Cao, W.; Zhu, Y. Improved estimation of rice aboveground biomass combining textural and spectral analysis of UAV imagery. *Precis. Agric.* **2019**, *20*, 611–629. [CrossRef]
21. Al-Saddik, H.; Laybros, A.; Billiot, B.; Cointault, F. Using image texture and spectral reflectance analysis to detect yellowness and esca in grapevines at leaf-level. *Remote Sens.* **2018**, *10*, 618. [CrossRef]
22. Guo, A.; Huang, W.; Dong, Y.; Ye, H.; Ma, H.; Liu, B.; Wu, W.; Ren, Y.; Ruan, C.; Geng, Y. Wheat yellow rust detection using uav-based hyperspectral technology. *Remote Sens.* **2021**, *13*, 123. [CrossRef]
23. Khanal, S.; Fulton, J.; Shearer, S. An overview of current and potential applications of thermal remote sensing in precision agriculture. *Comput. Electron. Agric.* **2017**, *139*, 22–32. [CrossRef]
24. Mahlein, A.K.; Alisaac, E.; Al Masri, A.; Behmann, J.; Dehne, H.W.; Oerke, E.C. Comparison and combination of thermal, fluorescence, and hyperspectral imaging for monitoring fusarium head blight of wheat on spikelet scale. *Sensors* **2019**, *19*, 2281. [CrossRef] [PubMed]
25. Zarco-Tejada, P.J.; Camino, C.; Beck, P.S.A.; Calderon, R.; Hornero, A.; Hernández-Clemente, R.; Kattenborn, T.; Montes-Borrego, M.; Susca, L.; Morelli, M.; et al. Previsual symptoms of Xylella fastidiosa infection revealed in spectral plant-trait alterations. *Nat. Plants* **2018**, *4*, 432–439. [CrossRef] [PubMed]
26. Poblete, T.; Camino, C.; Beck, P.S.A.; Hornero, A.; Kattenborn, T.; Saponari, M.; Boscia, D.; Navas-Cortes, J.A.; Zarco-Tejada, P.J. Detection of Xylella fastidiosa infection symptoms with airborne multispectral and thermal imagery: Assessing bandset reduction performance from hyperspectral analysis. *ISPRS J. Photogramm. Remote Sens.* **2020**, *162*, 27–40. [CrossRef]
27. Zhang, C.; Chen, W.; Sankaran, S. High-throughput field phenotyping of ascochyta blight disease severity in chickpea. *Crop Prot.* **2019**, *125*, 104885. [CrossRef]
28. Xie, C.; Yang, C. A review on plant high-throughput phenotyping traits using UAV-based sensors. *Comput. Electron. Agric.* **2020**, *178*, 105731. [CrossRef]

29. Zhang, J.; Pu, R.; Yuan, L.; Huang, W.; Nie, C.; Yang, G. Integrating remotely sensed and meteorological observations to forecast wheat powdery mildew at a regional scale. *IEEE J. Sel. Top. Appl. Earth Obs. Remote Sens.* **2017**, *7*, 4328–4339. [CrossRef]

30. Maimaitijiang, M.; Sagan, V.; Sidike, P.; Harting, S.; Fritschi, F.B. Soybean yield prediction from UAV using multimodal data fusion and deep learning. *Remote Sens. Environ.* **2019**, *237*, 111599. [CrossRef]

31. *Technical Specification on Remote Sensing Monitoring for Crop Diseases—Part 2: Wheat Powder Mildew*; Chinese Academy of Agricultural Sciences: Beijing, China, 2015.

32. Zhang, J.; Cheng, T.; Guo, W.; Xu, X.; Qiao, H.; Xie, Y.; Ma, X. Leaf area index estimation model for UAV image hyperspectral data based on wavelength variable selection and machine learning methods. *Plant Methods* **2021**, *17*. [CrossRef] [PubMed]

33. Haboudane, D.; Miller, J.R.; Pattey, E.; Zarco-Tejada, P.J.; Strachan, I.B. Hyperspectral vegetation indices and novel algorithms for predicting green LAI of crop canopies: Modeling and validation in the context of precision agriculture. *Remote Sens. Environ.* **2004**, *90*, 337–352. [CrossRef]

34. Gamon, J.; Penuelas, J.; Field, C. A narrow-waveband spectral index that tracks diurnal changes in photosynthetic efficiency. *Remote Sens. Environ.* **1992**, *41*, 35–44. [CrossRef]

35. Daughtry, C.S.T.; Walthall, C.L.; Kim, M.S.; Colstoun, E.B.; McMurtrey, J.E. Estimating corn leaf chlorophyll concentration from leaf and canopy reflectance. *Remote Sens. Environ.* **2000**, *74*, 229–239. [CrossRef]

36. Haboudane, D.; Miller, J.R.; Tremblay, N.; Zarco-Tejada, P.J.; Dextraze, L. Integrated narrow-band vegetation indices for prediction of crop chlorophyll content for application to precision agriculture. *Remote Sens. Environ.* **2002**, *81*, 416–426. [CrossRef]

37. Abdulridha, J.; Ampatzidis, Y.; Kakarla, S.C.; Roberts, P. Detection of target spot and bacterial spot diseases in tomato using UAV-based and benchtop-based hyperspectral imaging techniques. *Precis. Agric.* **2019**, *21*, 955–978. [CrossRef]

38. Penuelas, J.; Frédéric, B.; Filella, I. Semi-empirical indices to assess carotenoids/chlorophyll A ratio from leaf spectral reflectances. *Photosynthetica* **1995**, *31*, 221–230.

39. Gitelson, A.A.; Kaufman, Y.J.; Stark, R.; Rundquist, D. Novel algorithms for remote estimation of vegetation fraction. *Remote Sens. Environ.* **2002**, *80*, 76–87. [CrossRef]

40. Roujean, J.L.; Broen, F.M. Estimating PAR absorbed by vegetation from bidirectional reflectance measurements. *Remote Sens. Environ.* **1995**, *51*, 375–384. [CrossRef]

41. Gitelson, A.A.; Merzlyak, M.N.; Chivkunova, O.B. Optical properties and nondestructive estimation of anthocyanin content in plant leaves. *Photochem. Photobiol.* **2010**, *74*, 38–45. [CrossRef]

42. Mirik, M.; Michels, G.J.; Kassymzhanova-Mirik, S.; Elliott, N.C.; Catana, V.; Jones, D.B.; Bowling, R. Using digital image analysis and spectral reflectance data to quantify damage by greenbug (Hemitera: Aphididae) in winter wheat. *Comput. Electron. Agric.* **2006**, *51*, 86–98. [CrossRef]

43. Zarco-tejada, P.J.; Berjón, A.; López-lozano, R.; Miller, J.R.; Martín, P.; Cachorro, V.; González, M.R.; Frutos, A.D. Assessing vineyard condition with hyperspectral indices: Leaf and canopy reflectance simulation in a row-structured discontinuous canopy. *Remote Sens. Environ.* **2005**, *99*, 271–287. [CrossRef]

44. Merzlyak, M.N.; Gitelson, A.A.; Chivkunova, O.B.; Rakitin, V.Y. Non-destructive optical detection of pigment changes during leaf senescence and fruit ripening. *Physiol. Plant.* **1999**, *106*, 135–141. [CrossRef]

45. Filella, I.; Serrano, L.; Serra, J.; Peñuelas, J. Evaluating wheat nitrogen status with canopy reflectance indices and discriminant analysis. *Crop Sci.* **1995**, *35*, 1400–1405. [CrossRef]

46. De Castro, A.; Ehsani, R.; Ploetz, R.; Crane, J.H.; Abdulridha, J. Optimum spectral and geometric parameters for early detection of laurel wilt disease in avocado. *Remote Sens. Environ.* **2015**, *171*, 33–44. [CrossRef]

47. Gitelson, A.A.; Kaufman, Y.J.; Merzlyak, M.N. Use of a green channel in remote sensing of global vegetation from EOS-MODIS. *Remote Sens. Environ.* **1996**, *58*, 289–298. [CrossRef]

48. Liu, L.; Huang, W.; Pu, R.; Wang, J. Detection of internal leaf structure deterioration using a new spectral ratio index in the near-infrared shoulder region. *J. Integr. Agric.* **2014**, *13*, 760–769. [CrossRef]

49. Huete, A.R. A soil-adjusted vegetation index (SAVI). *Remote Sens. Environ.* **1988**, *25*, 295–309. [CrossRef]

50. Haralick, R.M.; Shanmugam, K.; Dinstein, I. Textural features for image classification. *IEEE Trans. Cybern.* **1973**, *SMC-3*, 610–621. [CrossRef]

51. Prasad, S.; Peddoju, S.K.; Ghosh, D. Multi-resolution mobile vision system for plant leaf disease diagnosis. *Signal Image Video Process.* **2016**, *10*, 379–388. [CrossRef]

52. Li, S.; Yuan, F.; Ata-UI-Karim, S.T.; Zheng, H.; Cheng, T.; Liu, X.; Tian, Y.; Zhu, Y.; Cao, W.; Cao, Q. Combining Color Indices and Textures of UAV-Based Digital Imagery for Rice LAI Estimation. *Remote Sens.* **2019**, *11*, 1763. [CrossRef]

53. Guo, J.; Zhang, J.; Xiong, S.; Zhang, Z.; Wei, Q.; Zhang, W.; Feng, W.; Ma, X. Hyperspectral assessment of leaf nitrogen accumulation for winter wheat using different regression modeling. *Precis. Agric.* **2021**, *22*, 1634–1658. [CrossRef]

54. Han, J.; Zhang, Z.; Cao, J.; Luo, Y.; Zhang, L.; Li, Z.; Zhang, J. Prediction of winter wheat yield based on multi-source data and machine learning in China. *Remote Sens.* **2020**, *12*, 236. [CrossRef]

55. Breiman, L. Random forest. *Mach. Learn.* **2001**, *45*, 5–32. [CrossRef]

56. Lu, B.; He, Y. Evaluating empirical regression, machine learning, and radiative transfer modelling for estimating vegetation chlorophyll content using bi-seasonal hyperspectral images. *Remote Sens.* **2019**, *11*, 1979. [CrossRef]

57. Graeff, S.; Link, J.; Claupein, W. Identification of powdery mildew (*Erysiphe graminis* sp. *tritici*) and take-all disease (*Gaeumanno-myces graminis* sp. *tritici*) in wheat (*Triticum aestivum* L.) by means of leaf reflectance measurements. *Cent. Eur. J. Biol.* **2006**, *1*, 275–288. [CrossRef]

58. Shi, Y.; Huang, W.; Zhou, X. Evaluation of wavelet spectral features in pathological detection and discrimination of yellow rust and powdery mildew in winter wheat with hyperspectral reflectance data. *J. Appl. Remote Sens.* **2017**, *11*, 026025. [CrossRef]

59. Zhao, J.; Fang, Y.; Chu, G.; Yan, H.; Hu, L.; Huang, L. Identification of leaf-scale wheat powdery mildew (*Blumeria graminis* f. sp. *Tritici*) combining hyperspectral imaging and an SVM classifier. *Plants* **2020**, *9*, 936. [CrossRef]

60. Rajendran, D.K.; Park, E.; Nagendran, R.; Hung, N.B.; Cho, B.K.; Kim, K.H.; Lee, Y.H. Visual analysis for detection and quantification of pseudomonas cichorii disease severity in tomato plants. *Plant Pathol. J.* **2016**, *32*, 300–310. [CrossRef]

61. Xie, C.; He, Y. Spectrum and image texture features analysis for early blight disease detection on eggplant leaves. *Sensors* **2016**, *16*, 676. [CrossRef]

62. Guo, A.; Huang, W.; Ye, H.; Dong, Y.; Ma, H.; Ren, Y.; Ruan, C. Identification of wheat yellow rust using spectral and texture features of hyperspectral images. *Remote Sens.* **2020**, *12*, 1419. [CrossRef]

63. Cao, W.; Qiao, Z.; Gao, Z.; Lu, S.; Tian, F. Use of unmanned aerial vehicle imagery and a hybrid algorithm combining a watershed algorithm and adaptive threshold segmentation to extract wheat lodging. *Phys. Chem. Earth Parts A/B/C* **2021**, *123*, 103016. [CrossRef]

64. Chandel, A.K.; Khot, L.R.; Yu, L. crop vigor and yield characterization using high-resolution aerial multispectral and thermal infrared imaging technique. *Comput. Electron. Agric.* **2021**, *182*, 105999. [CrossRef]

65. Francesconi, S.; Harfrouche, A.; Maesano, M.; Balestra, G.M. UAV-based thermal, RGB imaging and gene expression analysis allowed detection of fusarium head blight and gave new insights into the physiological responses to the disease in durum wheat. *Front. Plant Sci.* **2021**, *12*, 551. [CrossRef] [PubMed]

66. Calderón, A.; Navas-Cortés, J.A.; Lucena, C.; Zarco-Tejada, P.J. High-resolution airborne hyperspectral and thermal imagery for early detection of Verticillium wilt of olive using fluorescence, temperature and narrow-band spectral indices. *Remote Sens. Environ.* **2013**, *139*, 231–245. [CrossRef]

67. Han, Z.; Deng, L. Application driven key wavelengths mining method for aflatoxin detection using hyperspectral data. *Comput. Electron. Agric.* **2018**, *153*, 248–255. [CrossRef]

68. Zheng, Q.; Huang, W.; Cui, X.; Shi, Y.; Liu, L. New spectral index for detecting wheat yellow rust using sentinel-2 multispectral imagery. *Sensors* **2018**, *18*, 868. [CrossRef] [PubMed]

69. Jiang, X.; Zhen, J.; Miao, J.; Zhao, D.; Wang, J.; Jia, S. Assessing mangrove leaf traits under different pest and disease severity with hyperspectral imaging spectroscopy. *Ecol. Indic.* **2021**, *129*, 107901. [CrossRef]

70. Zhang, D.; Chen, G.; Zhang, H.; Jing, X.; Gu, C.; Weng, S.; Wang, Q.; Chen, Y. Integration of spectroscopy and image for identifying fusarium damage in wheat kernels using hyperspectral imaging. *Spectrochim. Acta Part A* **2020**, *236*, 118344. [CrossRef]

71. Meiforth, J.J.; Buddenbaum, H.; Hill, J.; Shepherd, J. Monitoring of canopy stress symptoms in New Zealand kauri trees analysed with AISA hyperspectral data. *Remote Sens.* **2020**, *12*, 926. [CrossRef]

72. Burnett, A.C.; Anderson, J.; Davidson, K.J.; Ely, K.S.; Lamour, J.; Li, Q.; Morrison, B.D.; Yang, D.; Rogers, A.; Serbin, S.P. A best-practice guide to predicting plant traits from leaf-level hyperspectral data using partial least squares regression. *J. Exp. Bot.* **2021**, *72*, 6175–6189. [CrossRef] [PubMed]

sensors

Article

Hyperspectral Estimation of Winter Wheat Leaf Area Index Based on Continuous Wavelet Transform and Fractional Order Differentiation

Changchun Li [1], Yilin Wang [1], Chunyan Ma [1,*], Fan Ding [1], Yacong Li [1], Weinan Chen [1], Jingbo Li [1,2] and Zhen Xiao [1]

1   School of Surveying and Land Information Engineering, Henan Polytechnic University, Jiaozuo 454000, China; lichangchun610@126.com (C.L.); 211904010019@home.hpu.edu.cn (Y.W.); 212004020067@home.hpu.edu.cn (F.D.); 211904020026@home.hpu.edu.cn (Y.L.); 211904020046@home.hpu.edu.cn (W.C.); lijingbo1024@163.com (J.L.); 212004020068@home.hpu.edu.cn (Z.X.)
2   Beijing Research Center for Information Technology in Agriculture, Beijing Academy of Agriculture and Forestry Sciences, Beijing 100097, China
*   Correspondence: mayan@hpu.edu.cn; Tel.: +86-187-9024-8606

Abstract: Leaf area index (LAI) is highly related to crop growth, and the traditional LAI measurement methods are field destructive and unable to be acquired by large-scale, continuous, and real-time means. In this study, fractional order differential and continuous wavelet transform were used to process the canopy hyperspectral reflectance data of winter wheat, the fractional order differential spectral bands and wavelet energy coefficients with more sensitive to LAI changes were screened by correlation analysis, and the optimal subset regression and support vector machine were used to construct the LAI estimation models for different growth stages. The precision evaluation results showed that the LAI estimation models constructed by using wavelet energy coefficients combined with a support vector machine at the jointing stage, fractional order differential combined with support vector machine at the booting stage, and wavelet energy coefficients combined with optimal subset regression at the flowering and filling stages had the best prediction performance. Among these, both flowering and filling stages could be used as the best growth stages for LAI estimation with modeling and validation R^2 of 0.87 and 0.71, 0.84 and 0.77, respectively. This study can provide technical reference for LAI estimation of crops based on remote sensing technology.

Keywords: winter wheat; leaf area index; fractional order differential; continuous wavelet transform; optimal subset regression; support vector machine

Citation: Li, C.; Wang, Y.; Ma, C.; Ding, F.; Li, Y.; Chen, W.; Li, J.; Xiao, Z. Hyperspectral Estimation of Winter Wheat Leaf Area Index Based on Continuous Wavelet Transform and Fractional Order Differentiation. *Sensors* **2021**, *21*, 8497. https://doi.org/10.3390/s21248497

Academic Editor: Jiyul Chang

Received: 11 November 2021
Accepted: 17 December 2021
Published: 20 December 2021

Publisher's Note: MDPI stays neutral with regard to jurisdictional claims in published maps and institutional affiliations.

1. Introduction

Leaf area index (LAI) is one of the important community structure parameters in ecosystem research, directly related to many ecological processes such as evapotranspiration, soil water balance, and net productivity. In addition, it is also an important spatial variable measuring wheat photosynthetic effective radiation, transmission, and an eco-environmental process model [1]. In addition, LAI is an important characteristic parameter describing the geometric structure of the wheat canopy. It can be used to quantitatively express the initial energy exchange on the canopy surface, directly reflect the energy, carbon dioxide, and physical environment of growth in the canopy diversification scale space, and reflect the dynamic characteristics and health status of wheat in the process of growth and development. Therefore, LAI estimation is essential for wheat growth monitoring and yield estimation. In the traditional LAI acquisition method, measurement was conducted on the field, which turns out to be destructive, time-consuming, and laborious, and not able to obtain LAI continuously in real time and on a large scale. Remote sensing has the factors of high temporal and spatial resolution and can be used to monitor a region quickly, widely, and periodically, which has now become the main means to estimate surface parameters and extract crop phenotypic parameters.

At present, many scholars have conducted research on crop LAI estimation using remote sensing technology and made some achievements. For example, Huang Jingfeng et al., Su Wei et al., and Fieuzal R et al. [2–4] calculated the position, amplitude, and amplitude of the red edge, respectively, by using the C_{HH} and L_{HH} band data of canopy red light band (680~760 nm), extracted lidar, and SAR parameters such as red edge area and vertical structure parameters, analyzed their correlation with LAI, and constructed LAI estimation models of rape, corn, and wheat. Liu Jun et al., Li Shumin et al., and Wang Laigang et al. [5–7] established LAI estimation models of maize and wheat at different growth stages by constructing a vegetation index and a spectral index, combining with measured LAI data and satellite remote sensing data, such as environmental satellite, MODIS, aster, and SPOT5, using mathematical statistics or the crop growth model PROSAIL, so as to realize LAI dynamic monitoring. By analyzing the current research situation, it could be seen that most of the existing studies are based on the original spectrum. They used the original spectrum to construct the correlation index, analyzed its relationship with crop LAI, and constructed the crop LAI remote sensing estimation model. Spectral differentiation technique can reflect the essential characteristics of crops by partially eliminating the influence of environment, such as atmospheric effect, vegetation shadow, and soil. In recent years, more and more attention has been paid to the monitoring of crop growth by spectral derivative technique and some research results have been obtained. For example, Wang Dengwei et al. [8] found that the first-order differential spectrum value of 750 nm is highly correlated with chlorophyll content. Chen Junying et al. [9] analyzed the characteristics of chlorophyll content and spectrum of rice and found that the first-order differential parameter of spectral reflectance of rice leaves has a strong correlation with chlorophyll content. Jiang Jinbao et al. [10–12] analyzed the correlation between different order spectral derivative indexes and wheat stripe rust and canopy chlorophyll content, optimized the spectral derivative indexes with a strong correlation, and constructed a wheat stripe rust monitoring model and canopy chlorophyll content estimation model, which achieved good results. Smith K L et al. [13] showed that the first-order differential ratio at 725 and 702 nm can be used to monitor the vegetation growth under gas leakage stress. Fractional order differentiation, as the generalization of integer order differentiation, was first proposed by Leibniz at the end of the 17th century and laid some theoretical foundation for its future development [14]. Fractional order differential techniques have unique advantages in signal analysis and processing and have been widely used recently in biological fields, physics, and engineering, among others [15,16]. In order to make full use of the spectral information acquired by the remote sensing platform, enhance the effectiveness of spectral response, and improve the accuracy of spectral modeling, it is necessary to preprocess the raw spectral data. Most previous studies have used integer-order differential transformations to process spectral data, which may cause partial loss of spectral information. The use of fractional order differential transform can realize the refinement of spectral information, which can effectively eliminate the influence of environmental background and deeply explore the potential information in the spectrum [17,18]. Jiang Ming et al. [19] studied the correlation between 0~2 order (interval 0.2 order) fractional differential spectral reflectance and heavy metal content in soil, obtained the correlation coefficients of each order fractional differential and soil heavy metals at each spectral sampling interval, and compared and analyzed the curve change law of correlation coefficients. Zhang Wenwen et al. [20] processed the spectral reflectance with the 11-order differential of 0~1 order (with an interval of 0.1 order) and analyzed the correlation between the differential value of each order and the measured Cu~(2+) content in corn leaves. The results showed that, compared with the common first-order differentiation, the fractional differentiation can highlight the correlation between the spectral reflectance of some bands and the Cu~(2+) content in leaves and expand the selection space of the characteristic band. Wang Jingzhe et al. [21] discussed the possibility of fractional differentiation technology in estimating the content of heavy metal chromium and organic carbon in desert soil using hyperspectral data. The research results showed that the accuracy and robustness of the model after fractional preprocessing are higher than that of integer differentiation. Li Changchun et al. [22] processed hyperspectral remote sensing data by factional differentiation, analyzed the correlation between fractional differentiation spectrum and wheat chlorophyll content, and constructed a chlorophyll content estimation model by using stepwise regression analysis, SVM, and artificial neural network. Spectral signal transformation can improve its sensitivity to crop LAI. Continuous wavelet transform (CWT) can effectively reduce noise, decompose spectral data, and extract more spectral positions and characteristic parameters [23]. At present, the wavelet coefficients extracted by CWT are effective in the inversion of heavy metals and phenotypic parameters in the crop canopy. For example, Chen Haoyu et al. [24] performed CWT on hyperspectral data to generate wavelet energy coefficient and established a BP neural network and SVM inversion model of soil organic matter content. Li Bao et al. [25] and Tan Xianming et al. [26] used univariate linear regression, SVM, and the partial least square method to construct the chlorophyll content estimation model for fresh peach leaves and maize canopy after continuous wavelet transform of spectral information, which was better than the traditional methods. Wang Yancang et al. [27] carried out CWT on hyperspectral data and constructed a quantitative inversion model of winter wheat leaf water content by using partial least square method.

This paper aims to construct and screen the best estimation models of LAI for different growth stages, with winter wheat as the research object. This study, firstly, preprocessed the canopy hyperspectral data by fractional order differential transform and continuous wavelet decomposition. Then, the fractional order differential spectral bands and wavelet energy coefficients with stronger correlation with the LAI based on correlation analysis were screened. Finally, the LAI estimation models for jointing, booting, flowering, and filling stages were constructed and evaluated by support vector machine and optimal subset regression, respectively. This study can provide theoretical and technical references for remote sensing estimation of crop LAI.

2. Materials and Methods

2.1. Overview of the Study Area and Experimental Design Scheme

The research area is located in the National Precision Agriculture Research and Demonstration Base in Xiaotangshan Town, Changping District, Beijing. It is flat and fertile tidal soil. The average altitude is approximately 36 m. The climate is mild with four distinct seasons, the average annual temperature is approximately 13 °C, the average annual rainfall is approximately 510 mm, and precipitation is mostly concentrated in summer, characterizing a typical warm temperate continental monsoon climate. Figure 1 shows the specific location of the study area.

The experimental area was 84 m in length from east to west and 32 m in length from north to south, with each plot measuring 6×8 m. There were 48 plots, 16 treatments, and 3 replications. The orthogonal experiment was conducted with different amounts of nitrogen fertilizer, different moisture contents, and different crop varieties. Four levels of N fertilizer were set: 0, 195, 390, and 585 kg/hm^2, respectively; three levels of water irrigation were set: rainfed (W1, no irrigation), normal water (W2, irrigation water 192 mm), and twice normal water (W3, irrigation water 384 mm); there are two crop varieties: Jing 9843 (J9843) and Zhong Mai 175 (ZM175). The sowing time was 7 October 2014, the planting density was 3.75 million plants/hm^2, the harvest date was on 11 June 2015, and the crops before sowing were corn.

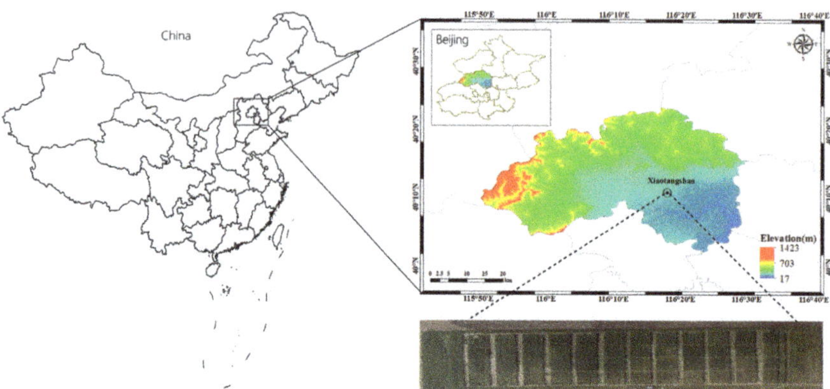

Figure 1. Schematic diagram of the study area and experimental design.

2.2. Data Acquisition

Canopy hyperspectral data and LAI data were measured at the jointing stage, booting stage, flowering stage, and filling stage of winter wheat in 2018 and 2019, respectively.

2.2.1. Canopy Hyperspectral Data Measurement

The canopy hyperspectral data were measured using an ASD Field SpecFR Pro 2500 spectrometer. The data collection was conducted during 10:00–14:00 BST on a clear day, with a sensor probe field of view of 25°, and placed vertically downward at approximately 1.0 m above the canopy. To eliminate the effect of environmental changes on the spectral measurements, the spectrometer was calibrated with a whiteboard before and after each measurement, and dark currents were collected every 5 min. Measurements were repeated 10 times on each plot and the average value was taken as the spectral reflectance of the canopy on the plot. Since the spectrometer has different spectral sampling intervals in different bands, after the spectral data acquisition, the spectral resampling interval was first set to 1 nm, and then the spectrum was smoothed using the Savitzky–Golay filter [28] of ViewSpec Pro software (Malvern Panalytical, Malvern, UK) to reduce the spectral noise and improve the signal-to-noise ratio of the spectral data.

2.2.2. LAI Data Measurement

LAI was measured using the LAI 2200 plant canopy analyzer. For wheat LAI measurement, three sample points were randomly selected for each plot. The LAI 2200 first performed backlit measurement in an open area to ensure that all LAI data obtained were accurate and valid. The LAI 2200 was then placed at the sample point parallel to the ridge and perpendicular to the ridge, respectively, with the probe placed close to the wheat plant. LAI values were measured 4 times, and the arithmetic mean value was taken as the LAI of the sample point and then the average of the LAI values at the three sample points was taken as the LAI value of the plot.

2.3. Data Processing Methods

2.3.1. Fractional Order Derivative Processing

Fractional order derivative is a fundamental mathematical operation with a wide range of applications in fields such as image enhancement processing and signal analysis [29,30]. Traditional integer order derivative will ignore some information related to crop physiological and biochemical parameters, which affects the accuracy, while fractional order derivative can effectively denoise and refine the local information of hyperspectral data to obtain more detailed information. The commonly used fractional order derivative includes three types as follows: Riemann–Liouville, Caputo, and Grünwald–Letnikov [31]. The paper uses the

Grünwald–Letnikov differentiation to process canopy hyperspectral data. The differential formula is as shown in Equation (1):

$$\frac{d^{\alpha} f(\lambda)}{d\lambda^{\alpha}} = f(\lambda) + (-\alpha)f(\lambda - 1) + \frac{(-\alpha)(-\alpha + 1)}{2} f(\lambda - 2) + \cdots + \frac{\Gamma(-\alpha + 1)}{m!\Gamma(-\alpha + 1)} f(\lambda - m) \tag{1}$$

In the formula, $\Gamma(\cdot)$ is the Gamma function, λ is the corresponding wavelength, m represents the difference between the upper and lower limits of the differential, α represents any order.

2.3.2. Continuous Wavelet Transform

Wavelet transform is called "mathematical microscope", which can decompose complex signals into wavelet signals of different scales (frequencies), with rich basis functions, good time-frequency localization, and multi-scale characteristics. There are two main types of wavelet transforms, that is, continuous wavelet transform (CWT) and discrete wavelet transform (DWT). The paper used CWT to decompose the hyperspectral data in order to obtain a series of wavelet energy coefficients at different scales. The wavelet coefficients contain two dimensions of decomposition scale and band. Therefore, the 1D hyperspectral data are converted to 2D wavelet energy coefficients by continuous wavelet transform. The calculation is shown in Equation (2):

$$W_f(a, b) = \int_{-\infty}^{+\infty} f(\lambda)\Psi_{a,b}(\lambda)d\lambda \tag{2}$$

$$\Psi_{a,b}(\lambda) = \frac{1}{\sqrt{a}}\Psi\left(\frac{\lambda - b}{a}\right) \tag{3}$$

where $f(\lambda)$ is the hyperspectral reflectance, λ is the spectral band in the range of 350–1350 nm, $\Psi_{a,b}$ denotes the wavelet basis function, a denotes the scale factor, b is the translation factor.

2.4. Modeling Methods

2.4.1. Optimal Subset Regression

Optimal subset regression is a method combining all alternative independent variables as a subset of the model for regression modeling. For a model with independent variables, the optimal subset can be used to build 2^{n-1} subset models, and the selection is performed to determine the best combination of independent variables for the model by using the maximum adjusted R^2 ($Adj \cdot R^2$) as the principle of independent variable selection [32].

$$Adj \cdot R^2 = 1 - \frac{RSS/(n - k - 1)}{TSS/(n - 1)} \tag{4}$$

$$RSS = \sum_{i=1}^{n} \left(y_i - \widehat{y}_i\right)^2 \tag{5}$$

$$TSS = \sum_{i=1}^{n} \left(y_i - \overline{y_i}\right)^2 \tag{6}$$

where RSS denotes the residual sum of squares, TSS denotes total sum of square difference, y_i is the measured value, $\widehat{y}_i$ is the estimated value, $\overline{y_i}$ is the estimated value of the measured value, n is the number of samples, k is the number of independent variables, and i is the sample identifier.

2.4.2. Support Vector Machines

Support vector machine (SVM) is a machine learning algorithm based on supervised learning and the principle of structural risk minimization. By projecting data into a high-dimensional space through a kernel function and finding the optimal hyperplane in the high-dimensional space, it better solves the dimensional catastrophe and overfitting problems, with good generalization ability and robustness. Therefore, it is usually used for pattern recognition, classification, and small sample regression analysis [33]. Support vector machines are more stable in training and are capable of obtaining more accurate results when used for small sample regression analysis. So, it can be used to find the optimum directly in the learning performance and complexity of the model based on limited data information, with a view to obtaining the best generalization capability.

2.5. Model Accuracy Evaluation

In the paper, 75% of the sample data was selected for model construction and the rest was used for model accuracy validation. The coefficient of determination (R^2), root mean squared error ($RMSE$), and standard root mean squared error ($nRMSE$) are selected as model accuracy evaluation indexes, and each evaluation indexes are calculated as follows.

$$R^2 = \frac{(\sum\limits_{i=1}^{n} y_i - \overline{y})^2}{(\sum\limits_{i=1}^{n} x_i - \overline{y})^2} \tag{7}$$

$$RMSE = \sqrt{\frac{\sum\limits_{i=1,j=1}^{n} (x_i - y_i)^2}{n}} \tag{8}$$

$$nRMSE = \sqrt{\frac{\sum\limits_{i=1,j=1}^{n} (x_i - y_i)^2}{n}} \Big/ \overline{y} \tag{9}$$

where x_i is the measured value, y_i is the estimated value, $\overline{y}$ is the mean value, i is the sample identifier, and n is the number of samples.

In general, a larger R^2 and a smaller $RMSE$ indicate better modeling results. In addition, $nRMSE \leq 10\%$ indicates that the consistency between measured and estimated values is excellent, $10\% < nRMSE \leq 20\%$ indicates that the consistency between measured and estimated values is good, $20\% < nRMSE \leq 30\%$ indicates that the consistency between measured and estimated values is moderate, and $nRMSE > 30\%$ indicates that the consistency between measured and estimated values is poor.

3. Results

3.1. Estimation of Wheat LAI Based on Fractional Order Differential Spectra

3.1.1. Correlation Analysis of Raw Spectra and Fractional Order Differential Spectra with Wheat LAI

The fractional order differentiation of the original hyperspectral data was performed using the Grünwald–Letnikov differentiation with order range 0–2 and step size 0.1, when the original spectrum, the first order differentiation spectrum, and the second order differentiation spectrum are represented. Twenty fractional order differential transformations were applied to the canopy raw spectra at the jointing stage, booting stage, flowering stage, and filling stage, respectively. The correlations between the raw spectra and differential spectra of each order and LAI were plotted for different fertility periods, and the results are shown in Figures 2 and 3. Meanwhile, the 10 differential spectra with a strong correlation were selected and plotted with the LAI correlation matrix, and the results are shown in Figure 4.

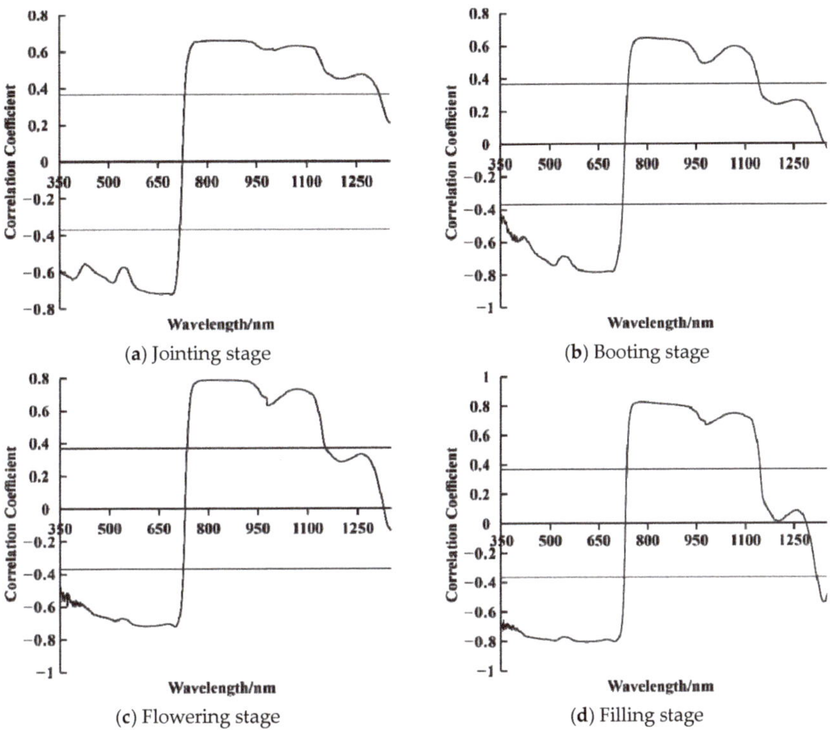

Figure 2. Correlation analysis of original spectrum and LAI at different growth stages.

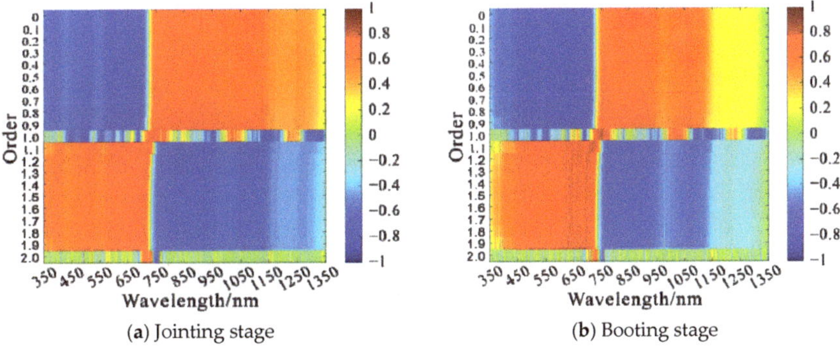

Figure 3. *Cont.*

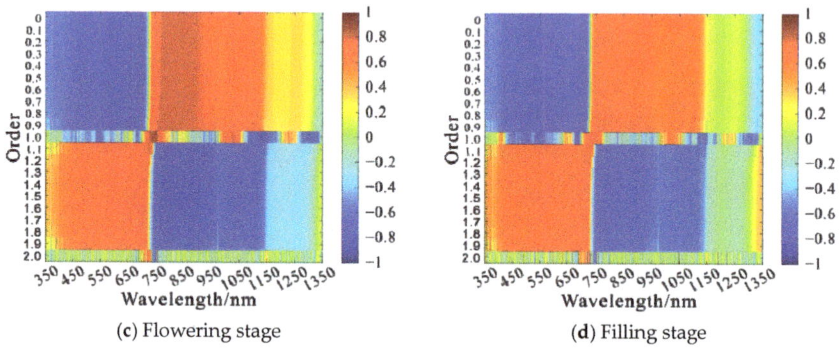

(c) Flowering stage   (d) Filling stage

Figure 3. Correlation analysis of fractional differentiation spectrum and leaf area index at different growth stages.

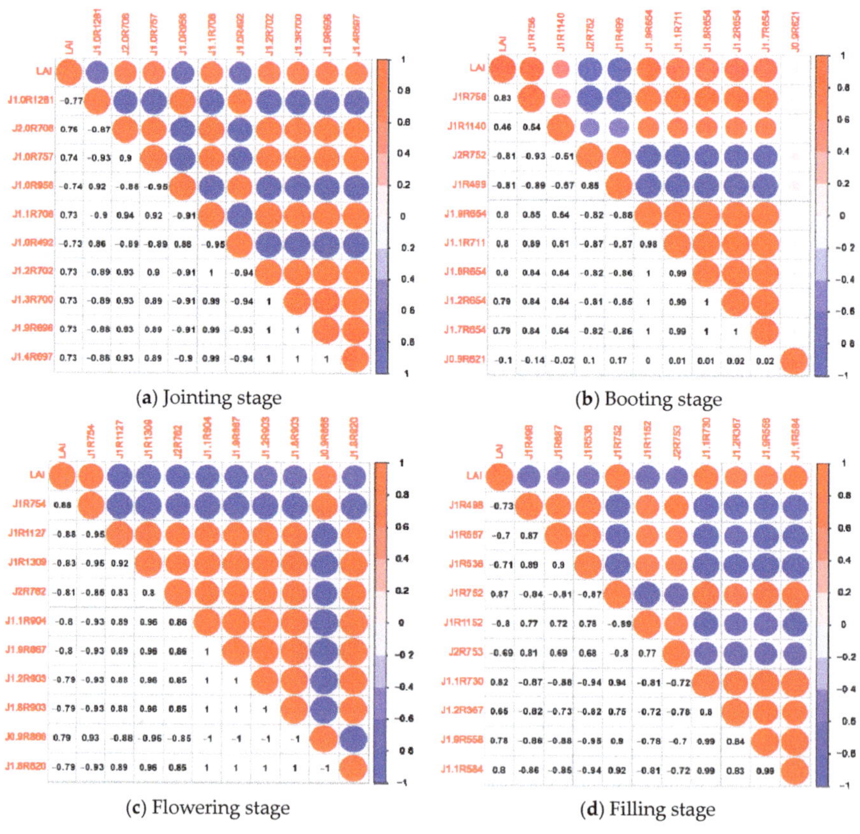

(a) Jointing stage   (b) Booting stage

(c) Flowering stage   (d) Filling stage

Figure 4. Correlation matrix diagram of selected fractional differentiation spectrum and leaf area index at different growth stages.

At the jointing stage, Figure 2a, using raw spectral reflectance with LAI to analyze correlatively, showed that it was significantly negatively correlated with LAI at the 0.01 level in the band range of 350–716 nm and significantly positively correlated with LAI at the 0.01 level in the band range of 733–1318 nm, with a maximum absolute value of the correlation coefficient $|\rho| = 0.72$. Correlation analysis using fractional order differential spectra and LAI and analysis of Figure 3a showed that the maximum value of the absolute value of the correlation coefficient $|\rho|$ between each order of differential spectra and LAI was above 0.72 and, when the order was 1, the maximum value of $|\rho|$ was up to 0.77. The number of spectral bands passing the 0.01 highly significant level test was above 946 except for the integer orders (1st and 2nd orders) and was up to 977 when the order was 1.1. The orders and bands where the 10 differential spectra with high correlation coefficients were located were, respectively, 1st order, 1281 nm, 2nd order, 708 nm, 1st order, 757 nm, 1st order, 956 nm, 1.1 order, 708 nm, 1st order, 492 nm, 1.2 order, 702 nm, 1.3 order, 700 nm, 1.9 order, 696 nm, 1.4 order, 697 nm, and their correlation with the correlation matrix of LAI was shown in Figure 4a.

At the booting stage, Figure 2b, using raw spectral reflectance with LAI to analyze correlatively, indicated it was significantly negatively correlated with LAI at the 0.01 level in the band range of 350–723 nm and significantly positively correlated with LAI at the 0.01 level in the band range of 741–1141 nm, with the maximum absolute value of the correlation coefficient $|\rho| = 0.78$. The correlation between fractional order differential spectra and LAI was analyzed. Figure 3b showed that the maximum absolute value of the correlation coefficient $|\rho|$ between each order of differential spectra and LAI was above 0.78 and, when the order was 1, the maximum value of $|\rho|$ could reach 0.83. Except for the integer order (1st and 2nd order), the number of spectral bands passing the 0.01 highly significant level test was above 738, and when the order was 0, 0.1, 0.2, 0.3, 0.4, and 0.5, the maximum number of bands was up to 775. The orders and bands where the 10 differential spectra with high correlation coefficients were located were as follows: 1st order, 756 nm, 1st order, 1140 nm, 2nd order, 752 nm, 1st order, 499 nm, 1.9 order, 654 nm, 1.1 order, 711 nm, 1.8 order, 654 nm, 1.2 order, 654 nm, 1.7 order, 654 nm, 0.9 order, 621 nm, and their correlation with the correlation matrix of LAI was shown in Figure 4b.

At the flowering stage, Figure 2c, using raw spectral reflectance with LAI to analyze correlatively, said that it was significantly negatively correlated with LAI at the 0.01 level in the band range of 350–723 nm and significantly positively correlated with LAI at the 0.01 level in the band range of 736–1154 nm, with the maximum absolute value of the correlation coefficient $|\rho| = 0.79$. The correlation between the fractional order differential spectra and LAI was analyzed. It was showed in Figure 3c that the maximum value of the absolute value of the correlation coefficient $|\rho|$ between each order of differential spectra and LAI was above 0.79 and, when the order was 1, the maximum value of $|\rho|$ could reach 0.88. Except for the integer order (1st and 2nd order), the number of spectral bands passing the 0.01 highly significant level test was above 757 and when the order was 0, 0.1, 0.2, 0.3, 0.4, the maximum number of bands could be 793. The orders and bands where the 10 differential spectra with high correlation coefficients were located were as follows: 1st order, 754 nm, 1st order, 1127 nm, 1st order, 1309 nm, 2nd order, 762 nm, 1.1 order, 904 nm, 1.9 order, 867 nm, 1.2 order, 903 nm, 1.8 order, 903 nm, 0.9 order, 866 nm, 1.8 order, 820 nm, and their correlation matrix with LAI was shown in Figure 4c.

At the filling stage, Figure 2d, using raw spectral reflectance with LAI to analyze correlatively, showed that it is significantly negatively correlated with LAI at the 0.01 level in the 350–727 and 1320–1350 nm band ranges and significantly positively correlated with LAI at the 0.01 level in the 736–1144 nm band range, with the maximum absolute value of the correlation coefficient $|\rho| = 0.83$. The correlation between fractional order differential spectra and LAI was analyzed. Figure 2d showed that the maximum value of the absolute value of the correlation coefficient $|\rho|$ between each order differential spectrum and LAI was above 0.72. When the order was 1, $|\rho|$ can reach up to 0.82. The number of spectral bands passing the 0.01 highly significant level test was above 766, except for integer orders (1st and 2nd orders), and up to 794 when the order was 0 and 0.1. The 10 differential spectra with high correlation coefficients were located in the order and band were as follows: 1st order, 498 nm, 1st order, 687 nm, 1st order, 538 nm, 1st order, 752 nm, 1st order, 1152 nm, 2nd order, 753 nm, 1st order, 730 nm, 1st order, 367 nm, 1st order, 558 nm, 1st order, 584 nm, and their correlation with LAI correlation matrix was shown in Figure 4d.

3.1.2. Construction of LAI Estimation Model Based on Optimal Subset Regression

During the construction of the LAI estimation model, it can be seen from the results of the optimal subset analysis (shown in Figure 5) that, at the jointing stage, five fractional order differential spectra, J2.0R708, J1.0R956, J1.1R708, J1.0R492, and J1.9R696, were selected as independent variables to construct the optimal subset regression model. the results of the subset analysis are shown in Figure 5a. At the booting stage, nine fractional order differential spectra, J1R756, J1R1140, J2R752, J1R499, J1.9 R654, J1.1R711, J1.8R654, J1.2R654, and J1.7R654, were selected as independent variables to construct the optimal subset regression model and the results of subset analysis are shown in Figure 5b. At the flowering stage, five fractional order differential spectra, J1R754, J1R1127, J1R1309, J1.9R867, and J1.8R820, were selected as independent variables to construct the optimal subset regression model and the results of the subset analysis are shown in Figure 5c. At the filling stage, three fractional order differential spectra, J1R687, J1R538, and J1R752, were selected as independent variables to construct the optimal subset regression model and the results of the subset analysis are shown in Figure 5d.

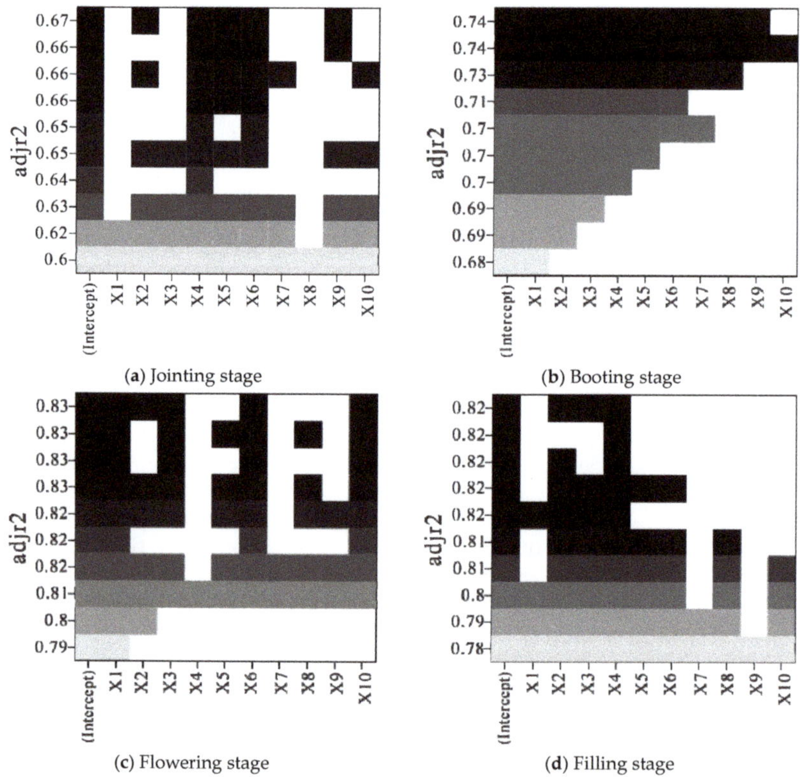

Figure 5. Optimal subset analysis of selected fractional differentiation spectrum for estimating LAI.

According to the optimal fractional order differential spectra preferentially selected in different growth stages, the models for LAI estimation at the jointing, booting, flowering, and filling stages were constructed using 75% of sample data based on the optimal subset regression method and 25% of sample data was used for model accuracy validation. The results of R^2, *RMSE*, and *nRMSE* in the modeling and validation of LAI estimation models are shown in Table 1.

Table 1. Optimal subset regression modeling results of LAI estimation based on fractional differentiation spectrum at different growth stages.

Growth Stages	Modeling Accuracy			Verification Accuracy		
	R^2	*RMSE* (μg/cm^2)	*nRMSE* (%)	R^2	*RMSE* (μg/cm^2)	*nRMSE* (%)
Jointing stage	0.72	0.47	12.65%	0.40	0.90	26.30%
Booting stage	0.67	1.75	61.86%	0.56	2.06	87.55%
Flowering stage	0.86	0.45	13.43%	0.63	0.74	21.61%
Filling stage	0.84	0.31	19.78%	0.70	0.62	41.69%

As shown in Table 1, the fractional order differential spectrum combined with optimal subset regression for LAI estimation had the best estimation effect at the flowering stage and the worst estimation effect at the booting stage, with *nRMSE* reaching 87.55%. The consistency between the estimated value and the measured value is particularly poor.

3.1.3. Construction of LAI Estimation Model Based on the Support Vector Machine

Select the first 10 fractional differential spectra with a strong correlation with LAI in each growth period as the independent variable, LAI as the dependent variable, and use 75% of the sample data to construct LAI estimation models during the jointing stage, booting stage, flowering stage, and filling stage under the method of SVM. Then, 25% of the sample data was used for accuracy verification. The results of modeling and verification R^2, *RMSE*, and *nRMSE* are shown in Table 2.

Table 2. Estimation of leaf area index based on fractional differentiation spectrum at different growth stages and modeling results of support vector machine.

Growth Stages	Modeling Accuracy			Verification Accuracy		
	R^2	*RMSE* (μg/cm^2)	*nRMSE* (%)	R^2	*RMSE* (μg/cm^2)	*nRMSE* (%)
Jointing stage	0.65	0.56	15.12%	0.65	0.60	17.09%
Booting stage	0.80	0.83	20.01%	0.76	0.82	23.32%
Flowering stage	0.87	0.47	13.94%	0.57	0.80	28.37%
Filling stage	0.83	0.38	21.24%	0.63	0.51	36.76%

As shown in Table 2, when combining fractional order differential spectra with SVM to estimate LAI, the accuracy of LAI estimation at the booting stage and flowering stage was comparable and the estimation at the flowering stage was slightly better than that at the booting stage.

It could be noticed from the comprehensive analysis of LAI estimation results for different growth stages, acquired by using fractional order differential spectra based on optimal subset regression and SVM methods, that at the jointing stage, the overall estimation effect was poor. Although the modeling R^2 of the optimal subset regression model reached 0.72, the R^2 of the model validation was only 0.40, which was a poor estimation effect, while the SVM model had a relatively good estimation effect with R^2 values of 0.65 and 0.60 in the modeling and validation of LAI estimation models, respectively. Overall, the LAI estimation performance at the booting stage was better than that at the jointing stage, the performance of the SVM model estimation was better compared with the optimal subset regression model, and the R^2 in the modeling and validation of the model reached 0.80 and 0.76, respectively. The LAI estimation performance at the flowering stage was the best overall. The estimation performances of optimal subset regression model and the SVM model were comparable and values of R^2 in the modeling and validation of the model reached 0.86, 0.63 and 0.87, 0.57, respectively. Compared with that at the flowering stage, the LAI estimation accuracy was slightly worse at the filling stage. The estimation results of the same optimal subset regression model and SVM model were comparable, with the values R^2 in the modeling and validation of the model reaching 0.84, 0.70 and 0.83, 0.63, respectively.

3.2. Estimation of Wheat LAI Based on Continuous Wavelet Transform

3.2.1. Analysis of Correlation between Wavelet Energy Coefficients and LAI

By using the second-order derivative of Gaussian function Mexican Hat as the wavelet basis of the continuous wavelet transform and applying the continuous wavelet transform to the canopy hyperspectral data of wheat at each growth stage, respectively, the wavelet energy coefficients at different scales is obtained. The correlation between LAI and wavelet energy coefficients at different growth periods was analyzed. The correlation graphs and correlation matrices between wavelet energy coefficients and LAI at different fertility periods were plotted, as shown in Figures 6 and 7.

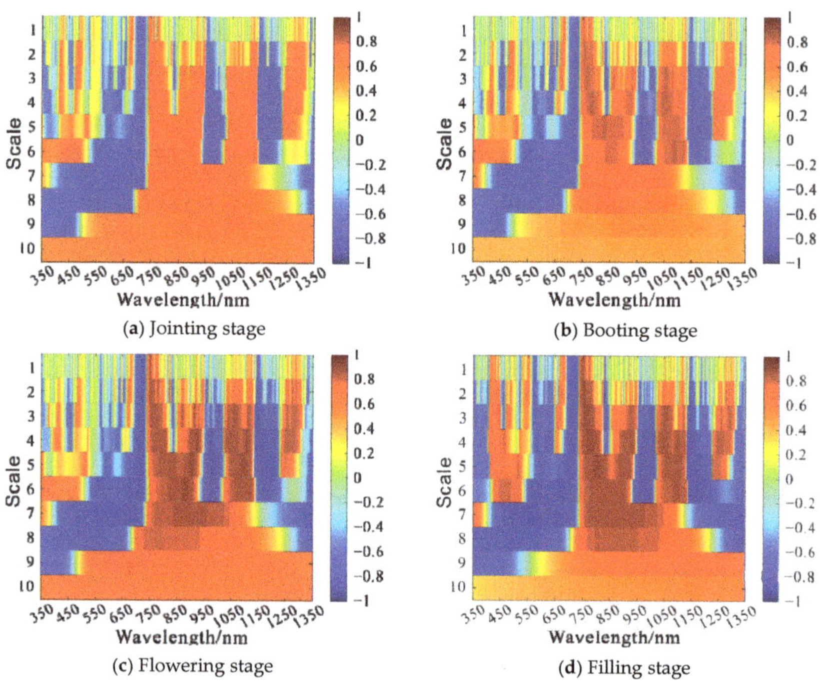

Figure 6. Correlation analysis of wavelet energy coefficient and leaf area index at different growth stages.

The results of the analysis of Figures 6 and 7 are as follows:

At the jointing stage, the correlation analysis was carried out by using the wavelet energy coefficient and LAI. As shown in Figure 6a, with the increase in the decomposition scale, there is first a rise on the absolute value of the correlation coefficient between wavelet energy coefficient and LAI $|\rho|$, then a drop. When the decomposition scale was 10, the maximum value of $|\rho|$ was above 0.70, and when the decomposition scale was 7, the maximum value of $|\rho|$ was up to 0.77. With the increase in the decomposition scale, the number of spectral bands passing the 0.01 highly significant level test gradually increased, and when the decomposition scale was 10, the number of bands was up to 1001. The decomposition scale and band where the 10 wavelet energy coefficients with high correlation coefficients were located were C7R725, C2R762, C3R600, C3R417, C6R722, C2R1150, C2R949, C3R766, C3R1276, C5R719, and their correlation matrix with LAI was shown in Figure 7a.

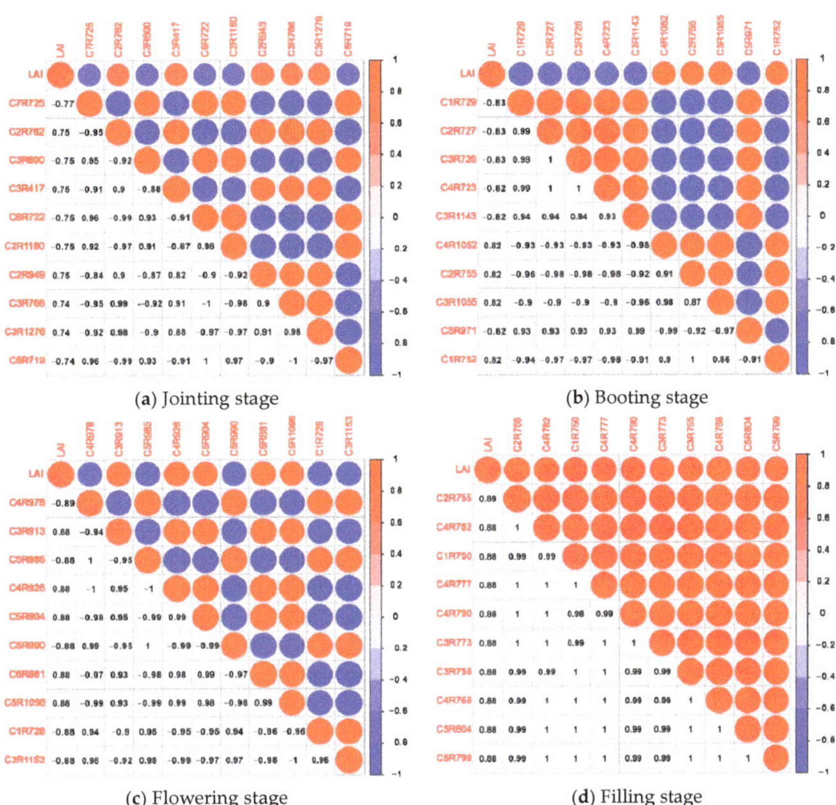

(**a**) Jointing stage

(**b**) Booting stage

(**c**) Flowering stage

(**d**) Filling stage

Figure 7. Correlation analysis of selected wavelet energy coefficient and leaf area index at different growth stages.

At the booting stage, the correlation analysis was carried out by using the wavelet energy coefficient and LAI. As shown in Figure 6b, with the increase in the decomposition scale, the absolute value of the correlation coefficient between wavelet energy coefficient and LAI $|\rho|$ decreased gradually. When the decomposition scale was 10, the maximum value of $|\rho|$ was above 0.72, and when the decomposition scale was 1, the maximum value of $|\rho|$ could reach 0.83. With the increase in the decomposition scale, the number of spectral bands passing the 0.01 highly significant level test gradually increases; when the decomposition scale was 10, the number of bands can reach up to 1001 bands. The decomposition scales and bands where the 10 wavelet energy coefficients with high correlation coefficients are located are C1R729, C2R727, C3R726, C4R723, C3R1143, C4R1052, C2R755, C3R1055, C5R971, C1R752, and their correlation matrix with LAI is shown in Figure 7b.

At the flowering stage, the correlation analysis was carried out by using the wavelet energy coefficient and LAI. As shown in Figure 6c, with the increase in the decomposition scale, the absolute value of the correlation coefficient between wavelet energy coefficient and LAI $|\rho|$ increased first and then decreased. When the decomposition scale was 10, the maximum value of $|\rho|$ was above 0.70, and when the decomposition scale was 4, the maximum value of $|\rho|$ could reach 0.89. With the increase in the decomposition scale, the number of spectral bands passing the 0.01 highly significant level test gradually increased, and when the decomposition scale was 10, the number of bands could reach 1001. The decomposition scales and bands where the 10 wavelet energy coefficients with high correlation coefficients were located were C4R978, C3R913, C5R985, C4R926, C5R904,

C5R990, C6R881, C5R1098, C1R728, C3R1153, and their correlation matrix with LAI was shown in Figure 7c.

At the filling stage, the correlation analysis was carried out by using the wavelet energy coefficient and LAI. As shown in Figure 6d, with the increase in the decomposition scale, the absolute value of the correlation coefficient $|\rho|$ between wavelet energy coefficients and LAI gradually decreased. When the decomposition scale was 10, the maximum value of $|\rho|$ was above 0.80, and when the decomposition scale was 2, the maximum value of $|\rho|$ could reach 0.89. With the increase in the decomposition scale, the number of spectral bands passing the 0.01 highly significant level test gradually increased, and when the decomposition scale was 6, the number of bands was up to 951. The decomposition scales and bands where the 10 wavelet energy coefficients with high correlation coefficients are located are C2R755, C4R782, C1R750, C4R777, C4R790, C3R773, C3R755, C4R768, C5R804, C5R799, and their correlation matrix with LAI was shown in Figure 7d.

3.2.2. Construction and Analysis of LAI Estimation Model Based on Optimal Subset Regression

The results of the optimal subset analysis (shown in Figure 8) in the construction of the LAI estimation model showed that, at the jointing stage, four wavelet energy coefficients, C3R600, C3R417, C6R722, and C5R719, were selected as independent variables to construct the optimal subset regression model, and the results of the subset analysis are shown in Figure 8a. At the booting stage, five wavelet energy coefficients, C1R729, C2R727, C3R726, C4R723, and C1R752, were selected as independent variables to construct the optimal subset regression model, and the results of the subset analysis are shown in Figure 8b. At the flowering stage, seven wavelet energy coefficients, C4R978, C4R926, C5R904, C5R990, C5R1098, C1R728, and C3R1153, were selected as independent variables to construct the optimal subset regression model, and the results of the subset analysis are shown in Figure 8c. At filling stage, the wavelet energy coefficient C2R755 was selected as the independent variable to construct the optimal subset regression model, and the results of the subset analysis are shown in Figure 8d.

According to the optimal fractional differential spectrum selected in different growth stages, the models for LAI estimation at the jointing, booting, flowering, and filling stages were constructed using 75% of sample data based on the optimal subset regression method and 25% of sample data were used for model accuracy validation. The results of R^2, *RMSE*, and *nRMSE* in the modeling and validation of LAI estimation models are shown in Table 3.

Table 3. Optimal subset regression modeling results of leaf area index estimation based on wavelet energy coefficient at different growth stages.

Growth Period	Modeling Accuracy			Verification Accuracy		
	R^2	*RMSE* (μg/cm^2)	*nRMSE* (%)	R^2	*RMSE* (μg/cm^2)	*nRMSE* (%)
Jointing stage	0.73	0.46	12.40%	0.59	0.71	19.47%
Booting stage	0.81	0.78	18.46%	0.56	0.95	23.64%
Flowering stage	0.87	0.43	12.84%	0.71	0.66	19.15%
Filling stage	0.84	0.31	19.91%	0.77	0.58	38.55%

As shown in Table 3, when combining wavelet energy coefficients with optimal subset regression to estimate LAI, the accuracy at the flowering stage was the highest; its modeling accuracy was (R^2 = 0.87, *RMSE* = 0.43μg/cm^2, *nRMSE* = 12.84%) and the verification accuracy was (R^2 = 0.71, *RMSE* = 0.66μg/cm^2, *nRMSE* = 19.15%).

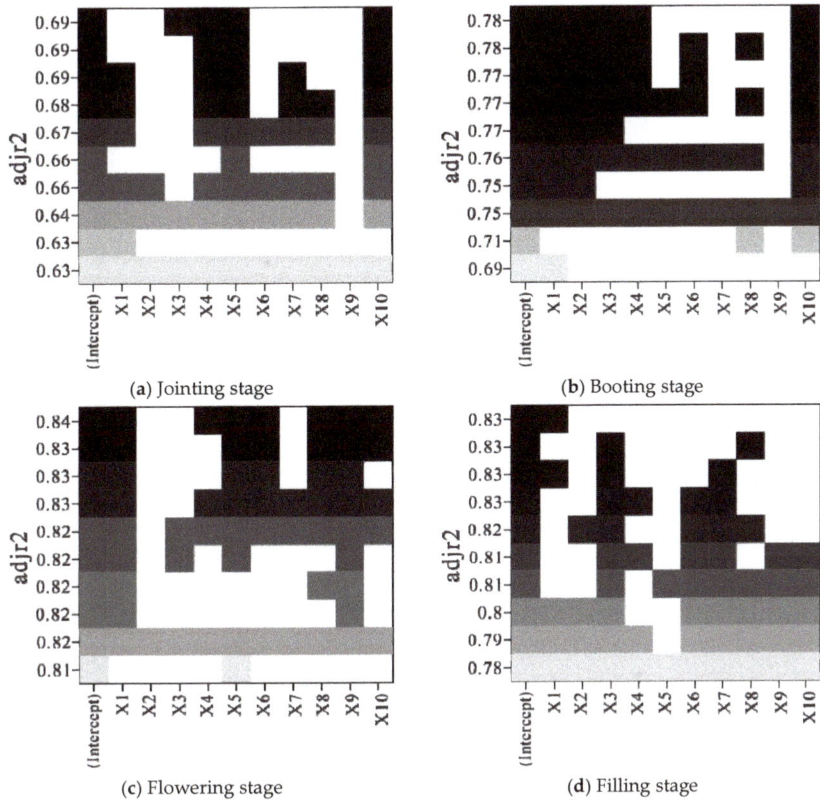

(a) Jointing stage

(b) Booting stage

(c) Flowering stage

(d) Filling stage

Figure 8. Optimal subset analysis of wavelet energy coefficient for estimating leaf area index at different growth stages.

3.2.3. Construction and Analysis of LAI Estimation Model Based on the Support Vector Machine

Select the first 10 fractional differential spectra with a strong correlation with LAI in each growth period as the independent variable, LAI as the dependent variable, and use 75% of the sample data to construct LAI estimation models during the jointing stage, booting stage, flowering stage, and filling stage under the method of SVM. Then, 25% of the sample data was used for accuracy verification. The results of modeling and verification R^2, *RMSE*, and *nRMSE* are shown in Table 4.

Table 4. Estimation of leaf area index based on wavelet energy coefficient at different growth stages and modeling results of support vector machine.

Growth Period	Modeling Accuracy			Verification Accuracy		
	R^2	*RMSE* (μg/cm^2)	*nRMSE* (%)	R^2	*RMSE* (μg/cm^2)	*nRMSE* (%)
Jointing stage	0.69	0.52	14.37%	0.74	0.58	16.83%
Booting stage	0.76	0.88	20.80%	0.63	0.87	23.40%
Flowering stage	0.87	0.42	15.47%	0.65	0.73	29.93%
Filling stage	0.90	0.39	11.47%	0.63	0.76	26.91%

Table 4 showed that, when combining wavelet energy coefficients with SVM to estimate the LAI number, the accuracy at the flowering stage and filling stage were comparable and their estimation effect was better than that at the jointing stage and booting stage.

In addition, their modeling R^2 was above 0.85, validation R^2 was greater than 0.60, and the modeled *nRMSE* was less than 20%.

It could be noticed from the comprehensive analysis of LAI estimation results for different growth stages, acquired by using wavelet energy coefficient based on optimal subset regression and SVM methods, that, compared with other growth stages, the overall estimation effect of LAI at the jointing stage was poor, and the estimation effect of optimal subset regression model and SVM model was equivalent, with the modeling and verification R^2 reaching 0.73, 0.59 and 0.69, 0.74, respectively. At the booting stage, the LAI estimation effect was better than that in jointing stage. Compared with SVM model, the optimal subset regression model had better estimation effect, with the modeling and verification R^2 of the model reaching 0.81 and 0.56, respectively. At the flowering and filling stages, LAI estimation effect was equivalent and the estimation effect of optimal subset regression model and SVM model was also equivalent; the modeling R^2 of the model reached 0.87, 0.84 and 0.87, 0.90, respectively, and the verification R^2 reached 0.71, 0.77 and 0.65, 0.63, respectively.

4. Discussion

By comprehensively analyzing the estimation results of wheat LAI at four growth stages using fractional differential spectrum and wavelet energy coefficient based on optimal subset regression and SVM method, it was noticed that the maximum values of R^2 in modeling and verification of the estimation model were 0.90 and 0.77, respectively, and the average values of R^2 in modeling and verification were 0.79 and 0.64, respectively, indicating generally that the LAI estimation performance was good, which is mainly due to the following reasons. First, the data source is hyperspectral data, which have the advantages of high spectral resolution and rich spectral information of features, able to be used to analyze the spectral characteristics of features from multiple angles and directions and comprehensively express the detailed information of crops, and are suitable for remote sensing estimation of crop phenotype parameters. Second, the difficulty of crop LAI estimation using hyperspectral data lies in the accurate extraction of LAI-sensitive spectral information from hyperspectral data; however, the data acquisition process was affected by factors such as environment and background, resulting in noise in the original spectral data that affected the extraction of sensitive information. After fractional order differentiation and wavelet transform processing, the noise effect can be effectively eliminated, the spectral information can be refined, and the crop LAI estimation effect can be effectively improved, which is consistent with the conclusions of research by Li Changchun et al. [22], Fang Shenghui et al. [34], and Yao Shengnan et al. [35].

The result of comprehensive analysis on the effect of LAI estimation at different growth stages showed that, at the jointing stage, the maximum values of R^2 in modeling and verification were only 0.73 and 0.74, which was relatively poor compared with that at other growth stages. The values of R^2 in the modeling and validation of the LAI estimation model at the flowering stage and the filling stage were slightly different, with the average values of 0.87, 0.85 and 0.64, 0.68, respectively. The modeled *nRMSE* was slightly different, with the average values of 13.92 and 18.10%, respectively, both of which were less than 20%; however, at the filling stage, the maximum value of verified *nRMSE* was 41.69% and the average value was 35.98%, while at the flowering stage, the maximum value of verified *nRMSE* was only 29.93% and the average value was 24.66%, indicating that the LAI estimation effect at the flowering stage was slightly better than that at the filling stage, which is mainly because wheat growth reaches its peak at the flowering stage and then LAI starts to decrease. In addition, LAI is relatively sensitive to the canopy spectrum.

The result of comprehensive analysis on the LAI estimation effects of different methods at different growth stages showed that, at the jointing stage, the estimation effect of the optimal subset regression model based on wavelet energy coefficient was equivalent to that of the SVM model, and the R^2 values in modeling and verification were 0.73, 0.59 and 0.69, 0.74, respectively. At the booting stage, the LAI estimation accuracy of the SVM model based on fractional differential spectrum was the highest, and the R^2 values in modeling and verification reached 0.80 and 0.76, respectively. At the flowering stage, the LAI estimation accuracy of the optimal subset regression model based on wavelet energy coefficient was the highest, and the R^2 values in the modeling and verification were 0.87 and 0.71, respectively. At the filling stage, the LAI estimation accuracy of the optimal subset regression model based on wavelet energy coefficient was the highest, and the R^2 values in the modeling and verification reached 0.84 and 0.77, respectively. It provided the best scheme for wheat LAI estimation at different growth stages.

For four growth periods, the effect of LAI estimation by optimal subset regression models and SVM models based on fractional order differential spectra and wavelet energy coefficients was analyzed. Compared with fractional differential spectrum, the optimal subset regression model and SVM model based on wavelet energy coefficient have better LAI estimation effect; the maximum values of R^2 in modeling and verification were 0.90 and 0.77, respectively, and the average values were 0.81 and 0.66, respectively. This was mainly because, as a new spectral processing method, wavelet transform can effectively reduce noise and decompose spectral data, mine spectral hidden information, extract more sensitive information, and effectively improve the accuracy and generalization ability of the model, which was consistent with the research conclusions of Ebrahimi et al. [23] and Cheng et al. [36].

5. Conclusions

In this paper, the collected hyperspectral data of wheat canopy at different growth stages were processed by fractional order differentiation and continuous wavelet transform. According to the correlation analysis results, the fractional order differential spectrum and wavelet energy coefficients with strong correlation were selected. Combined with the optimal subset regression and SVM method, LAI estimation models of wheat at different growth stages were constructed, respectively. Based on the results of modeling and validation accuracy assessment to screen the optimal estimation model of LAI at each growth stage and the fertility period with the best LAI estimation, this study can provide a theoretical and technical reference for LAI estimation of crops based on remote sensing technology. However, the LAI estimation methods used in the paper have some limitations and need to be further improved in the following two aspects.

The optimal subset regression and support vector machine algorithms used in the paper are both empirical regression models, which ignore the radiative transfer processes of vegetation canopy and atmosphere. The next study needs to parameterize the remote sensing radiative transfer processes to further improve the stability of the model.

Due to the limitation of the experimental conditions, only the sample data of one experimental area were used in this study, and the influence of experimental data by locality was not considered. The subsequent study can verify the prediction effect of the model by increasing the sample size, crop species, and sample years, which is very meaningful for the further transfer of the model to practical applications.

Author Contributions: Conceptualization, C.L.; data curation, C.M.; methodology, Y.W.; software, F.D. and Y.W.; formal analysis, Y.L.; writing—original draft preparation, Y.W.; writing—review and editing, C.L.; validation, F.D.; investigation, W.C. and Z.X.; visualization, J.L.; funding acquisition, C.L. All authors have read and agreed to the published version of the manuscript.

Funding: This study was supported by the Natural Science Foundation of China (41871333), the Scientific and Technological Innovation Team of Universities in Henan Province (22IRTSTHN008).

Institutional Review Board Statement: Not applicable.

Informed Consent Statement: Not applicable.

Data Availability Statement: Since the datasets were acquired through field collection by the Beijing Research Center for Information Technology in Agriculture, all data cannot be shared due to legal, ethical, and privacy restrictions.

Acknowledgments: The authors thank the Beijing Agricultural Information Technology Research Center for the provided dataset. We thank all the authors for their support and all the reviewers who participated in this review.

References

1. Sonnentag, O.; Talbot, J.; Chen, J.M.; Roulet, N.T. Using direct and indirect measurements of leaf area index to characterize the shrub canopy in an ombrotrophic peatland. *Agric. For. Meteorol.* **2007**, *144*, 200–212. [CrossRef]
2. Huang, J.; Wang, Y.; Wang, F.; Liu, Z. Red edge characteristics and leaf area index estimation model using hyperspectral data for rape. *Trans. Chin. Soc. Agric. Eng.* **2006**, *22*, 22–26.
3. Su, W.; Zhan, Q.; Zhang, M.; Wu, D.; Zhang, R. Estimation method of crop leaf area index based on airborne LiDAR data. *Trans. Chin. Soc. Agric. Mach.* **2016**, *47*, 272–277.
4. Fieuzal, R.; Baup, F. Estimation of leaf area index and crop height of sunflowers using multi-temporal optical and SAR satellite data. *Int. J. Remote Sens.* **2016**, *37*, 2780–2809. [CrossRef]
5. Liu, J.; Pang, X.; Li, Y.; Du, L. Inversion study on leaf area index of summer maize using remote sensing. *Trans. Chin. Soc. Agric. Mach.* **2016**, *47*, 309–317.
6. Li, S.; Li, H.; Sun, D.; Zhou, L. Estimation of regional leaf area index by remote sensing inversion of PROSAIL canopy spectral model. *Spectrosc. Spect. Anal.* **2009**, *29*, 2725–2729.
7. Wang, L.; Tian, Y.; Zhu, Y.; Yao, X.; Zheng, G.; Cao, W. Estimation of winter wheat leaf area index by fusing different spatial and temporal resolution remote sensing data. *Trans. Chin. Soc. Agric. Eng.* **2012**, *28*, 117–124.
8. Wang, D.; Huang, C.; Zhang, W.; Ma, Q.; Zhao, P. Relationships Analysis between Cotton Chlorophyll Content, Chlorophyll Density and Hyperspectral Data. *Cotton Sci.* **2008**, *36*, 368–371. [CrossRef]
9. Chen, J.; Tian, Q.; Shi, R. Study on Simulation of Rice Leaf's Chlorophyll Concentration via the Spectrum. *Remote Sens. Inform.* **2005**, *6*, 12–16. [CrossRef]
10. Jiang, J.; Chen, Y.; Huang, W. Using Hyperspectral Derivative Indices to Diagnose Severity of Winter Wheat Stripe Rust. *Opt. Tech.* **2007**, *33*, 620–623. [CrossRef]
11. Jiang, J.; Chen, Y.; Huang, W. Using Hyperspectral Remote Sensing to Estimate Canopy Chlorophyll Density of Wheat under Yellow Rust Stress. *Spectrosc. Spect. Anal.* **2010**, *30*, 2243–2247.
12. Jiang, J.; Chen, Y.; Huang, W. Using Hyperspectral Derivative Index to Monitor Winter Wheat Disease. *Spectrosc. Spect. Anal.* **2007**, *27*, 2475–2479. [CrossRef]
13. Smith, K.L.; Steven, M.D.; Colls, J.J. Use of hyperspectral derivative ratios in the red-edge region to identify plant stress responses to gas leaks. *Remote Sen. Environ.* **2004**, *92*, 207–217. [CrossRef]
14. António, M.L.; José, A.T.M. A Review of Fractional Order. *Entropies* **2020**, *22*, 1374.
15. Lopes, A.M.; Machado, J.T. Fractional Order Models of Leaves. *J. Vib. Control* **2014**, *20*, 998–1008. [CrossRef]
16. Tarasov, V.E. On history of mathematical economics: Application of fractional calculus. *Mathematics* **2019**, *7*, 509. [CrossRef]
17. Huang, G.; Xu, L.; Pu, Y. Summary of research on image processing using fractional calculus. *Appl. Res. Comput.* **2012**, *29*, 8.
18. Xu, J.; Feng, X.; Guan, L.; Wang, S.; Hu, Q. Fractional Differential Application in Reprocessing Infrared Spectral Data. *Control Instrum. Chem. Ind.* **2012**, *39*, 347–351.
19. Jiang, M.; Guo, Y.; Qian, J.; Ding, M. Effect of Fractional Differential on Soil Hypermetallic Hyperspectral Data at Different Sampling Intervals. *Bull. Surv. Mapp.* **2018**, *10*, 37–40+45.
20. Zhang, W.; Yang, K.; Xia, T.; Liu, C.; Sun, T. Correlation Analysis on Spectral Fractional-order Differential and the Content of Heavy Metal Copper in Corn Leaves. *Sci. Technol. Eng.* **2017**, *17*, 33–38.
21. Wang, J.; Ding, J.; Zhang, D.; Liu, W. Estimation of Desert Soil Organic Carbon Content Based on Hyperspectral Data Preprocessing with Fractional Differential. *Trans. Chin. Soc. Agric. Eng.* **2016**, *32*, 161–169.
22. Li, C.; Shi, J.; Ma, C.; Cui, Y.; Wang, Y.; Li, Y. Estimation of chlorophyll content in winter wheat based on wavelet transform and fractional differential. *Trans. Chin. Soc. Agric. Mach.* **2021**, *52*, 172–182.
23. Ebrahimi, H.; Rajaee, T. Simulation of groundwater level variations using wavelet combined with neural network, linear regression and support vector machine. *Glob. Planet Chang.* **2017**, *148*, 181–191. [CrossRef]
24. Chen, H.; Yang, G.; Han, X.; Liu, X.; Liu, F.; Wang, N. Hyperspectral inversion of soil organic matter content based on continuous wavelet transform. *J. Agric. Sci. Technol.-Iran.* **2021**, *23*, 132–142.
25. Li, B.; Wang, M.; Wang, G.; Hu, Y.; Li, W.; Liu, Y.; Xu, J. Detection of chlorophy content of peach leaves based on hyperspectral technology. *Eng. Sur. Mapp.* **2018**, *27*, 6.
26. Tan, X.; Wang, Z.; Zhang, J.; Wang, B.; Yang, F.; Yang, W. Estimation of maize canopy chlorophyll density under drought stress based on continuous wavelet transform. *Agric. Res. Arid Areas* **2021**, *39*, 155–161.
27. Wang, Y.; Zhang, X.; Jin, Y.; Gu, X.; Feng, H.; Wang, C. Quantitative retrieval of water content in winter wheat leaves based on continuous wavelet transform. *J. Triticeae Crop.* **2020**, *40*, 503–509.
28. Savitzky, A.; Golay, M. Smoothing and Differentiation of Data by Simplified Least Squares Procedures. *Anal. Chem.* **1964**, *36*, 1627–1639. [CrossRef]

29. Wang, C.; Lan, L.; Zhou, S. Adaptive fractional differential and its application to image texture enhancement. *J. Chongqing Univ.* **2011**, *34*, 32–37.

30. Hong, Y.; Chen, Y.; Yu, L.; Liu, Y.; Liu, Y.; Zhang, Y.; Yi, L.; Hang, C. Combining Fractional Order Derivative and Spectral Variable Selection for Organic Matter Estimation of Homogeneous Soil Samples by VIS–NIR Spectroscopy. *Remote Sens.* **2018**, *10*, 479. [CrossRef]

31. Zhao, Q.; Ge, X.; Ding, J.; Wang, J.; Zhang, Z.; Tian, M. Combination of fractional order differential and machine learning algorithm for spectral esbimation of soil organic carbon content. *Laser Optoelectron. Prog.* **2020**, *57*, 253–261.

32. Cai, W.; Zhao, S.; Wang, Y.; Peng, F. Estimation of winter wheat residue cover using spectral and textural information from Sentinel-2 data. *J. Remote Sens.* **2020**, *24*, 1108–1119.

33. Liu, L.; Si, J.; Lin, Q. A fast terminal sliding mode control of permanent magnet synchronous motor with varible parameters predicted by SVM. *J. Xian Jiaotong Univ.* **2021**, *55*, 53–60.

34. Fang, S.; Yue, Y.; Liang, Q. Retieval of cholrophyll content using continuous wavelet analysis across a range of vegetation species. *Geomat. Inform. Sci. Wuhan Univ.* **2015**, *40*, 296–302.

35. Yao, S.; Jiang, J.; Shi, X.; Wang, W.; Meng, H. Hyperspectral estimation of canopy chlorophyll content in soybean under natural gas micro leakage stress. *Geogr. Geo-Inform. Sci.* **2019**, *35*, 22–27.

36. Cheng, T.; Rivard, B.; Sánchez-Azofeifa, A.; Féret, J.; Jacquemoud, S.; Ustin, S.L. Predicting leaf gravimetric water content from foliar reflectance across a range of plant species using continuous wavelet analysis. *J. Plant Physiol.* **2012**, *169*, 1134–1142. [CrossRef]

Article

Assessment of Vineyard Canopy Characteristics from Vigour Maps Obtained Using UAV and Satellite Imagery

Javier Campos, Francisco García-Ruíz and Emilio Gil *

Department of Agro Food Engineering and Biotechnology, Universitat Politècnica de Catalunya, Esteve Terradas, 8, 08860 Castelldefels, Spain; javier.campos@upc.edu (J.C.); fco.jose.garcia@upc.edu (F.G.-R.)
* Correspondence: emilio.gil@upc.edu; Tel.: +34-93-5521099

Abstract: Canopy characterisation is a key factor for the success and efficiency of the pesticide application process in vineyards. Canopy measurements to determine the optimal volume rate are currently conducted manually, which is time-consuming and limits the adoption of precise methods for volume rate selection. Therefore, automated methods for canopy characterisation must be established using a rapid and reliable technology capable of providing precise information about crop structure. This research providedregression models for obtaining canopy characteristics of vineyards from unmanned aerial vehicle (UAV) and satellite images collected in three significant growth stages. Between 2018 and 2019, a total of 1400 vines were characterised manually and remotely using a UAV and a satellite-based technology. The information collected from the sampled vines was analysed by two different procedures. First, a linear relationship between the manual and remote sensing data was investigated considering every single vine as a data point. Second, the vines were clustered based on three vigour levels in the parcel, and regression models were fitted to the average values of the ground-based and remote sensing-estimated canopy parameters. Remote sensing could detect the changes in canopy characteristics associated with vegetation growth. The combination of normalised differential vegetation index (NDVI) and projected area extracted from the UAV images is correlated with the tree row volume (TRV) when raw point data were used. This relationship was improved and extended to canopy height, width, leaf wall area, and TRV when the data were clustered. Similarly, satellite-based NDVI yielded moderate coefficients of determination for canopy width with raw point data, and for canopy width, height, and TRV when the vines were clustered according to the vigour. The proposed approach should facilitate the estimation of canopy characteristics in each area of a field using a cost-effective, simple, and reliable technology, allowing variable rate application in vineyards.

Keywords: vineyard; pesticide application; variable rate application; unmanned aerial vehicle; satellite; nanosatellite

Citation: Campos, J.; García-Ruíz, F.; Gil, E. Assessment of Vineyard Canopy Characteristics from Vigour Maps Obtained Using UAV and Satellite Imagery. *Sensors* **2021**, *21*, 2363. https://doi.org/10.3390/s21072363

Academic Editor: Jiyul Chang

Received: 9 March 2021
Accepted: 26 March 2021
Published: 29 March 2021

Publisher's Note: MDPI stays neutral with regard to jurisdictional claims in published maps and institutional affiliations.

1. Introduction

The European Green Deal, recently launched by the European Commission [1], is designed to deal with climate and environment-related challenges, attempting to develop sustainable responses. Among the several topics included in the Green Deal, agricultural activities and all aspects related to food production are addressed in the European Farm to Fork Strategy to ensure a reasonable, healthy, and environment-friendly food system. This strategy includes the impacting measures linked with the use of a plant protection product (PPP), given its negative effects on air and water quality, soil degradation, food safety, and human health.

One of the most important challenges considered in the Farm to Fork strategy is the objective to reduce the overall use and risk of chemical pesticides by 50% and the use of more hazardous pesticides by 50% by 2030. This objective is particularly important for orchard fruits and vineyards. Vineyards, while accounting for only 7% of the agricultural land area in the European Union consume 48% of the total active ingredients [2]. The crops at fruit orchards and vineyards, similar to all 'three-dimensional' (3D) crops, are characterised by the variation in their canopy characteristics and the heterogeneity in a parcel [3], making it difficult to achieve safe and optimal pesticide application.

The latest improvements in the available technology and its adaptation to these types of crops have resulted in remarkable achievements in both the reduction in the total amount of PPP and an increased control of the losses and, consequently, in the reduction in environmental contamination. These advancements involved the use of an accurate method to identify, characterise, and quantify the amount of pesticide to be sprayed, which is considered as the most important factor related to the success of the pesticide application process. Miranda-Fuentes et al. [4] demonstrated the effects of different methods for crown characterisation in isolated olive trees on the obtained results, concluding that irrespective of the selected method for canopy evaluation, some minimum requirements in terms of accuracy must be ensured to apply the most suitable amount of pesticide. Pesticide dose and dose expression were demonstrated as two factors affected by the canopy characteristics of citrus plantations [5], showing that a large vegetation implies major differences in the canopy deposition and coverage. Drift values in spray application in apple plantations were also directly related to crop foliage characteristics [6], with a fully foliated canopy resulting in a 25-times less drift than the one obtained in a dormant canopy stage. A similar conclusion was drawn by Grella et al. [7], who showed that the crop canopy structure plays a role in determining the drift values at both apple and vineyard plantations, particularly focusing on the crop type, training system, and growth stage.

Canopy characterisation at fruit and vineyard plantations has been commonly discussed in recent years, with numerous proposed methodologies ranging from simple manual processes [8,9] to those using sophisticated devices, such as LiDAR [4,10,11], ultrasonic sensors [12–14], unmanned aerial vehicles (UAVs) [15–17], and satellites [18–20]. The advantages and disadvantages of all methodologies have been probed, making it difficult to select the most accurate one.

UAVs and satellite imagery, classified as remote sensing methodologies, have been promoted in the last few years [21,22] as remarkable techniques for canopy characterisation. Differences in their management, accuracy, economical cost, and other important factors have been extensively discussed, yielding various advantages and disadvantages— directly related to the targeted crop and conditions—of both methods. Although satellite image acquisition of large areas saves considerable time, it has a low and inadequate resolution for precision viticulture [23,24]. The effectiveness of Sentinel-2 imagery and high-resolution UAV aerial images was evaluated [25], concluding that the resolution of the satellite imagery was insufficient for their direct use for describing vineyard variability. In contrast, Di Gennaro et al. [26] demonstrated the effectiveness and high resolution of Sentinel-2 imagery in the canopy characterisation process at vineyard plantations. Recent advancements in UAV-related research have led to a wide range of UAV applications for monitoring vineyard performance, such as rate of canopy development, canopy structure spatial variability, and disease incidence [27–31]. Similar to manned aircraft and satellite-based remote sensing, UAVs are convenient in terms of simple flight preparation and flexible operations [25], independently of their technical specifications (fixed-wing or multirotor); however, they are more effective for small and medium-sized vineyards [32,33]. According to Ouyang et al. [34], the operational flexibility of UAVs allows the timely assessment of canopy management outcomes, compared with manned aircraft and satellite remote sensing.

The variable rate application (VRA) of a PPP in 3D crops represents an important step forward in the sustainable use of pesticides, allowing accurate spray deposition and reduction in the drift loss by adjusting the optimal amount of the PPP applied to the canopy structure. This technology can be implemented using two different methods. The first one is adjusting the working parameters of the spray process based on the canopy characteristics measured 'on the go' using electronic devices [35–38]. The second is using previously generated canopy maps by manual or remote sensing measurements and their transformation into prescription maps using dedicated tools [39–41]. The second option based on canopy maps requires the implementation of an accurate process for canopy characterisation, and its extension and implementation on a large scale in commercial parcels are directly related to the degree of automation and ease of the process [42], being the development of canopy maps the most influencing process.

There is a need to develop an automated method for canopy characterisation that can consequently promote the implementation of the VRA process for sustainable PPP management in vineyards. Other studies on canopy characterisation in orchards and fruit crops have been conducted based on the use of the normalised differential vegetation index (NDVI) as the main parameter obtained utilising remote sensing platforms and its relationship with principal canopy dimensions [43].

The general objective of this research was to investigate the potential relationships among manual field measurements and remote-sensing-based methods (UAVs and satellite imagery) for canopy characterisation at commercial vineyard parcels. The following specific objectives were addressed:

- To compare the fitting results of linear regression models between manual canopy characterisation and both aerial platforms, considering different spatial and spectral resolutions.
- To investigate the effect of plant-by-plant versus clustered data on the precision and accuracy of canopy characteristics determination.
- To propose the most successful method for obtaining reliable prescription maps to be implemented in the VRA process.

2. Materials and Methods

2.1. Study Site

The research was conducted in the Alt Penedès region, one of the most important wine production areas in Catalonia, Spain. A total of five commercial vineyards (Table 1 and Figure 1) of four different varieties (Chardonnay, Merlot, Cabernet Sauvignon, and Macabeu) were included in the study. Four plots (A–D) were located in Torrelavit (Barcelona, Spain), and experiments were conducted in 2018 and 2019; the fifth plot (E) was located in El Plà del Penedès (Barcelona, Spain) and was only used for data collection in 2018. All selected vineyards were trained in a double cordon spur pruning system with green pruning when the shoot length exceeded 10 cm. All vines were in full production, non-irrigated, and with ages ranging from 21 to 31 years. The terrain slope was 5–10% for plots A–D and 0% for plot E. Furthermore, the soil was regularly harrowed to control weeds in rows and under vines.

Table 1. Main characteristics of selected vineyard plots.

Plot	Variety	Row Spacing (m)	Vine Spacing (m)	Area (ha)	X Coord. (m)	Y Coord. (m)	Ref. System
A	Chardonnay	2.2	1.2	2.35	392,194	4,587,999	
B	Merlot	2.2	1.2	2.97	392,234	4,587,843	ETRS 89
C	C. Sauvignon	2.2	1.2	1.53	391,856	4,588,055	UTM31
D	C. Sauvignon	2.2	1.2	3.14	391,744	4,588,107	
E	Macabeu	2.8	1.2	4.90	391,265	4,584,841	

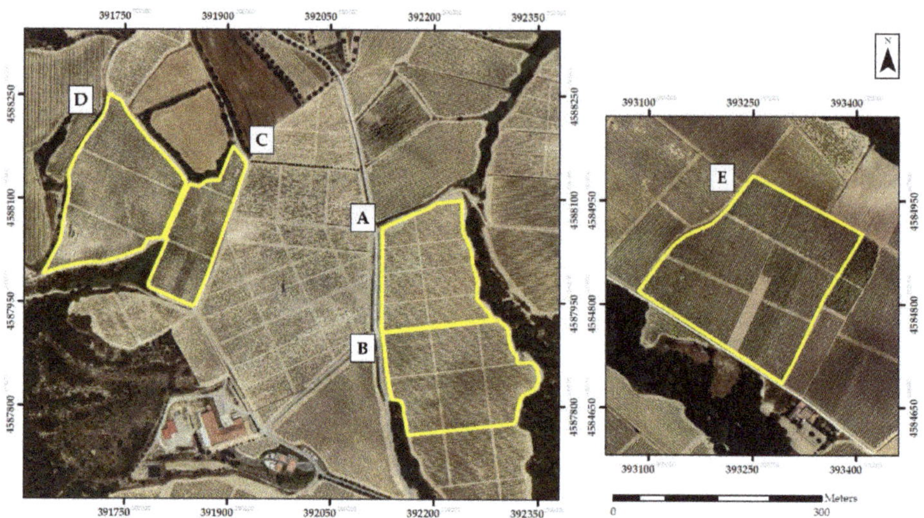

Figure 1. Orthophotomaps of studied vineyard plots.

2.2. Field Sampling Design

To select a representative and unbiased subset of vines and conduct the field measurements (sampling vines), a multi-stage (nested) systematic uniform random (SUR) sampling design was established. This type of sampling is more efficient than simple random sampling (with or without replacement), easy to implement, and particularly appropriate when the population is heterogeneous [44,45]. SUR sampling allows distributing the sampling locations uniformly over the entire surface of a plot, thus ensuring a known probability of selection for the entire population [46,47]. To implement multi-stage SUR sampling in each vineyard, a predefined set of sampling periods (m) was used to divide the entire population based on its structure. In this study, vine period refers to the number of plants between sampling vines, and row period is defined as the number of plants between the rows sampled. At each level, every m-th unit in the population is selected, and the position of the first sampling vine is chosen with a random start [44,46]. A random start is an integer between 1 and m (Figure 2). The number of sampling vines per plot was selected based on the characteristics of the plot (shape, area, length, orientation of rows, and vine spacing). The characteristics of the SUR sampling for each plot are listed in Table 2.

Table 2. Characteristics of SUR sampling for each vineyard plot in 2018 and 2019.

Plot	2018					2019				
	Rnd. Starting Row	Rnd. Starting Vine	Row Period	Vine Period	Total Samp. Vines	Rnd. Starting Row	Rnd. Starting Vine	Row Period	Vine Period	Total Samp. Vines
A	3	6	5	30	70	2	10	5	30	72
B	2	17	7	35	50	3	7	7	35	52
C	3	6	4	35	34	2	10	5	30	32
D	5	20	7	25	56	3	4	7	25	58
E	3	8	9	23	68	-	-	-	-	-

Figure 2. Scheme of SUR sampling used in plot A in 2018. Identification system for sampling vines in field.

Each sampling location consisted of a single vine (1.2 m canopy row assigned) and was appropriately identified in each plot with a coloured tape to allow easy identification in the field during the various seasons to maintain the sampling vines at the different measurement dates. Furthermore, to identify the sampling locations subsequently from aerial images, the selected vines were physically marked on the ground (between crop rows) with two white lime marks (Figure 2) to identify the start and end of each sampled vine.

2.3. Manual Canopy Characterisation

Manual field measurements were conducted coincident with three different canopy stages—*beginning of flowering* (BBCH 59), *berries pea size* (BBCH 75), and *beginning of ripening* (BBCH 81)—according to the BBCH monograph. The BBCH scale is a system for a uniform coding of phenologically similar growth stages of all mono- and dicotyledonous plant species [48].

Canopy characterisation for each of the sampling vines consisted of measuring the most representative parameters (canopy height and width). Manual measurements were conducted using a regular measuring tape following the EPPO standard [49]. Each measurement included 95% of the canopy, excluding protruding branches [50]. In each sampling vine, three measurements were performed by two different surveyors. The final value was calculated from the average of the corresponding six measurements per vine. Subsequently, the leaf wall area (LWA) [51] and tree row volume (TRV) [50,52,53] were calculated, being the officially recognised parameters for pesticide dose expression [49].

2.4. Aerial Platforms and Multispectral Sensors Used

2.4.1. UAV-Based Image Acquisition

This section presents the methodology for collecting and processing the images captured using a UAV as proposed by Campos et al. [17,42]. A UAV hexacopter (model: CondorBeta, Dronetools SL, Sevilla, Spain) loaded with a multispectral camera (model: RedEdge, Micasense, Seattle, WA, USA) flew over the vineyards. The camera was equipped with five spectral bands: red (R) centred at 668 nm with a bandwidth of 10 nm, green (G) centred at 560 nm with a bandwidth of 20 nm, blue (B) centred at 475 nm with a bandwidth of 20 nm, red edge (RE) centred at 717 nm with a bandwidth of 10 nm, and near-infrared (NIR) centred at 840 nm with a bandwidth of 40 nm. Focal length was 5.5 mm and sensor resolution 1280 × 960 pixels (width × height).

Flights were conducted 95 m above ground level at a cruise flight speed of 6 m s^{-1}. Overlapping zones were adjusted to 80% in the flight sense and 60% in the transverse sense. Flights were executed on the same dates and crop stages as described for manual canopy characterisation.

From the spectral images obtained using the Micasense RedEdge, an orthophotomap with a ground sample distance of 6.48 cm pixel^{-1} was obtained. Agisoft Metashape (Agisoft LLC, St. Petersburg, Russia) was the software used for photogrammetric processes. Each orthophotomap was radiometrically calibrated using calibration plates as greyscale standards (22%, 32%, 44%, and 51% reflectance), which were placed close to the area where the UAV took off and landed. These plates were placed on the ground inside the vineyard with the objective of ensuring several frames in which the both the vine canopy and the reflectance standards are present together as the UAV flies over the defined area. From each spectral band, the 12-bit digital value in each calibration panel was extracted. A power function was used to transform each pixel in the image to its corresponding reflectance value for each of the orthophotomaps. Georeferencing of the five mosaics resulting from each spectral band of the multispectral sensor was performed using fixed ground control points in the study area. The position of the natural ground control points was accurately recorded using a global navigation satellite system with real-time kinematic (RTK) correction (model: GPS1200+, Leica Geosystems AG., Heerbrugg, Switzerland). This georeferencing process was conducted only during the first flight (BBCH 65) in each season. For the remaining flights (BBCH 75 and 81), the same ground control points were maintained. The photogrammetric calculation process yielded an RMSE of 18.62 cm.

2.4.2. Satellite-Based Image Acquisition

Satellite-based images were obtained from PlanetScope (PS), a commercial constellation of nanosatellites consisting of more than 130 triple CubeSat miniature satellites (<5 kg) called as Dove (Planet Labs Inc., San Francisco, CA, USA). Although PS operates under a commercial license, many of its products are open-access for research purposes. Dove satellites are equipped with a line scanner imaging sensor with four spectral bands in the blue (455–515 nm), green (500–590 nm), red (590–670 nm), and NIR (780–860 nm) regions, providing high-resolution imagery (3 m spatial resolution) with an approximately daily revisit time. Cloud-free, orthorectified, and scaled top of atmosphere radiance level 3B images [54] were acquired from the study areas in 2018 and 2019 to maximally match the dates of manual canopy characterisation to allow comparison of the two methods. Each image from PS covered approximately 192 km^2, and one single frame captured the entire study area on each acquisition date.

2.5. Image Analysis

2.5.1. Canopy Vigour Map Generation

For the UAV-based imagery, the process followed by Campos et al. was utilised [17,42]. The first step for image analysis was the calculation of a vegetation index that expresses the vigour of the vines at each stage of the growth season. Several indices were considered, but given its extensive knowledge among viticulturists, and the considerable literature existing characterizing vineyards by the NDVI [17–19,22,25], it was finally chosen for this research. The NDVI [55] has been proven to be closely correlated to biomass development and crop stress [56–58]. The NDVI was calculated as a combination of the R and NIR bands (Equation (1)) (Figure 3b):

$$\text{NDVI} = \frac{\text{NIR} - \text{R}}{\text{NIR} + \text{R}} \tag{1}$$

As the vineyards were planted in rows, vineyard-only pixels were segmented from an image by applying an NDVI threshold to eliminate the undesired elements, such as weeds, shadows, and soil. In all flights, the NDVI threshold was changed and established manually based on a visual inspection of the image. The pixels below and above the selected threshold were considered noise and classified as a '0' and considered vineyard pixels and coded as '1', respectively. The result was a binary mask image containing vineyard-only pixels (Figure 3c).

Combining the original NDVI images and the vineyard-only masks, the vineyard rows were masked out. In the corresponding newly created images, the non-canopy pixels became '0', whereas the vineyard canopy pixels retained their original NDVI value (Figure 3d). Subsequently, an inverse distance weighting interpolation was performed to generate a continuous NDVI map (Figure 3e), which was finally classified into homogeneous vigour areas.

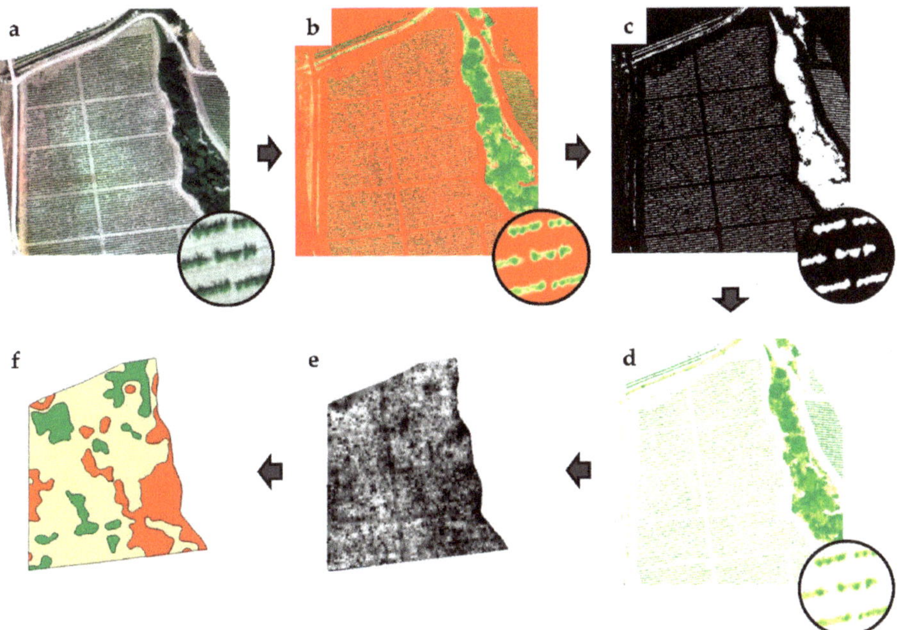

Figure 3. Analysis of workflow to obtain clustered vigour maps: (**a**) radiometrically calibrated multiband image, (**b**) NDVI image, (**c**) binary mask of vineyard-only pixels, (**d**) NDVI vineyard-only pixels, (**e**) continuous NDVI map, (**f**) clustered vigour map (red: low vigour, yellow: medium vigour, green: high vigour).

Finally, for classification purposes, the interpolated NDVI images were divided into quintiles (P20, P40, P60, and P80). NDVI values lower than P20, between P20 and P80, and higher than P80 were categorised as low, medium, and high vigour, respectively. This resulted in a three-class vigour level (high, medium, and low) (Figure 3f). The above-mentioned entire process is illustrated in Figure 3.

A similar approach was followed in the case of satellite imagery, where the NDVI was calculated using bands 3 (red) and 4 (NIR) from the four-band product delivered by PS (Equation (1)). Because of the low spatial resolution (pixel size was larger than the distance between the vineyard rows), segmentation between the canopy and the background elements (weeds, shadows, and soil) was not possible. The raw NDVI images were classified into three vigour levels (high, medium, and low) following the quintile rules previously explained.

2.5.2. Extraction of Information of Sampling Vines

The manually measured vines had to be identified in each orthophotomap. Therefore, a multiband RGB image was generated for each plot and flying date to enhance the visualisation of the white lime marks defining the beginning and ending of each sampling vine. A rectangular polygon guided by both lime marks (Figure 4a) was manually generated using QGIS software [59]. Combining the only-vineyard pixel mask (Figure 4b) with the polygon layer of each sampling vine, the mask was clipped, keeping only the pixels (logic 0 and 1) within the rectangles of interest (Figure 4c). Finally, a polygonisation process was performed to obtain the sampling vine-only polygon contours (Figure 4d).

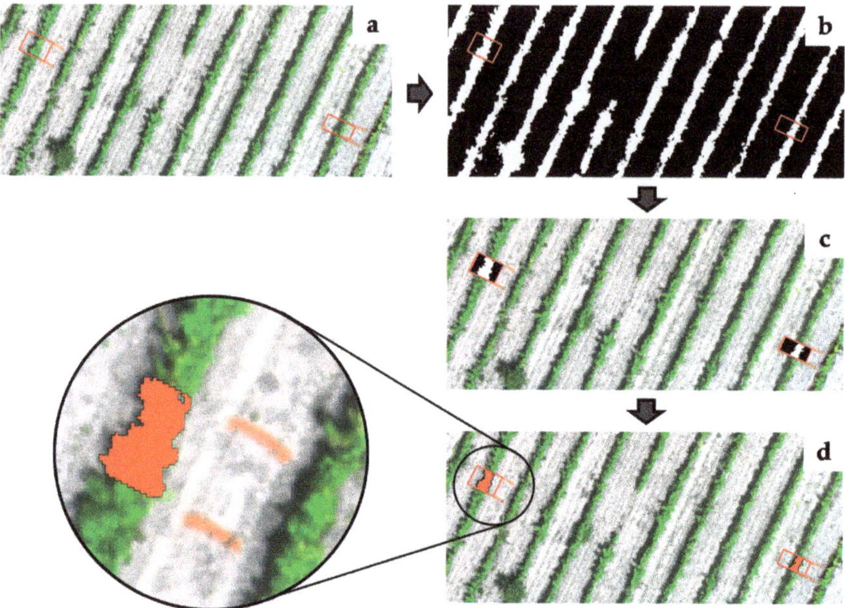

Figure 4. Overview of entire process from lime marks in ground to obtain sampling vine-only polygon contours: (**a**) polygons defining sampling vines, (**b**) binary mask, (**c**) sampling vine binary mask, (**d**) sampling vine-only polygon contours.

From each sampling vine, the following information was obtained:

- Raw NDVI mean: It was calculated as the mean of all pixels contained inside the sampling vine-only polygon contour. It was obtained for the UAV-based imagery ($NDVI_D$) and satellite-based imagery ($NDVI_S$).
- Clustered vigour: Each sampling vine was assigned to a vigour class (high, medium, and low vigour) based on the three zones previously defined. This information was obtained as a categorical variable. It was determined for UAV-based imagery (C_vigour_D) and satellite-based imagery (C_vigour_S).
- Polygon-projected area (Prj_area_D): To calculate the projected area of each sampling vine, the area of each polygon contour defining the vines was calculated using the field calculator tool in QGIS software [59]. This variable was calculated only for UAV-based imagery because the canopy and background in satellite-based imagery could not be segmented.
- Sampling category (Edge_pnt): Based on the geographic coordinates (ETRS89 UTM31), each sampling vine was classified as an edge point depending on its position in the plot. The sampling vines located within the inner buffer of 3 m from the plot border were considered as edge points. This information was obtained as a categorical variable (YES/NO).

2.6. Data Management

To conduct the planned comparisons of the different methods, an organised database was generated. Additionally, following Campos et al. [17], a new variable ($NDVI_D \times Prj\_area_D$) was introduced in the database, which was obtained by combining the $NDVI_D$ and the Prj_area_D. Table 3 lists the variables included in the database.

Table 3. Database fields.

Database Variables	Units	Type of Acquisition	Type of Data	Example of Data
Plot		-	Categorical	A
Vineyard variety		-	Categorical	Chardonnay
Year		-	Categorical	2019
BBCH		-	Categorical	75
Sampling vine		-	Numerical	35
$NDVI_D$		UAV	Numerical	0.72
C_vigor_D		UAV	Categorical	Medium
Prj_area_D	m^2	UAV	Numerical	0.17
$NDVI_D \times Prj\_area_D$		UAV	Numerical	0.12
H_M	m	Manual	Numerical	0.78
W_M	m	Manual	Numerical	0.33
LWA_M	$m^2\,ha^{-1}$	Manual	Numerical	7090.91
TRV_M	$m^3\,ha^{-1}$	Manual	Numerical	1170.00
$NDVI_S$		Satellite	Numerical	0.54
C_vigor_S		Satellite	Categorical	Medium
Edje_pnt		Satellite	Categorical	NO

Following the database generation, Spearman's rank correlation [60] analysis was executed to determine the relationship between remote sensing-based information and canopy structural measurements performed in the field. The remote-sensing-based variable that correlated (higher Spearman's ρ) the most with any of the canopy characteristics manually obtained was further selected for a deeper analysis. It is important to note that in the case of satellite imagery data, the statistical analysis was performed considering two different scenarios. In the first scenario, all the points were included regardless of their classification as edge points. The second scenario only considered data points that were not at the edge of the field plots. In the case of UAV-based imagery, where single vines can be clearly detected, the above process was not required.

The datasets of both remote sensing platforms were analysed following two different scenarios: considering every single data point as an individual value and using an aggregation (clusters) mode. The first evaluation of the obtained data was conducted considering the raw values for all single data points generated by the three different measurement procedures (UAV, satellite, and manual measurements), hereafter referred as single point data (SPD) analysis. For every single point identified, the ground-measured and remote-sensing-estimated canopy parameters were evaluated, and the potential relationships were analysed. This first proposed evaluation method allowed to a pixel-based conversion of the NDVI to any canopy parameter measured for every evaluated canopy stage and every single parcel. The remarkable discontinuity in the contiguous pixels in the maps generated from the raw pixel values (Figure 3e) is a technical limitation for the VRA of the inputs. A common technique used to solve this problem is to classify the raw values into a determined number of zones or clusters, which are treated as homogeneous management areas (Figure 3f). Furthermore, the main purpose of this research was the development of practical and useful canopy maps for the VRA of pesticides. Considering the above, the relationships among the averages of the canopy height, width, TRV, and LWA and the average remote sensing-based vegetation indices for all three different zones in every parcel classified as low, medium, and high canopy vigour zones were analysed (Figure 3f). This analysis is referred as aggregated data (AD) analysis.

Statistical analysis of all involved parameters and both proposed methods was performed to determine the potential application of linear regression. In all cases, a detailed comparison of each pair of variables was performed to obtain the most suitable linear regression model. To ensure the normality assumption, the variables were Ln-transformed. If normality was not satisfied, the linear model was rejected. This process was executed using the RStudio software [61].

3. Results

3.1. General View of Measured Parameters

The main descriptive statistical parameters obtained in the canopy characterisation are shown in Figures 5 and 6. The results obtained after the manual measurements (canopy height and canopy width) and the corresponding calculated parameters (TRV and LWA) present a logical development process of the canopy with the season variation (Figure 5). Based on the data, compared to starting point BBCH 59, the canopy dimensions increased by 1.5 times up to BBCH 75. This increase was due to the rapid growth rate of the green structures occurring between bud burst and the end of flowering under normal climatic conditions. After BBCH 75, stabilisation of the canopy development was observed, and the main parameters were maintained at similar levels. The results obtained using the two aerial platforms (UAV and satellite) exhibit differences (Figure 6). At all canopy stages, the NDVI statistical ranges (difference between the maximum and minimum values) obtained with the UAV were wider than those obtained using the satellites. Amplitudes of 0.38, 0.35, and 0.61 were obtained using the UAV in the first, second, and third canopy stages, respectively, whereas the corresponding amplitudes determined using the satellites were 0.1, 0.34, and 0.27, respectively.

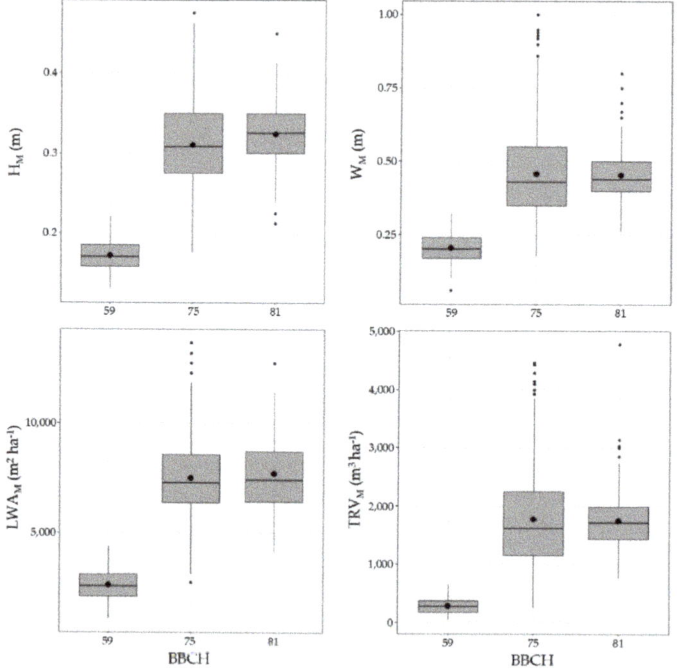

Figure 5. Box plots of principal descriptive statistical parameters for manual measurements. ● mean values.

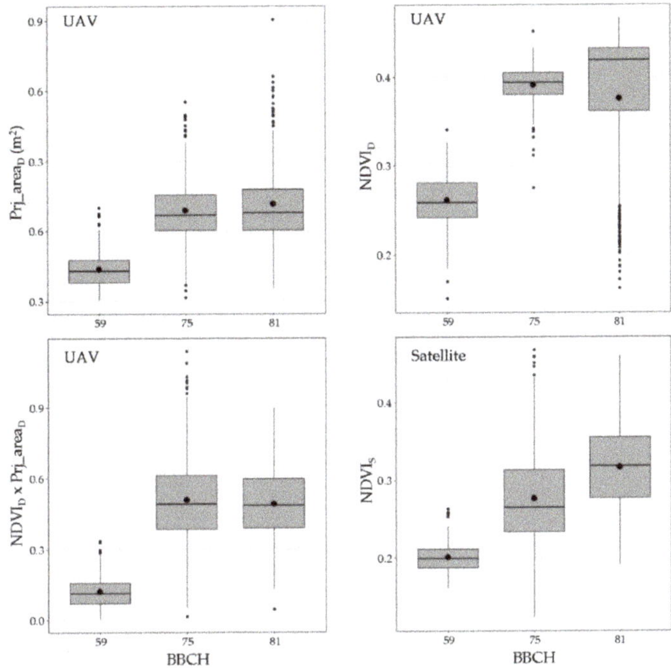

Figure 6. Box plots of principal descriptive statistical parameters for remote sensing variables. ● mean values.

3.2. Manual Data vs. UAV Variables

3.2.1. Data Correlation

To determine the spectral parameter with the most suitable correlation with any vegetative parameter (canopy height, width, TRV, or LWA), Spearman rho values were analysed. Table 4 summarises Spearman's rho correlation matrix for the UAV variables.

Table 4. Spearman's rho correlation matrix for variables obtained using UAV.

	(1)	(2)	(3)	(4)	(5)	(6)	(7)
(1). $NDVI_D$	1						
(2). Prj_area_D	0.44	1					
(3). $NDVI_D \times Prj\_area_D$	0.66	0.95	1				
(4). H_M	0.47	0.84	0.82	1			
(5). W_M	0.45	0.83	0.81	0.81	1		
(6). LWA_M	0.62	0.75	0.79	0.96	0.71	1	
(7). TRV_M	0.53	0.86	0.86	0.95	0.92	0.91	1

The correlation values of the four canopy parameters and the NDVI ranged from 0.45 to 0.62, and those of the projected area were larger, ranging from 0.75 to 0.86. Additionally, $NDVI_D \times Prj\_Area_D$ was also compared with the canopy characteristics, and the obtained values ranged from 0.79 to 0.86.

An in-depth analysis of only the best correlated parameters in Table 4 indicates that the TRV is the most remarkable canopy parameter in terms of the correlation with the information obtained using the UAV. The projected area (Prj_Area_D) and the combination of the NDVI and the projected area ($NDVI_D \times Prj\_Area_D$) are the two most remarkable parameters compared to the TRV, with a rho value of 0.86 in both cases, suggesting that both are strong correlations [62]. Additionally, the UAV-based extraction of the projected area

(Prj_Area$_D$) is very strongly correlated with the canopy width (rho value of 0.83). This can be expected because the projected area varies owing to the changes in the vegetation width while maintaining the canopy length equal to the plantation distance. However, this parameter still presents a very strong correlation [62] with the canopy height (rho value of 0.84). Similar results are obtained with NDVI$_D$ × Prj_Area$_D$, exhibiting strong correlations with the canopy height and width (rho values of 0.82 and 0.81, respectively). The previous analysis and the main objective of this research, i.e., to determine the most remarkable relationships among the spectral parameters obtained using aerial platforms and canopy characterisation values, are considered. Accordingly, NDVI$_D$ × Prj_Area$_D$ is found as the most remarkable parameter. Therefore, in the following sections, detailed analysis and evaluation of this relationship are presented.

3.2.2. Linear Regression Model

Considering the SPD dataset, the linear regression models between NDVI$_D$ × Prj_Area$_D$ and all manually measured parameters describing the canopy characteristics (canopy height, canopy width, LWA, and TRV) were evaluated. The data were Ln-transformed to ensure the normality assumption for the residues. The only variable that followed this normality assumption and yielded suitable residual plots for the model ($p > 0.05$ in the Kolmogorov–Smirnov test) was the TRV (Figure 7). Considering the results obtained after normality evaluation, for the remainder evaluated variables (canopy height, canopy width, and LWA), the intended linear regression models were rejected ($p < 0.05$ in the Kolmogorov–Smirnov test).

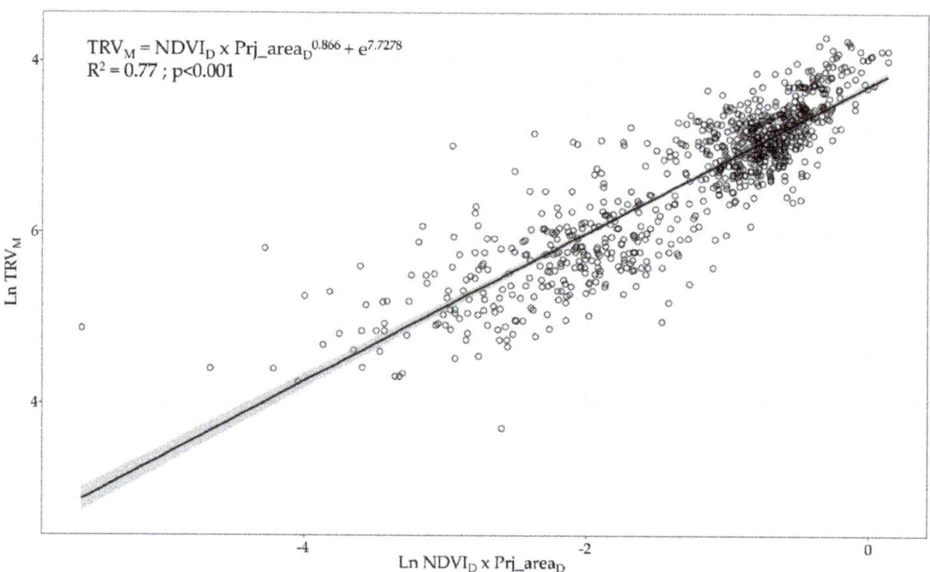

Figure 7. Linear regression model ($R^2 = 0.77$, $p < 0.001$) evaluating relationship between Ln TRV$_M$ and Ln NDVI$_D$ × Prj_area$_D$. Grey band shows 95% confidence interval.

When the normality evaluation was performed using the AD dataset, all studied variables (canopy height, canopy width, LWA, and TRV) followed the normality assumption of the residues ($p > 0.05$ in the Kolmogorov–Smirnov test). The linear regression models built as combinations of NDVI$_D$ × Prj_Area$_D$ and all manually measured parameters describing the canopy characteristics yielded high coefficients of determination: R^2 of 0.93 for the canopy height, 0.84 for the canopy width, 0.91 for the LWA, and 0.94 for the TRV (Figure 8).

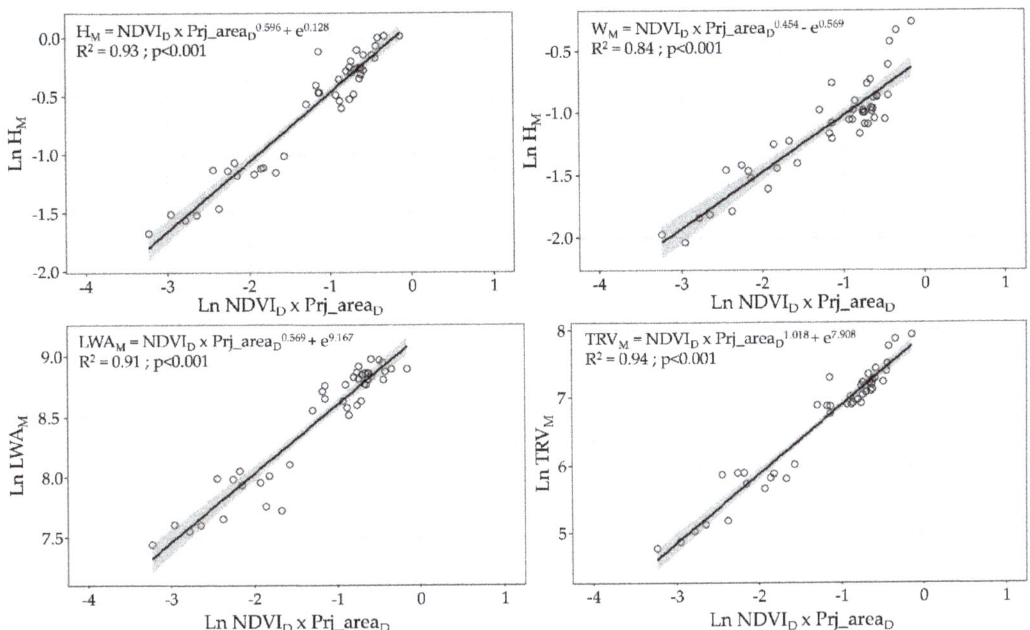

Figure 8. Linear regression models evaluating relationship among Ln NDVI$_D$ x Prj_area$_D$ and transformed canopy structural parameters (Ln H$_M$, Ln W$_M$, Ln LWA$_M$, and Ln TRV$_M$). Grey bands present 95% confidence interval.

As shown in Figure 8, when the data are grouped by vigour zones, the correlation values among all analysed variables are improved. Therefore, considering the potential use of this technique for the implementation of the VRA process, the model shown in Figure 8 seems the most appropriate for determining the optimal volume rate considering the canopy characteristics [8,41].

3.3. Manual Data vs. Satellite Variables

3.3.1. Data Correlation

To evaluate the correlations between the NDVI$_S$ and all vegetative parameters manually obtained (canopy height, canopy width, LWA, or TRV), Spearman's rho correlation matrices were analysed in the case of the satellite dataset. The results of the Spearman's rho values of the satellite variables, considering or rejecting the edge points, are listed in Table 5.

Table 5. Spearman's rho values for variables obtained using satellite imagery considering and rejecting edge points.

	H$_M$	W$_M$	LWA$_M$	TRV$_M$
NDVI$_S$ considering edge points	0.49	0.52	0.40	0.52
NDVI$_S$ rejecting edge points	0.66	0.67	0.57	0.68

Based on the results summarised in Table 5, the spectral values are affected by the border effect. The pixels located close to the edge of a parcel seem to be contaminated by adjacent elements (mainly roads), reducing the spectral value of the pixels. However, the border effect did not impact the manually measured structural parameters of the canopy. Consequently, the spectral values obtained using the satellite imagery present a certain border effect; thus, an in-depth data analysis was conducted without edge points.

Considering similar relationships among the $NDVI_S$ and the evaluated structural parameters (Table 5), all variables were included in the following linear regression analysis.

3.3.2. Linear Regression Model

Considering the SPD dataset, the linear regression models between the $NDVI_S$ and all manually measured parameters describing the canopy characteristics were evaluated. The data were Ln-transformed to ensure the normality assumption of the residues. The only variable that followed this normality assumption and showed suitable residual plots for the model ($p > 0.05$ in the Kolmogorov–Smirnov test) was the canopy width (Figure 9). Based on the results of the normality evaluation, for the remainder evaluated variables (canopy height, LWA, and TRV), the planned linear regression models were excluded ($p < 0.05$ in the Kolmogorov–Smirnov test).

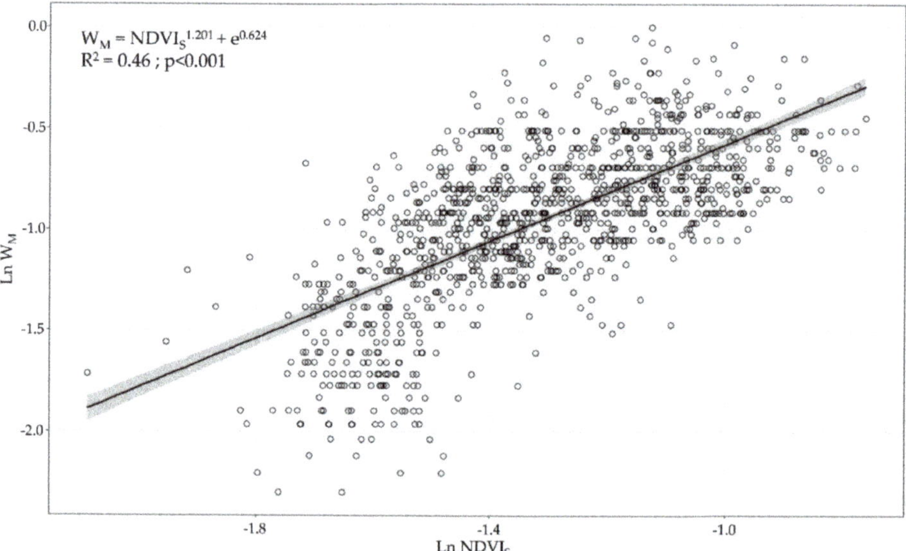

Figure 9. Linear regression model ($R^2 = 0.46$, $p < 0.001$) evaluating relationship between Ln W_M and Ln $NDVI_S$. Grey band shows 95% confidence interval.

However, when the satellite data were clustered (i.e., ADA) and the same data analysis was performed, the variables following the normality assumption of the residues ($p > 0.05$ in the Kolmogorov–Smirnov test) were the canopy height, canopy width, and TRV. For the LWA, the regression model was rejected because the normality assumption was not achieved ($p < 0.05$ in the Kolmogorov–Smirnov test). The linear regression models built as combinations of the $NDVI_S$ and height, width, and TRV yielded coefficients of determination R^2 of 0.48 for the canopy height, 0.51 for the canopy width, and 0.50 for the TRV (Figure 10).

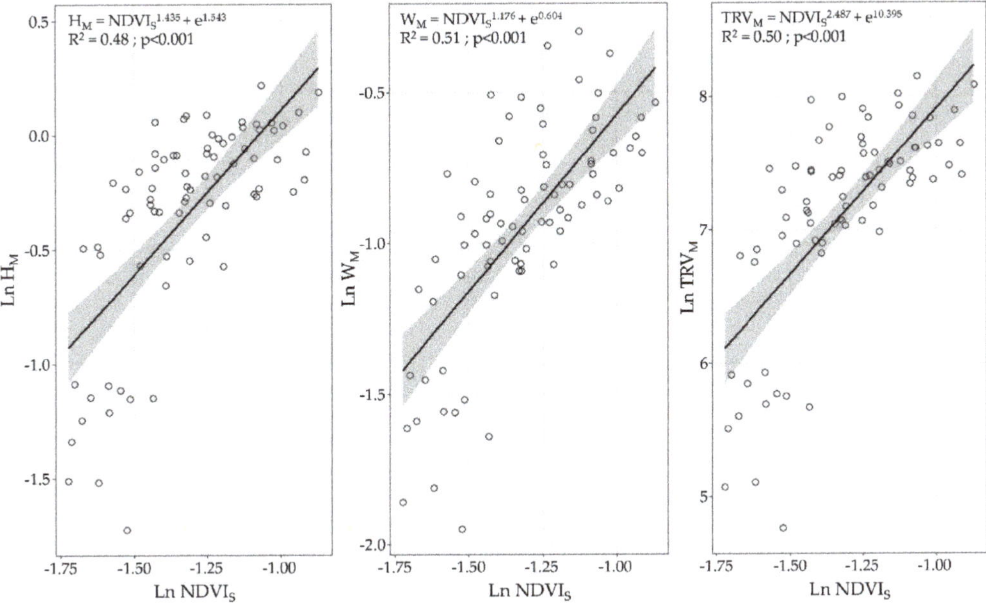

Figure 10. Linear regression models evaluating relationships among Ln $NDVI_S$ and transformed canopy structural parameters (Ln H_M, Ln W_M, and Ln TRV_M). Grey bands present 95% confidence interval.

4. Discussion

The three methods used to measure and characterise vine development during the season are proven to detect the different patterns associated with the vegetation growth, as observed in Figures 5 and 6. From BBCH 59 to BBCH 75, a relevant increase in the parameters is detected, coinciding with the rapid shoot growth of the vines in the first stages of development. This was also observed in other studies in various crops and using different spectral sensors, such as Sentinel 2, Landsat, and RapidEye [58,63–66]. Around BBCH 75, the vine structure and the canopy architecture are modified by several management operations (i.e., shoot positioning, trimming, hedging, and leaf thinning) to maintain vegetative and fruiting balance. Consequently, there are no observable differences between BBCH 75 and BBCH 81 in the parameters assessing the canopy structure. This suggests that an alternative time for data capture can be used to better describe the rapid changes in the vegetation in the first stages after the first leaves unfold. Starting from BBCH 59, a second measurement around the end of flowering (BBCH 69) can represent the evolution of the canopy structure more realistically. In comparison, a single measurement after BBCH 75 is sufficient to point out the characteristics of the vegetation until harvesting. The latter is also confirmed when the sample distribution around the regression lines is analysed (Figures 7 and 8), with a clear cluster of data points corresponding to advanced development stages (high values of the canopy height, width, LWA, and TRV). In comparison, there is a lack of points in the lower end of the x and y axes in all plots. The effect of the forcing canopy architecture in the vineyard was also observed in the temporal evolution of the LWA obtained from the field measurements in BBCH 60, 61, 69, 75, 77, 79, and 81 [42].

Remote sensing data acquired using UAV and satellite platforms were used to expedite the canopy characterisation process, which is a key procedure when determining rational PPP doses and application volumes adapted to the canopy status. However, the accuracy differed between the platforms as well as between the raw and clustered data point analysis.

First, when comparing the linear regression models, a better fitting (higher coefficient of determination) was obtained with the UAV-based data than when using the satellite information. It is also remarkable to note that at least one of the four measured variables (canopy height, canopy width, LWA, and TRV) showed significant linear regression with the remotely sensed data of both aerial platforms. UAV-based NDVI $\times$ ProjArea$_D$ yielded considerably high coefficients of determination above 0.84, which presents a new area of development for UAV technology as a tool for canopy characterisation to enable VRA principles. Other studies using high-resolution NDVI maps have found significant differences between vines belonging to different vigour classes [42,65,67]; however, none of them have related or modelled the structural characteristics of vines and the NDVI values in viticulture. Ampatzidis et al. [43] obtained a strong relationship (R^2 = 0.65) between the canopy height and the NDVI from a UAV survey of citrus groves, which have a more complex structure than vertical shoot positioned-trained vines. Satellite-based results presented significant relationships ($p < 0.001$) with the canopy height, canopy width, and TRV with coefficients of determination of approximately 0.45, reflecting the importance of ensuring sufficient spatial, spectral, and temporal resolution to reduce noise in the data. In line with this, when comparing the NDVI values from both platforms, a clear difference between the NDVI statistical ranges (difference between the highest and lowest values) was found when using UAV-based data (0.38, 0.35, and 0.61) and satellite-based data (0.1, 0.34, and 0.27, respectively). These differences were caused by the differences in the sensor spectral characteristics, spatial resolution (from a few centimetres to several meters per pixel), and spectral mixing. UAV-based imagery allows canopy segmentation and partially removes vegetation, soil, and shadow spectral mixing at the pixel level [33,58]. This discrimination is impossible when using imagery at 3 m $\times$ pixel^{-1}, causing a diminishing effect on the NDVIs from vines as compared to the pure canopy spectral signals acquired from the high-resolution UAV-based images. Matese et al. [33] and Gatti et al. [65] found similar results using RapidEye satellite with a 5 m $\times$ pixel^{-1} spatial resolution (NDVI statistical ranges of 0.15 and 0.13, respectively) due to the important spectral mixing effects. Devaux et al. [66] used temporal NDVI information from Sentinel 2 satellite imagery to track vineyard growth during the season, which provided a methodology to determine the approximate dates for conducting vine structure management operations. Similarly, some consistency is found with other published results when comparing the maximum level reached by satellite-based NDVIs in the period of maximum vegetation development (July in the Northern Hemisphere). Based on the data in Figure 6, the average NDVI value for the satellite-based data is 0.27, with a maximum of 0.47. This is comparable with the per plot average NDVI reported by Devaux et al. [66] (0.4) and Gatti et al. [65] (0.44) in vineyards trained in vertical shoot positioning and grown without cover crop. Moreover, using Sentinel 2 imagery at 10 m $\times$ pixel^{-1} spatial resolution, NDVI values of 0.5 were reported by the above studies as well as by Johnson et al. [18,68]. Discrepancies between the NDVI values can be attributed to the differences in the technical specifications of the radiometric sensors used in the studies and the variation in the crop management and geographic area considered. Another important effect related to the spatial resolution of remote sensing platforms that must be considered is the ability to characterise crops in the early stages of a season when the shoot length ranges from 10 to 30 cm but the canopy density is minimal. Satellite-based imagery includes a higher proportion of unwanted elements per pixel than the area occupied by the vine canopy and require a minimum shoot and leaf development for remote assessment, as observed in the data point dispersion in Figures 9 and 10. The border effect on the spectral images was detected as another factor related to the spatial resolution of the platform. Pixels close to the border zones in a parcel are considerably affected by the adjacent pixels belonging to the intended zone (mainly roads), reducing the estimated NDVI values. This fact becomes more important as the spatial resolution of the platform is increased and thus is much more important in the case of satellites than in the case of UAVs [69].

Second, a comparison of the raw point data (SPD) and clustered data (AD) demonstrated that, in general, an improvement was achieved when the samples were analysed in terms of averages based on the vigour class in the field. Although this method limited the possibility of converting a grid image into a canopy characteristic grid map (cell-by-cell), it was reliable for characterising classified maps into the three classes. From a practical perspective, this is in alignment with the method by which remote sensing companies offer canopy maps. Specifically, they perform post-processing of raw NDVI maps to yield management maps with filtered and homogeneous areas, which facilitates prescription management. Concurrently, when performing spraying applications in the variable rate mode, the machine has to change the pressure and the nozzle flow rate depending on the prescription area where it is located at each moment. Volume rate prescription areas should ensure a minimum size to avoid continuous changes in the sprayer's operational parameters, and this is well achieved when prescription maps are clustered into two or three categories.

5. Conclusions

This study focuses on the development of linear regression models to predict the structural characteristics of vegetation in vineyards using aerial remote sensing. Using a rapid and cost-effective technology to monitor a canopy on a high temporal and spatial basis is key to estimate the changes in the canopy volume and density and to adapt the PPP dose with increased rationality and sustainability.

The developed methodology achieved robust characterisation (R^2 higher than 0.84 in all cases) of the TRV, LWA, canopy height, and canopy width using the vegetation indices obtained from UAV images when the remote sensing data were classified into three vigour classes. This enables a reliable determination of the canopy characteristics, allowing the generation of PPP prescription maps defining the different vigour zones, which can be completely adapted for implementation of the VRA process [17,42]. The practical applicability of the proposed methodology is limited by the number of available maps along the season, encountered difficulties and long time required to generate the canopy vigour maps, and high price of UAV services in comparison with satellite-based options.

Satellite technology was investigated to overcome these limitations. Experiments yielded statistically significant linear relationships ($R^2 > 0.48$ in all cases) between the NDVI and the canopy parameters (TRV, canopy height, and canopy width). These results, together with the higher temporal resolution and lower prices compared to those of UAVs, suggested the potential benefits of using satellite-based imagery for the VRA process based on zonal vigour variability.

Irrespective of the aerial platform evaluated (UAV or satellite) and considering the final objective of a practical implementation of the VRA of a PPP, the benefits of AD management compared with SPD evaluation were demonstrated. Pesticide distribution based on canopy vigour zones [17,70] will allow significant reduction in the use of a PPP, which is in alignment with the recently published Farm to Fork strategy [71].

Thus, future studies should focus on improving canopy characterisation considering the pixel size of satellite imagery, adaptation of field measurements for validation, and full automation of the entire process.

Author Contributions: Methodology, J.C. and E.G.; Investigation, J.C.; Data Curation, J.C. and F.G.-R.; Writing—Original Draft Preparation, all members; Writing—Review and Editing, E.G.; Project Administration, E.G. and J.C. All authors have read and agreed to the published version of the manuscript.

Funding: This research was developed with partial funding from the FI-AGAUR grant from the Generalitat de Catalunya (2017 FI_B 00893).

Institutional Review Board Statement: Not applicable.

Informed Consent Statement: Not applicable

Data Availability Statement: Not applicable.

Acknowledgments: The authors would like to thank to Miguel Torres, Jean Leon and Andreu Pinyol for their support in the field trials.

Conflicts of Interest: The authors declare no conflict of interest.

References

1. COM. Communication from the Commission to the European Parliament, the European Council, the European Economic and Social Committee and the Committee of the Regions: European Green Deal COM/2019/640. 2019. Available online: https://eur-lex.europa.eu/legal-content/EN/TXT/DOC/?uri=CELEX:52019DC0640&from=EN (accessed on 10 December 2020).
2. EUROSTAT. *The Use of Plant Protection Products in the European Union. Data 1992–1999*; Office for Official Publications of the European Communities: Luxembourg, 2002; p. 14.
3. Gil, E.; Arnó, J.; Llorens, J.; Sanz, R.; Llop, J.; Rosell-Polo, J.R.; Gallart, M.; Escolà, A. Advanced Technologies for the Improvement of Spray Application Techniques in Spanish Viticulture: An Overview. *Sensors* **2014**, *14*, 691–708. [CrossRef] [PubMed]
4. Miranda-Fuentes, A.; Llorens, J.; Gamarra-Diezma, J.L.; Gil-Ribes, J.A.; Gil, E. Towards an Optimized Method of Olive Tree Crown Volume Measurement. *Sensors* **2015**, *15*, 3671–3687. [CrossRef]
5. Garcerá, C.; Fonte, A.; Salcedo, R.; Soler, A.; Chueca, P. Dose Expression for Pesticide Application in Citrus: Influence of Canopy Size and Sprayer. *Agronomy* **2020**, *10*, 1887. [CrossRef]
6. Praat, J.-P.; Maber, J.; Manktelow, D. The Effect of Canopy Development and Sprayer Position on Spray Drift from a Pipfruit Orchard. *NZPP* **2000**, *53*, 241–247. [CrossRef]
7. Grella, M.; Gallart, M.; Marucco, P.; Balsari, P.; Gil, E. Ground Deposition and Airborne Spray Drift Assessment in Vineyard and Orchard: The Influence of Environmental Variables and Sprayer Settings. *Sustainability* **2017**, *9*, 728. [CrossRef]
8. Pergher, G.; Petris, R. Canopy structure and deposition efficiency of vineyard sprayers. *J. Agric. Eng.* **2007**, *38*, 31–38. [CrossRef]
9. Gil, E.; Llorens, J.; Landers, A.; Llop, J.; Giralt, L. Field validation of DOSAVIÑA, a decision support system to determine the optimal volume rate for pesticide application in vineyards. *Eur. J. Agron.* **2011**, *35*, 33–46. [CrossRef]
10. Rosell, J.R.; Llorens, J.; Sanz, R.; Arnó, J.; Ribes-Dasi, M.; Masip, J.; Escolà, A.; Camp, F.; Solanelles, F.; Gràcia, F. Obtaining the three-dimensional structure of tree orchards from remote 2D terrestrial LIDAR scanning. *Agric. For. Meteorol.* **2009**, *149*, 1505–1515. [CrossRef]
11. Llorens, J.; Gil, E.; Llop, J.; Escolà, A. Ultrasonic and LIDAR Sensors for Electronic Canopy Characterization in Vineyards: Advances to Improve Pesticide Application Methods. *Sensors* **2011**, *11*, 2177–2194. [CrossRef]
12. Escolà, A.; Planas, S.; Rosell, J.R.; Pomar, J.; Camp, F.; Solanelles, F.; Gracia, F.; Llorens, J.; Gil, E. Performance of an Ultrasonic Ranging Sensor in Apple Tree Canopies. *Sensors* **2011**, *11*, 2459–2477. [CrossRef]
13. Vitali, M.; Tamagnone, M.; La Iacona, T.; Lovisolo, C. Measurement of grapevine canopy leaf area by using an ultrasonic-based method. *OENO One* **2013**, *47*, 183–189. [CrossRef]
14. Gamarra-Diezma, J.L.; Miranda-Fuentes, A.; Llorens, J.; Cuenca, A.; Blanco-Roldán, G.L.; Rodríguez-Lizana, A. Testing Accuracy of Long-Range Ultrasonic Sensors for Olive Tree Canopy Measurements. *Sensors* **2015**, *15*, 2902–2919. [CrossRef] [PubMed]
15. Mathews, A.J.; Jensen, J.L.R. Visualizing and Quantifying Vineyard Canopy LAI Using an Unmanned Aerial Vehicle (UAV) Collected High Density Structure from Motion Point Cloud. *Remote Sens.* **2013**, *5*, 2164–2183. [CrossRef]
16. Mathews, A.J. A Practical UAV Remote Sensing Methodology to Generate Multispectral Orthophotos for Vineyards: Estimation of Spectral Reflectance Using Compact Digital Cameras. *IJAGR* **2015**, *6*, 65–87. [CrossRef]
17. Campos, J.; Llop, J.; Gallart, M.; García-Ruíz, F.; Gras, A.; Salcedo, R.; Gil, E. Development of canopy vigour maps using UAV for site-specific management during vineyard spraying process. *Precis. Agric.* **2019**, *20*, 1136–1156. [CrossRef]
18. Johnson, L.F.; Roczen, D.E.; Youkhana, S.K.; Nemani, R.R.; Bosch, D.F. Mapping vineyard leaf area with multispectral satellite imagery. *Comput. Electron. Agric.* **2003**, *38*, 33–44. [CrossRef]
19. Martinez-Casasnovas, J.; Agelet-Fernandez, J.; Arnó, J.; Ramos, M. Analysis of vineyard differential management zones and relation to vine development, grape maturity and quality. *Span. J. Agric. Res.* **2012**, *10*, 326–337. [CrossRef]
20. Khan, A.; Khan, U.; Waleed, M.; Khan, A.; Kamal, T.; Marwat, S.N.K.; Maqsood, M.; Aadil, F. Remote Sensing: An Automated Methodology for Olive Tree Detection and Counting in Satellite Images. *IEEE Access* **2018**, *6*, 77816–77828. [CrossRef]
21. Karakizi, C.; Oikonomou, M.; Karantzalos, K. Spectral Discrimination and Reflectance Properties of Various Vine Varieties from Satellite, UAV and Proximate Sensors. *Int. Arch. Photogramm. Remote Sens. Spat. Inf. Sci.* **2015**, *40*, 31–37. [CrossRef]
22. Borgogno-Mondino, E.; Lessio, A.; Tarricone, L.; Novello, V.; de Palma, L. A Comparison between Multispectral Aerial and Satellite Imagery in Precision Viticulture. *Precis. Agric.* **2018**, *19*, 195–217. [CrossRef]
23. Anastasiou, E.; Balafoutis, A.; Darra, N.; Psiroukis, V.; Biniari, A.; Xanthopoulos, G.; Fountas, S. Satellite and Proximal Sensing to Estimate the Yield and Quality of Table Grapes. *Agriculture* **2018**, *8*, 94. [CrossRef]
24. Sassu, A.; Gambella, F.; Ghiani, L.; Mercenaro, L.; Caria, M.; Pazzona, A.L. Advances in Unmanned Aerial System Remote Sensing for Precision Viticulture. *Sensors* **2021**, *21*, 956. [CrossRef] [PubMed]
25. Khaliq, A.; Comba, L.; Biglia, A.; Ricauda Aimonino, D.; Chiaberge, M.; Gay, P. Comparison of Satellite and UAV-Based Multispectral Imagery for Vineyard Variability Assessment. *Remote Sens.* **2019**, *11*, 436. [CrossRef]
26. Di Gennaro, S.F.; Dainelli, R.; Palliotti, A.; Toscano, P.; Matese, A. Sentinel-2 Validation for Spatial Variability Assessment in Overhead Trellis System Viticulture Versus UAV and Agronomic Data. *Remote Sens.* **2019**, *11*, 2573. [CrossRef]
27. Albetis, J.; Jacquin, A.; Goulard, M.; Poilvé, H.; Rousseau, J.; Clenet, H.; Dedieu, G.; Duthoit, S. On the Potentiality of UAV Multispectral Imagery to Detect *Flavescence dorée* and Grapevine Trunk Diseases. *Remote Sens.* **2019**, *11*, 23. [CrossRef]

28. De Castro, A.; Jiménez-Brenes, F.M.; Torres-Sánchez, J.; Peña, J.M.; Borra-Serrano, I.; López-Granados, F. 3-D Characterization of Vineyards Using a Novel UAV Imagery-Based OBIA Procedure for Precision Viticulture Applications. *Remote Sens.* **2018**, *10*, 584. [CrossRef]

29. Kerkech, M.; Hafiane, A.; Canals, R. Vine disease detection in UAV multispectral images using optimized image registration and deep learning segmentation approach. *Comput. Electron. Agric.* **2020**, *174*. [CrossRef]

30. Pádua, L.; Marques, P.; Hruška, J.; Adão, T.; Peres, E.; Morais, R.; Sousa, J.J. Multi-Temporal Vineyard Monitoring through UAV-Based RGB Imagery. *Remote Sens.* **2018**, *10*, 1907. [CrossRef]

31. Romero, M.; Luo, Y.C.; Su, B.F.; Fuentes, S. Vineyard water status estimation using multispectral imagery from an UAV platform and machine learning algorithms for irrigation scheduling management. *Comput. Electron. Agric.* **2018**, *147*, 109–117. [CrossRef]

32. Andújar, D.; Moreno, H.; Bengochea-Guevara, J.M.; de Castro, A.; Ribeiro, A. Aerial imagery or on-ground detection? An economic analysis for vineyard crops. *Comput. Electron. Agric.* **2019**, *157*, 351–358. [CrossRef]

33. Matese, A.; Toscano, P.; Di Gennaro, S.F.; Genesio, L.; Vaccari, F.P.; Primicerio, J.; Belli, C.; Zaldei, A.; Bianconi, R.; Gioli, B. Intercomparison of UAV, Aircraft and Satellite Remote Sensing Platforms for Precision Viticulture. *Remote Sens.* **2015**, *7*, 2971–2990. [CrossRef]

34. Ouyang, J.; De Bei, R.; Fuentes, S.; Collins, C. UAV and Ground-Based Imagery Analysis Detects Canopy Structure Changes After Canopy Management Applications. *OENO One* **2020**, *54*, 1093–1103. [CrossRef]

35. Giles, D.K.; Delwiche, M.J.; Dodd, R.B. Sprayer control by sensing orchard crop characteristics: Orchard architecture and spray liquid savings. *J. Agric. Eng. Res.* **1989**, *43*, 271–289. [CrossRef]

36. Gil, E.; Escolà, A.; Rosell, J.R.; Planas, S.; Val, L. Variable rate application of plant protection products in vineyard using ultrasonic sensors. *Crop Prot.* **2007**, *26*, 1287–1297. [CrossRef]

37. Jeon, H.Y.; Zhu, H.; Derksen, R.; Ozkan, E.; Krause, C. Evaluation of ultrasonic sensor for variable-rate spray applications. *Comput. Electron. Agric.* **2011**, *75*, 213–221. [CrossRef]

38. Gil, E.; Llorens, J.; Llop, J.; Fàbregas, X.; Escolà, A.; Rosell, J.R. Variable rate sprayer. Part 2–Vineyard prototype: Design, implementation, and validation. *Comput. Electron. Agric.* **2013**, *95*, 136–150. [CrossRef]

39. Siegfried, W.; Viret, O.; Huber, B.; Wohlhauser, R. Dosage of plant protection products adapted to leaf area index in viticulture. *Crop Prot.* **2007**, *26*, 73–82. [CrossRef]

40. Garcerá, C.; Fonte, A.; Moltó, E.; Chueca, P. Sustainable Use of Pesticide Applications in Citrus: A Support Tool for Volume Rate Adjustment. *Int. J. Environ. Res. Public Health* **2017**, *14*, 715. [CrossRef]

41. Gil, E.; Campos, J.; Ortega, P.; Llop, J.; Gras, A.; Armengol, E.; Salcedo, R.; Gallart, M. DOSAVIÑA: Tool to calculate the optimal volume rate and pesticide amount in vineyard spray applications based on a modified leaf wall area method. *Comput. Electron. Agric.* **2020**, *160*, 117–130. [CrossRef]

42. Campos, J.; Gallart, M.; Llop, J.; Ortega, P.; Salcedo, R.; Gil, E. On-Farm Evaluation of Prescription Map-Based Variable Rate Application of Pesticides in Vineyards. *Agronomy* **2020**, *10*, 102. [CrossRef]

43. Ampatzidis, Y.; Partel, V.; Meyering, B.; Albrecht, U. Citrus rootstock evaluation utilizing UAV-based remote sensing and artificial intelligence. *Comput. Electron. Agric.* **2019**, *164*, 104900. [CrossRef]

44. Wulfsohn, D. Sampling techniques for plants and soil. In *Advanced Engineering Systems for Specialty Crops: A Review of Precision Agriculture for Water, Chemical, and Nutrient Application, and Yield Monitoring*; Upadhyaya, S.K., Giles, D.K., Haneklaus, S., Schnug, E., Eds.; Johan Heinrich von Thunen-Institut: Braunschweig, Germany, 2010; Special Issue 340; pp. 3–30.

45. Wulfsohn, D.; Aravena Zamora, F.; Potin Téllez, C.; Zamora Lagos, I.; Garcia-Fiñana, M. Multilevel systematic sampling to estimate total fruit number for yield forecasts. *Precis. Agric.* **2012**, *13*, 256–275. [CrossRef]

46. Mayhew, T.M. Taking tissue samples from the placenta: An illustration of principles and strategies. *Placenta* **2008**, *29*, 1–14. [CrossRef]

47. Mayhew, T.M.; Lucocq, J.M. From gross anatomy to the nanomorphome: Stereological tools provide a paradigm for advancing research in quantitative morphomics. *J. Anat.* **2015**, *226*, 309–321. [CrossRef] [PubMed]

48. Meier, U. BBCH-Monograph. In *Growth Stages of Plants-Entwicklungsstadien von Planzen–Estadios de las Plantas-Développement des Plantes*; Blackwell Wissenschaftsverlag: Berlin, Germany; Vienna, Austria, 1997; p. 622.

49. EPPO. PP 1/239 (3) Dose Expression for Plant Protection Products. EPPO Bulletin. 2021. Available online: https://onlinelibrary.wiley.com/doi/full/10.1111/epp.12704 (accessed on 10 December 2020).

50. Manktelow, D.W.L.; Praat, J.P. The Tree-Row-Volume Spraying System and its Potential use in New Zealand. In Proceedings of the NZ Plant Protection Conference, Lincoln, New Zealand, 18–21 August 1997; pp. 119–124.

51. Walklate, P.; Cross, J. An examination of Leaf-Wall-Area dose expression. *Crop. Prot.* **2012**, *35*, 132–134. [CrossRef]

52. Byers, R.E.; Hickey, K.D.; Hill, C.H. Base gallonage per acre. *Va. Fruit* **1971**, *60*, 19–23.

53. Byers, R.E.; Lyons, C.G., Jr.; Yoder, K.S.; Horsburgh, R.L.; Barden, J.A.; Donohue, S.J. Effect of apple tree size and canopy density on spray chemical deposit. *Hortscience* **1984**, *19*, 93–94.

54. Wilson, N.; Greenberg, J.; Jumpasut, A.; Collison, A.; Weichelt, H. *Absolute Radiometric Calibration of Planet Dove Satellites, Flocks 2p & 2e*; Planet: San Francisco, CA, USA, 2017.

55. Rouse, J.W.; Haas, R.H.; Schell, J.A.; Deering, D.W. Monitoring vegetation systems in the Great Plains with ERTS, Third ERTS Symposium, NASA SP-351. *NASA Spec. Publ.* **1973**, *I*, 309–317.

56. Dobrowski, S.Z.; Ustin, S.L.; Wolpert, J.A. Remote estimation of vine canopy density in vertically shoot-positioned vineyards: Determining optimal vegetation indices. *Aust. J. Grape Wine Res.* **2002**, *8*, 117–125. [CrossRef]

57. Johnson, L.F. Temporal stability of an NDVI-LAI relationship in a Napa Valley vineyard. *Aust. J. Grape Wine Res.* **2003**, *9*, 96–101. [CrossRef]

58. Sun, L.; Gao, F.; Anderson, M.C.; Kustas, W.P.; Alsina, M.M.; Sanchez, L.; Sams, B.; McKee, L.; Dulaney, W.; White, W.A.; et al. Daily mapping of 30 m LAI and NDVI for grape yield prediction in California vineyards. *Remote Sens.* **2017**, *9*, 317. [CrossRef]

59. QGIS Development Team. QGIS Geographic Information System. Open Source Geospatial Foundation. 2016. Available online: http://qgis.osgeo.org (accessed on 10 September 2020).

60. Dodge, Y. Spearman Rank Correlation Coefficient. In *The Concise Encyclopedia of Statistics*; Springer: New York, NY, USA, 2008. [CrossRef]

61. RStudio Team. *RStudio: Integrated Development for R*; RStudio, PBC: Boston, MA, USA, 2020. Available online: http://www.rstudio.com/ (accessed on 10 September 2020).

62. Prions, S.; Haerling, K.A. Making sense of methods and measurement: Spearman-Rho ranked-ordered coefficient. *Clin. Simul. Nurs.* **2014**, *10*, 535–536. [CrossRef]

63. Montero, F.J.; Meliá, J.; Brasa, A.; Segarra, D.; Cuesta, A.; Lanjeri, S. Assessment of vine development according to available water resources by using remote sensing in La Mancha, Spain. *Agric. Water Manag.* **1999**, *40*, 363–375. [CrossRef]

64. Cunha, M.; Marçal, A.R.S.; Rodrigues, A. A Comparative Study of Satellite and Ground-Based Vineyard Phenology. In *Imagin[e, G] Europe*; Manakos, I., Kalaitzidis, C., Eds.; IOS Press: Amsterdam, The Netherlands, 2010; pp. 68–77. [CrossRef]

65. Gatti, M.; Garavani, A.; Vercesi, A.; Poni, S. Ground-truthing of remotely sensed within-field variability in a cv. Barbera plot for improving vineyard management. *Aust. J. Grape Wine Res.* **2017**, *23*, 399–408. [CrossRef]

66. Devaux, N.; Crestey, T.; Leroux, C.; Tisseyre, B. Potential of Sentinel-2 Satellite Images to Monitor Vine Fields Grown at a Territorial Scale. *OENO One* **2019**, *53*. [CrossRef]

67. Acevedo-Opazo, C.; Tisseyre, B.; Guillaume, S.; Ojeda, H. The potential of high spatial resolution information to define within-vineyard zones related to vine water status. *Precis. Agric.* **2008**, *9*, 285–302. [CrossRef]

68. Johnson, L.F.; Bosch, D.F.; Williams, D.C.; Lobitz, B.M. Remote sensing of vineyard management zones: Implications for wine quality. *Appl. Eng. Agric.* **2001**, *17*, 557–560. [CrossRef]

69. Sozzi, M.; Kayad, A.; Marinello, F.; Taylor, J.; Tisseyre, B. Comparing Vineyard Imagery Acquired from Sentinel-2 and Unmanned Aerial Vehicle (UAV) Platform. *OENO One* **2020**, *54*, 189–197. [CrossRef]

70. Michaud, M.; Watts, K.C.; Percival, D.C.; Wilkie, K.I. Precision pesticide delivery based on aerial spectral imaging. *Can. J. Biosyst. Eng.* **2008**, *50*, 9–15.

71. COM. Communication from the Commission to the European Parliament, the Council, the European Economic and Social Committee and the Committee of the Regions: A Farm to Fork Strategy for a Fair, Healthy and Environmentally-Friendly Food System COM/2020/381 Final. 2020. Available online: https://eur-lex.europa.eu/legal-content/EN/TXT/?uri=CELEX:52020DC0381 (accessed on 10 December 2020).

Article

Comparison of Spaceborne and UAV-Borne Remote Sensing Spectral Data for Estimating Monsoon Crop Vegetation Parameters

Jayan Wijesingha *, Supriya Dayananda, Michael Wachendorf and Thomas Astor

Grassland Science and Renewable Plant Resources, Universität Kassel, Steinstraße 19,
D-37213 Witzenhausen, Germany; supriyad24@gmail.com (S.D.); mwach@uni-kassel.de (M.W.);
thastor@uni-kassel.de (T.A.)
* Correspondence: jayan.wijesingha@uni-kassel.de or gnr@uni-kassel.de; Tel.: +49-561-804-1245

Abstract: Various remote sensing data have been successfully applied to monitor crop vegetation parameters for different crop types. Those successful applications mostly focused on one sensor system or a single crop type. This study compares how two different sensor data (spaceborne multispectral vs unmanned aerial vehicle borne hyperspectral) can estimate crop vegetation parameters from three monsoon crops in tropical regions: finger millet, maize, and lablab. The study was conducted in two experimental field layouts (irrigated and rainfed) in Bengaluru, India, over the primary agricultural season in 2018. Each experiment contained $n = 4$ replicates of three crops with three different nitrogen fertiliser treatments. Two regression algorithms were employed to estimate three crop vegetation parameters: leaf area index, leaf chlorophyll concentration, and canopy water content. Overall, no clear pattern emerged of whether multispectral or hyperspectral data is superior for crop vegetation parameter estimation: hyperspectral data showed better estimation accuracy for finger millet vegetation parameters, while multispectral data indicated better results for maize and lablab vegetation parameter estimation. This study's outcome revealed the potential of two remote sensing platforms and spectral data for monitoring monsoon crops also provide insight for future studies in selecting the optimal remote sensing spectral data for monsoon crop parameter estimation.

Keywords: monsoon crops; leaf area index; leaf chlorophyll concentration; crop water content; multispectral; hyperspectral

Citation: Wijesingha, J.; Dayananda, S.; Wachendorf, M.; Astor, T. Comparison of Spaceborne and UAV-Borne Remote Sensing Spectral Data for Estimating Monsoon Crop Vegetation Parameters. *Sensors* **2021**, *21*, 2886. https://doi.org/10.3390/s21082886

Academic Editor: Jiyul Chang

Received: 15 March 2021
Accepted: 16 April 2021
Published: 20 April 2021

Publisher's Note: MDPI stays neutral with regard to jurisdictional claims in published maps and institutional affiliations.

1. Introduction

The global cropland area is predicted to decline by 1.8–2.4% by 2030 due to conversion of arable croplands to mostly built-up landcover, and 80% of this land cover change is expected to occur in Asia and Africa [1]. Bengaluru is one of the megacities (over 10 million population) in southern India [2], which has already lost 62% of the vegetated area, while the urban area increased by 125% between 2001 and 2011 [3]. Agricultural production has intensified (i.e., high nitrogen (N) fertiliser usage, drip irrigation), and the cropping pattern has changed to meet the increasing food demand for the growing population. Between 2006 and 2012, the cropping pattern in Bengaluru changed from high water use paddy cultivation to dry land cereals and pulses (e.g., maize, finger millet, lablab). According to the state-level statistics, maize and finger millet crop yield increased by 4 to 6% annually, while pulse yield (including lablab) soared by 15% [4].

Increasing crop production using available arable lands while sustainably managing resources (e.g., water, soil) and reducing climate change is challenging [5]. Thus, near-real-time crop status monitoring could be a way forward to manage available resources and reduce inputs (i.e., precision agriculture). However, crop monitoring approaches need to be adapted to distinct crop types, in different growth stages (phenology), and under different agricultural practices. Remote sensing (RS) is one of the primary tools for crop monitoring [6]. RS facilitates contactless data collection over a given crop area using reflected electromagnetic energy, enabling the characterisation of an area's spatiotemporal information. The development of RS data collection and analysis techniques helps to achieve accurate models to estimate crop parameters.

Various sensor platforms (i.e., terrestrial, airborne, and spaceborne) have been employed to collect data about cropping areas and estimate crop growth and health parameters through different modelling approaches [7]. Generally, the reflected electromagnetic energy from the plant changes according to the physiological and the structural condition of crops and the surrounding environment [8]. Both multi- and hyperspectral sensors have been utilised from different platforms to capture these varying reflected energies. Hyperspectral sensors capture reflected energy at many narrow spectral bands (usually more than 30 bands). In comparison, multispectral sensor data contains fewer spectral bands with larger bandwidth [9]. Due to the higher spectral sensitivity of the hyperspectral data, there is a significant potential to capture a wider variety of different physiological and structural crop traits [8]. To make the clear comparison of the spectral resolution difference of the RS data for crop trait estimation, it is necessary to obtain RS data with similar spatial resolution. However, most studies which compared the spectral resolution sensitivity (hyperspectral vs. multispectral) for crop trait estimation were based on different spatial resolution; for example [10] employed field spectroscopy data as hyperspectral data with point observation and satellite data as multispectral data with 10 m spatial resolution for estimation of maize crop traits.

Empirical (statistical) models (both parametric and non-parametric) or physical models (e.g., radiative transfer model inversion) have been employed to estimate crop parameters using spectral data [11]. The empirical models inspect the association between in-situ measured target crop vegetation parameter and spectral reflectance data collected from RS. The reflectance data or their transformations (e.g., first derivative) or vegetation index (VI) developed from many wavebands were the inputs for the empirical models. A linear regression model is one of the standard parametric empirical modelling methods which estimates crop traits by utilising single waveband reflectance data or VI data as input [12]. In contrast, all—or only the essential—waveband reflectance data (original and transformed) and a multitude of VI data can be used as inputs for non-parametric empirical modelling with, e.g., machine learning methods (i.e., random forest, Gaussian process) [13]. Since both parametric and non-parametric models are data driven methods, a comparison of these methods for estimation of crop traits using RS spectral data can always provide capabilities of different modelling methods [14].

Many crop vegetation parameters that indicate growth and health status have been estimated using RS spectral data, e.g., leaf area index (LAI), leaf chlorophyll content (LCC), and canopy water content (CWC) [7]. LAI (m^2/m^2) is the leaf area per unit ground area, an essential plant biophysical variable to understand growth, health, and yield [15]. When considering other photosynthetically-active plant parts besides the leaves, it is called the green area index or plant area index [8]. Crop LAI estimation using RS reflectance data and empirical modelling approaches (both parametric and non-parametric) have shown promising results, but also considerable variation in prediction quality (coefficient of determination (R^2) ranges from 0.36 to 0.97) [16].

The LCC (both chlorophyll a and b) is a crop biochemical indicator for photosynthetic capacity, environmental stress, and N status of leaves [17,18]. LCC ($\mu g/cm^2$) is referred to as leaf-level quantification, while the multiplication of LCC with LAI is considered canopy chlorophyll content (CCC-g/cm^2). Spectral reflectance from the green to near-infrared region shows a strong relationship with LCC values [8]. According to available literature, LCC can be estimated with a maximum relative error of less than 20% from both multi- and hyperspectral sensors [19,20].

Quantification of CWC (g/m^2) attempts to identify crop water stress by estimating the quantity of water per unit area of the ground surface [21]. Water absorption regions (970 nm and 1200 nm) of the spectral reflectance data have been employed to estimate

CWC using RS spectral data [21–23]. However, few studies were able to accurately estimate ($R^2 > 0.7$) maize crop CWC using linear regression models with VI derived from wavebands from the green, red-edge and near-infrared regions [24,25]. Conversely, the crop CWC has not yet been estimated using full spectral data to uncover the full potential of hyperspectral information.

Successful estimation of crop vegetation parameters with RS spectral data has been demonstrated for various crop types such as wheat, rice, barley, and maize [7,26,27]. However, RS data application has not been examined for crops like finger millet and lablab, which are major monsoon crops in the tropical region (e.g., Bengaluru, Southern India). Furthermore, few studies have compared different remote sensing platforms (e.g., in-situ vs airborne vs spaceborne) and sensors (multispectral vs hyperspectral) for crop vegetation parameters estimation [16,28]. Thus, this study sought to fill the identified research and knowledge gap for RS for monsoon crop monitoring. The primary objective of this study is to evaluate two different RS spectral data types (420–970 nm) with a similar spatial resolution (~1 m), namely spaceborne multispectral (WorldView3–8 bands) and unmanned aerial vehicle (UAV) borne hyperspectral (Cubert–126 bands) for estimating three crop vegetation parameters (LAI, LCC, and CWC) from three crop types (finger millet, maize, and lablab) under different agricultural treatments (irrigation and fertiliser). The specific sub-objectives of this study were:

- To build crop-specific parametric and non-parametric models to estimate crop vegetation parameters
- To evaluate the developed vegetation parameter estimation models against (a) the spectral sensitivity of the RS data (multispectral vs hyperspectral), (b) modelling method (parametric and non-parametric), and (c) crop type (finger millet, maize, and lablab)
- To explore how crop-wise vegetation parameter estimation is affected by agricultural treatment (irrigation and fertiliser)

2. Materials and Methods

2.1. Study Site and Experimental Design

This study was performed in an experimental station on the premises of the University of Agricultural Science (UAS), Bengaluru, Karnataka state, India (12°58′20.79″ N, 77°34′50.31″ E, 920 m.a.s.l). The climate of the study area is a tropical savanna climate with 29.2 °C mean annual temperature. The south-west monsoon rain between June to October contributes substantially to the mean total annual rainfall of 923 mm. The dominant soil types in the area are Kandic Paleustalfs and Dystric Nitisols.

Two experimental layouts were established with two water treatments: drip irrigated (I) (controlled according to available precipitation), and rainfed (R) (Figure 1). The experiment was conducted in the 2018 Kharif season (July–October). In each experimental layout, four repetitions of finger millet (*Eleusine coracana* L.) (cultivar ML-365), maize (*Zea mays* L.) (cultivar NAH1137), and lablab (*Lablab purpureus* L.) (cultivar HA3) were cultivated with three different N fertiliser treatments (low, medium, and high) (36 blocks within one experimental layout). At the high fertiliser level, the recommended dosage of N fertiliser (50 kg N ha^{-1}, 150 kg N ha^{-1}, and 25 kg N ha^{-1}, respectively, for finger millet, maize and lablab) was applied. A reduced amount was applied at medium fertiliser treatment (58%, 56%, and 53% of the recommended N dosage, respectively, for finger millet, maize, and lablab). No N fertiliser was applied in the low-level fertiliser treatment for the three crop types. Phosphorous (P) and potassium (K) fertiliser were applied at the time of sowing at different levels following the recommended doses for the respective crop types [29].

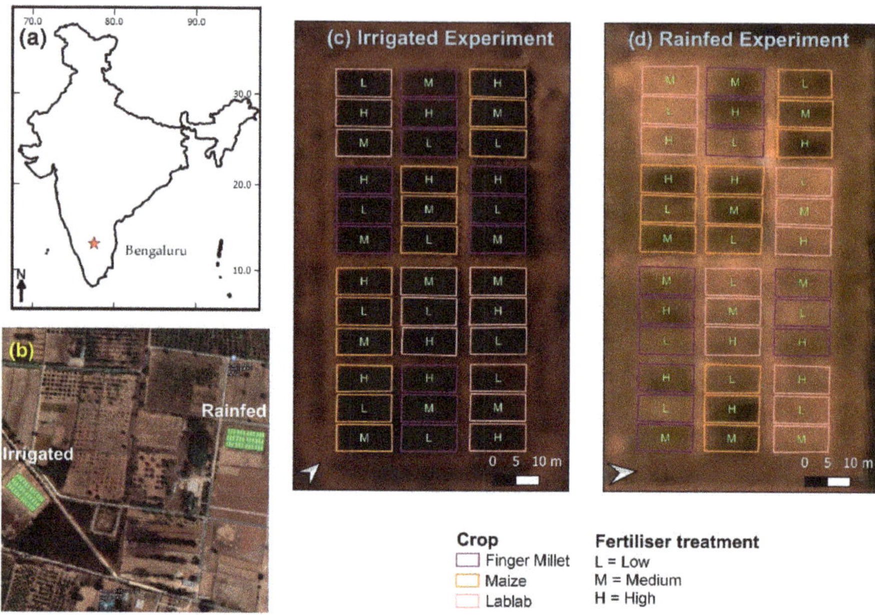

Figure 1. (**a**) Bengaluru, India; (**b**) Overview of the two experiment sites overlaid with Google satellite layer; (**c**) irrigated experiment layout, and (**d**) rainfed experiment layout with true colour composite Cubert hyperspectral image (Red = 642 nm, Green = 550 nm, Blue = 494 nm).

A single crop block was 6 m by 12 m, and the crop blocks were designed in a randomised block design. Each block was divided into two parts for destructive sampling (i.e., CWC) and non-destructive sampling (i.e., LAI, LCC). Field-level data collection and RS data collection campaigns were conducted between 29–31 October 2018. The phenological stages of the crops at the time of the field campaign are summarised in Table 1.

Table 1. Phenological stages of the crops when the remote sensing and in-situ data were collected. Based on Table A3 from [29].

Crop	Phenological Stage (Days after Sowing)	
	Irrigated Experiment	**Rainfed Experiment**
Finger millet	Inflorescence emergence (87)	Inflorescence emergence (79)
Lablab	Ripening (83)	Development of fruit (78)
Maize	Development of fruit (87)	Development of fruit (79)

2.2. In-Situ Field Data

Block-level LAI and LCC data were collected as non-destructive measurements. LAI was measured using an LI-COR LAI-2000 plant canopy analyser (LI-COR Inc., Lincoln, NE, USA). One single LAI measurement consisted of a three-time repetition of one above-canopy measurement followed by four below-canopy measurements between two crop rows. [30]. All LAI measurements were performed after 16:00 when the sun was at the horizon. LCC was measured using a handheld SPAD-502 Plus Chlorophyll meter (Konica Minolta, Osaka, Japan). The device measures the absorbances of the leaf in red and near-infrared regions. The device retrieves an arbitrary, unitless, numerical 'SPAD' value (SV) based on absorbance values. Four plants were randomly selected in each block, and three measurements per plant from the last fully developed leaf were taken. The block-level SV was computed as the average of all 12 measured SVs. According to [31], the consensus regression Equation (1) was applied to convert the SV into LCC in μg/cm^2:

$$LCC \left(\mu g \, cm^{-2} \right) = \frac{(99 \times SV)}{(144 - SV)} \tag{1}$$

After LAI and LCC measurements, destructive biomass sampling was conducted. From each block, two plants were removed, and above-ground fresh biomass weight was recorded. A subsample was dried using a sun dryer (maximum temperature was 75 °C) until no further weight loss was found (approx. 3 days). Based on dried sample weight, total dry biomass weight was computed. According to the sampled plant area, fresh biomass content (kg/m^2) and dry biomass content (kg/m^2) were determined. The canopy water content (CWC) was computed (Equation (2)) using fresh and dry biomass contents [22]:

$$CWC \left(kg \, m^{-2} \right) = fresh \, biomass \, content - dry \, biomass \, content \tag{2}$$

2.3. Remote Sensing Data

RS datasets acquired from two platforms and sensor systems were utilised in this study: (a) multispectral WorldView3 satellite data, and (b) hyperspectral Cubert UHD data mounted on a UAV.

2.3.1. WorldView 3 Data

A WorldView-3 multispectral satellite scene from 26 October 2018 was used as satellite RS data. The satellite image contained eight multispectral bands between 397 nm to 1039 nm, covering the visible and near-infrared regions of the electromagnetic spectrum (Table 2). The image's spatial resolution is 1.24 m [32,33].

Table 2. WorldView-3 multispectral image's bands and their effective bandwidths [32].

Band Name	Centre Wavelength (nm)	Effective Bandwidth (nm)
Coastal blue (CB)	427.4	40.5
Blue (BL)	481.9	54.0
Green (GR)	547.1	61.8
Yellow (YE)	604.3	38.1
Red (RD)	660.1	58.5
Red-edge (RE)	722.7	38.7
Near-infrared 1 (N1)	824.0	100.4
Near-infrared 2 (N2)	913.6	88.9

The fast line-of-sight atmospheric analysis of spectral hypercubes (FLAASH) method in ENVI 5.0 software (Harris Geospatial Solutions Inc., Broomfield, CO, USA) was applied to pre-process the satellite image using the image's metadata [34]. The pre-processed image pixel contained atmospherically-corrected surface reflectance values. However, the coastal blue (CB) band from WorldView3 data was not incorporated for the crop parameter vegetation modelling due to substantial influence from atmospheric scattering. Additionally, six vegetation indices (VIs) were calculated (Table 3). These VIs were chosen from published literature due to their proven potential to estimate LAI, LCC, and CWC [12,23,35] and compatibility with WorldView3 wavebands.

Table 3. Vegetation indices (VI) and their equations for WordView-3 (WV3) and Cubert (CUB) images. WV3 band names: GR: green, RD: red, RE: red-edge, N1: near-infrared 1. CUB bands are indicated by wavelength (ρ_{xxx}) in nanometres. (NDVI: normalised difference vegetation index, DATT4: The 4th VI introduced by [36], MTVI: modified triangular vegetation index, REIP: red-edge inflexion point, and WI: water index)

VI	Formula for WV3 Bands	Formula for CUB Bands	Reference
$NDVI_{800,670}$	$\frac{N1-RD}{N1+RD}$	$\frac{\rho_{800}-\rho_{670}}{\rho_{800}+\rho_{670}}$	[37]
$NDVI_{750,550}$	$\frac{N1-GR}{N1+GR}$	$\frac{\rho_{750}-\rho_{550}}{\rho_{750}+\rho_{550}}$	[38]
DATT4	$\frac{RD}{GR \times RE}$	$\frac{\rho_{670}}{\rho_{550} \times \rho_{706}}$	[36]
MTVI	$1.2\left[1.2\left(N1-GR\right)-2.5\left(RD-GR\right)\right]$	$1.2\left[1.2\left(\rho_{802}-\rho_{550}\right)-2.5\left(\rho_{670}-\rho_{550}\right)\right]$	[39]
REIP	$700+40\left[\frac{\left(\frac{RD+RE}{2}\right)-RE}{N1+RE}\right]$	$700+40\left[\frac{\left(\frac{\rho_{670}+\rho_{782}}{2}\right)-\rho_{702}}{\rho_{742}+\rho_{702}}\right]$	[40]
WI	$\frac{N1}{N2}$	$\frac{\rho_{902}}{\rho_{970}}$	[23]

2.3.2. Cubert Hyperspectral Data

A custom-made octocopter equipped with the Cubert Hyperspectral FireFleye S185 SE (Cubert GmbH, Ulm, Germany) snapshot camera was utilised as a UAV-borne imaging system. The hyperspectral camera is a 2D imager with a multi-point spectrometer. The camera has 450–998 nm spectral sensitivity and contains 138 spectral bands with a 4-nm sampling interval. The bands' full width at half maximum value is 4.8 nm at 450 nm and 25.6 nm at 850 nm. The spectral image is 50 by 50 pixels in size, and the camera focal length is 12 mm. Additionally, the camera has a panchromatic sensor that provides images with 1000 by 990 pixels [41,42].

The UAV-borne hyperspectral images were acquired on 29–30 October 2018 in both irrigated and rainfed experimental sites between 11:30–14:00 under clear sky conditions. At each site, the UAV-borne dataset was collected at 100-m flying height. According to the flying height, the ground sampling distance of the UAV dataset was 1.0 m. All flight missions were configured to keep 80% overlap (forward and side), and the UAV was flown with 2 ms^{-1} horizontal speed. Before each UAV flight, the camera was radiometrically calibrated to obtain surface reflectance values using a white calibration panel [43,44]. For georeferencing, the UAV images, 1-m^2 ground control points (black and white wooden crosses) were laid on the ground before the flights, and the positions of points were measured using a Trimble global navigation satellite system.

A workflow described by [43] was applied to produce a digital ortho-mosaic from single UAV-borne hyperspectral images using Agisoft PhotoScan Professional version 1.4.1 (64 bit) software (Agisoft LLC, St. Petersburg, Russia). Due to noise in the spectral bands between 450–470 nm, the final ortho-mosaic contained only 126 spectral bands (470–970 nm). Six VI images were computed in addition to the spectral band images (Table 3).

2.4. Model-Building Workflow for Crop Vegetation Parameter Estimation

From the WorldView-3 satellite dataset (WV3) and the UAV-borne hyperspectral dataset (CUB), mean values were extracted from the non-destructively sampled portions of the plots for (a) vegetation indices (VIs), and (b) all spectral wavebands (WBs). A 2-m internal buffer to the plot was applied to avoid edge effects. To estimate the crop parameters (LAI, LCC, and FMY) for each crop type, (a) parametric modelling (linear regression-LR) was conducted using VIs, and (b) non-parametric modelling (random forest regression-RFR) was performed with selected WBs based on feature importance analysis.

The relationship between the estimator (e.g., VI) and the dependent variable (e.g., LAI) was built using a linear equation (straight line) in the LR models. Before the LR model was built, a crop-wise Pearson correlation coefficient (r) was computed between the crop vegetation parameter and the VIs. A single LR model using the highest correlated VI was built to estimate crop-wise vegetation parameters.

RFR is one of the most prominent non-parametric regression algorithms that has been frequently applied for crop parameter modelling with RS data [13]. It is an ensemble modelling approach that employs decision trees and bagging [45]. This ensemble tree-based architecture supports the handling of a multitude of correlated variables [46]. The most influential bands were identified using the Boruta feature selection algorithm to reduce the computational intensity and overfitting. Boruta is an iterative process: in each iteration, features with a lower contribution to the accuracy were removed, and new random variables were introduced, thereby selecting essential variables for the model [47]. From the Boruta feature selection method, specific WBs from CUB and WV3 data were selected for crop-wise vegetation parameters. The selected WBs were utilised to build RFR models. Based on [48], one-third of the number of estimators was set as 'the number of drawn candidate variables in each split-(*mtry*)' hyperparameter value in each RFR model. The other hyperparameter, 'the number of trees in the forest' and 'the minimum number of observations in a terminal node–(*node size*)', were kept as 500 and 5, respectively, for all RFR models.

Additionally, the importance of the selected wavelengths was determined using the actual impurity reduction (AIR) importance value [49]. The AIR is a Gini importance value that was corrected for bias. Based on AIR values, the most important waveband for estimating each crop-wise vegetation parameter could be identified.

All the modelling procedures were executed using the 'mlr3' library and its extensions in the R programming language [50,51]. The 'ranger' library was employed inside the 'mlr3' library to build RFR models [52], and the 'Boruta' library was utilised for the feature selection step [47]. In total, 12 models for each crop vegetation parameter were developed (i.e., 2 modelling methods [LR and RFR] × 2 RS datasets [CUB and WV3] × 3 crops [FM, MZ, and LB]).

Due to limited data (24 data records per crop), cross-validation (CV) was applied in the model-building workflow. In CV, 12 models were trained and validated as follows: one data point from the irrigated site and one data point from the rainfed site were left out each time for validation, and the remaining 22 points (11 from the irrigated site and 11 from the rainfed site) were utilised for training the model. Based on the predicted vs actual values in the validation phase, the root means squared error (RMSE) was computed Equation (3). To standardise the RMSE values, normalised RMSE (nRMSE) was calculated by dividing RMSE from the range of the corresponding crop parameter value (the difference between the minimum and maximum values) Equation (4). The coefficient of determination (R^2) [53] was computed based on actual and predicted values Equation (5). Based on the distribution of the nRMSE and R^2 values, the crop-wise best model was identified for each crop vegetation parameter. Moreover, each model's predictive capability was examined using normalised residual values Equation (6) against two water and three fertiliser treatments. Positive or negative normalised residual values indicate overestimated or underestimated values, respectively:

$$RMSE = \sqrt{\frac{1}{n}\sum_{i=1}^{n}(y_i - \hat{y}_i)^2} \tag{3}$$

$$nRMSE = \frac{RMSE}{(max(y) - min(y))} \times 100 \tag{4}$$

$$R^2 = \left[1 - \frac{\sum_{i=1}^{n}(y_i - \hat{y}_i)^2}{\sum_{i=1}^{n}(y_i - \overline{y}_i)^2}\right] \tag{5}$$

$$normalised\ residual\ value = \frac{\hat{y} - y}{\hat{y} + y} \tag{6}$$

where y is the actual crop vegetation parameter, $\hat{y}$ is the predicted parameter, $\overline{y}$ is the average value of the actual parameter, and n is the number of samples.

3. Results

3.1. Crop Vegetation Parameter Data

The descriptive statistics of the crop-wise parameter data are presented in Table A1. The LAI values ranged between 0.4–3.2 m^2/m^2 for finger millet, 0.2–3.5 m^2/m^2 for lablab, and 1.0–3.0 m^2/m^2 for maize in both the irrigated and rainfed sites. For all three crops, the irrigated field always showed considerably higher LAI values than the rainfed site. According to the crop-wise two-way analysis of variance (ANOVA) test for LAI values, significant differences ($p < 0.001$) in LAI between irrigation treatments (I and R) were found for all three crops. N fertiliser (low, medium, and high) did not significantly affect ($p > 0.1$) LAI for any of the crops. However, there was a significant effect of interaction between irrigation and N fertiliser for lablab LAI ($p = 0.03$), with the combinations of N fertiliser and irrigation increasing average LAI.

The highest average LCC was found in irrigated maize (76.4 $\mu g/cm^2$), while the rainfed finger millet had the lowest average LCC (10.2 $\mu g/cm^2$) (Table A1). Like LAI, irrigation significantly positively affected ($p < 0.001$) LCC for all three crops. Fertilizer only significantly affected maize LCC ($p = 0.03$) positively. In contrast, there was a significant effect from the interaction between irrigation and N fertiliser for both finger millet ($p = 0.01$) and maize ($p = 0.05$) for LCC, with N fertiliser combinations irrigation increasing.

The highest CWC was found for maize (average CWC = 1.5 and 0.9 kg/m^2 for irrigated and rainfed) (Table A1), whereas lablab had the lowest CWC (0.7 kg/m^2 and 0.08 kg/m^2 for irrigated and rainfed experiments, respectively). According to the ANOVA test, CWC was significantly affected by irrigation for finger millet, lablab, and maize ($p < 0.001$). Besides, the CWC for finger millet revealed a significant positive effect of fertiliser ($p = 0.05$) and the interaction between water and fertiliser ($p = 0.01$).

Crop-wise LAI was strongly correlated with CWC (r = 0.85, 0.78, and 0.74 for finger millet, maize, and lablab, respectively). Similarly, crop-wise LCC was also positively correlated with both LAI (r = 0.81, 0.64, and 0.60) and CWC (r = 0.60, 0.62 and 0.54) for finger millet, maize, and lablab, respectively.

3.2. Spectral and Vegetation Index Data

The pattern of the spectral reflectance curves from the two experimental sites (I and R) exhibited a substantial difference for both RS datasets (Figure 2). In the irrigated plots, both CUB and WV3 spectral curves followed a typical healthy vegetation spectral reflectance curve. However, in the rainfed data, both CUB and WV3 reflectance data deviated in the visible region of the spectrum from healthy vegetation spectral curve due to higher soil spectral signals (Figure 1c).

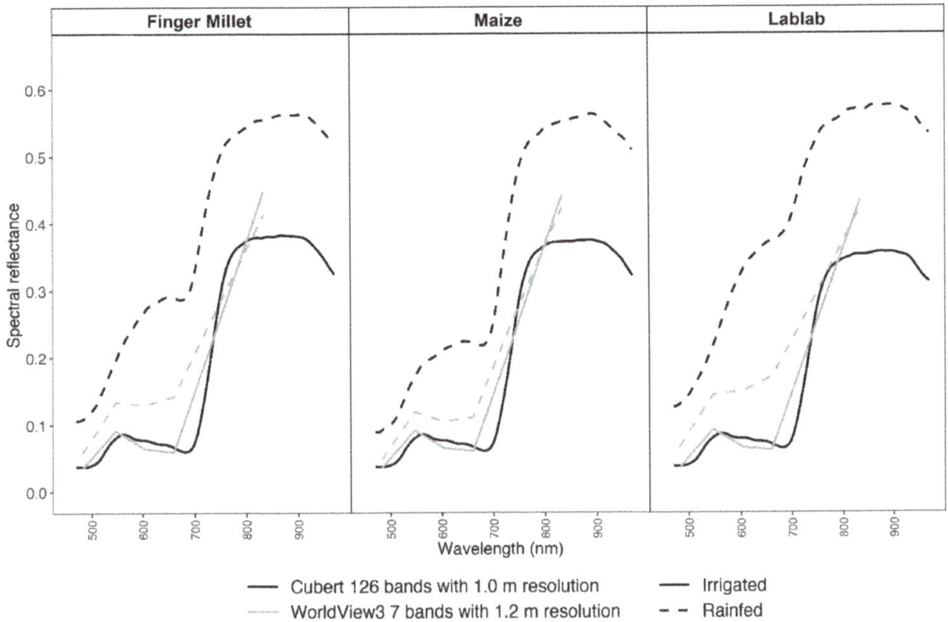

Figure 2. Average spectral reflectance data for millet, lablab, and maize from Cubert (black) and WorldView3 (grey) data for irrigated (solid line) and rainfed (dashed line) experiments.

The crop-wise VI significantly differed ($p < 0.001$) between the two RS data types (CUB and WV3) as well as between the irrigation treatments (Figure 3). However, the WV3 water index (WI) was the only index that did not show a substantial difference ($p > 0.3$) between the irrigation treatments.

3.3. Crop Vegetation Parameter Estimation with Linear Regression

LR models with VI were employed to estimate crop-wise vegetation parameters using two RS datasets. A total of six models were built separately (3 crops $\times$ 2 RS datasets). LAI estimation from CUB and WV3 data showed similar results for all three crops (Table 4). All six models obtained $R^2_{cv} \geq 0.73$. CUB VI for LAI estimation achieved nRMSE$_{cv}$ of 15.7%, 14.9%, and 15.6% for finger millet, maize, and lablab, respectively (Table 5). Likewise, nRMSE$_{cv}$ of 16.1%, 15.9%, and 16.0% were obtained for finger millet, maize, and lablab LAI estimation, respectively, using WV3 data. NDVI$_{800_670}$ was the best VI for LAI estimation using CUB data for all three crops. For WV3 data, NDVI$_{800_670}$ was the best for maize LAI estimation, while REIP was the best VI for finger millet and lablab LAI.

The VI-based LR models for estimating LCC showed lower R^2_{cv} values (Table 4). For finger millet, LCC estimation models with CUB VI data (nRMSE$_{cv}$ = 18.0%) performed better than with WV3 VI data (nRMSE$_{cv}$ = 21.0%). Maize LCC estimation models resulted in the highest normalised error and the lowest R^2_{cv} values. From the two RS datasets, WV3 VI performed better than CUB VI for maize LCC estimation. In contrast, lablab LCC estimation models from both RS datasets showed similar performances (nRMSE$_{cv}$ = 23.3% and R^2_{cv} = 0.37). Of the tested VIs, NDVI (NDVI$_{750_550}$, NDVI$_{800_670}$) and DATT4 were the most highly correlated with LCC for both RS datasets.

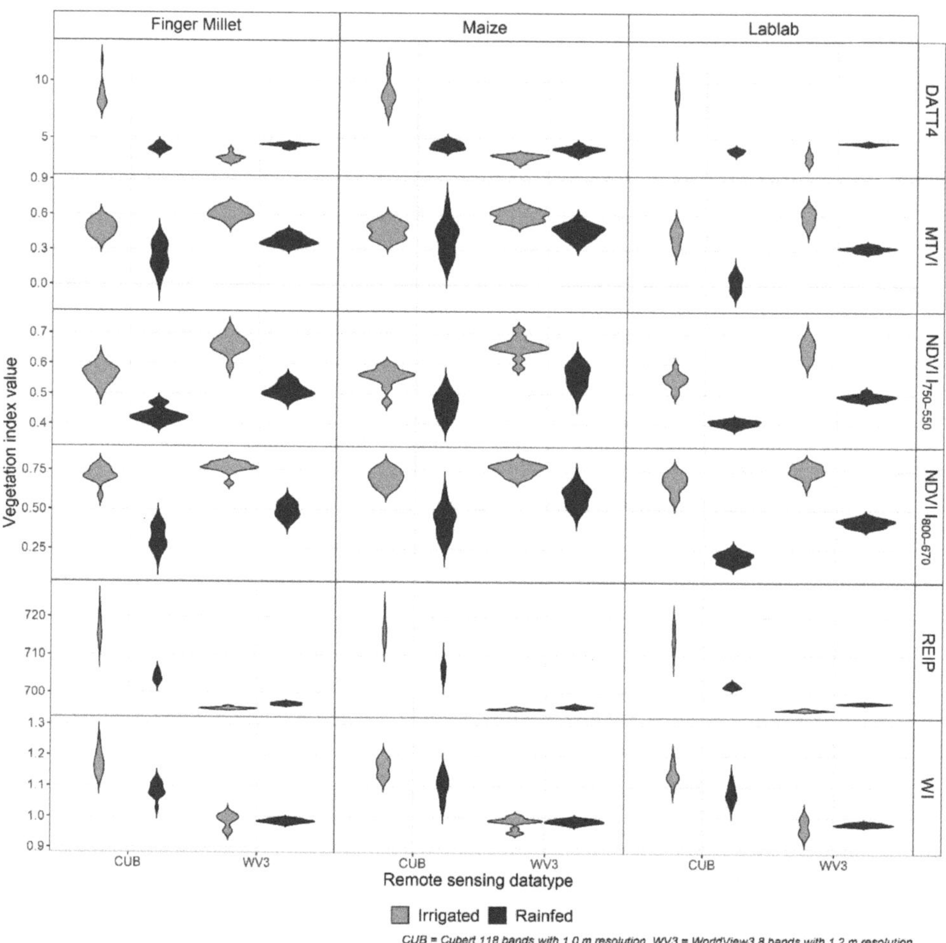

Figure 3. Distribution of crop-wise vegetation indices (VI) for finger millet, lablab, and maize from Cubert (CUB) and WorldView3 (WV3) data from irrigated (grey) and rainfed (black) experiments. (NDVI: normalised difference vegetation index, DATT4: The 4th VI introduced by Datt (1998), MTVI: modified triangular vegetation index, REIP: red-edge inflexion point, and WI: water index).

Crop-wise CWC estimation from VI from two RS datasets obtained less than 20% nRMSE$_{cv}$ (Table 4). The nRMSE$_{cv}$ values for CWC estimation with CUB data were 19.5%, 19.9%, and 16.3% for finger millet, maize, and lablab, respectively, while nRMSE$_{cv}$ values for finger millet, maize, and lablab were 19.9%, 17.0%, and 15.6%, respectively, for CWC estimation with WV3 data. The NDVI indices from CUB resulted in the best CWC estimation for all three crop types, while WV3-based NDVI$_{750_550}$, REIP, and DATT4 were strongly correlated with CWC values, respectively, from finger millet, lablab, and maize.

Table 4. Summary of the crop parameter estimation model results from linear regression (LR) using the best-correlated vegetation index (VI). Bold values indicate the lowest nRMSEcv values among the two remote sensing datasets for each crop type. All the reported linear regression models with the best vegetation index showed *p*-value less than 0.05. (LAI: leaf area index, LCC: leaf chlorophyll content, CWC: canopy water content, CUB: Cubert, WV3: WorldView3, r: Pearson correlation coefficient between VI and crop-wise vegetation parameter, R^2_{cv}: coefficient of determination from cross-validation, $nRMSE_{cv}$: normalised root means squares error from cross-validation).

Parameter	Crop	RS Data	LR Model with VIs		
			Best Vegetation Index	**r**	**R^2_{cv}**	**$nRMSE_{cv}$ (%)**
LAI **(m^2/m^2)**	Finger millet	CUB	$NDVI_{800\_670}$	0.88	0.74	**15.7**
		WV3	REIP	0.88	0.74	16.1
	Lablab	CUB	$NDVI_{800\_670}$	0.90	0.77	**15.6**
		WV3	REIP	0.90	0.77	15.9
	Maize	CUB	$NDVI_{800\_670}$	0.90	0.77	**14.9**
		WV3	$NDVI_{800\_670}$	0.89	0.73	16.0
LCC **($\mu g/cm^2$)**	Finger millet	CUB	DATT4	0.83	0.63	**18.0**
		WV3	$NDVI_{750\_550}$	0.76	0.50	21.0
	Lablab	CUB	$NDVI_{750\_550}$	0.67	0.37	**23.3**
		WV3	$NDVI_{750\_550}$	0.66	0.36	23.4
	Maize	CUB	$NDVI_{800\_670}$	0.59	0.21	24.1
		WV3	DATT4	0.61	0.26	**23.3**
CWC **(kg/m^2)**	Finger millet	CUB	$NDVI_{750\_550}$	0.73	0.44	**19.5**
		WV3	$NDVI_{750\_550}$	0.73	0.43	19.9
	Lablab	CUB	$NDVI_{800\_670}$	0.77	0.53	16.3
		WV3	REIP	0.81	0.58	**15.6**
	Maize	CUB	$NDVI_{800\_670}$	0.68	0.36	19.9
		WV3	DATT4	0.76	0.51	**17.0**

Table 5. Selected wavebands from Boruta feature selection algorithms for each crop vegetation parameter (LAI: leaf area index, LCC: leaf chlorophyll content, CWC: canopy water content) from two remote sensing datasets. Cubert bands are indicated as the band wavelength (ρ_{xxx}) in nanometres.

Parameter	Crop	Selected Wavebands from Cubert Data	Selected Wavebands from WorldView3 Data
LAI **(m^2/m^2)**	Finger Millet	$\rho_{522}, \rho_{526}, \rho_{582}, \rho_{642}, \rho_{694}, \rho_{702}, \rho_{706}, \rho_{722}, \rho_{730}, \rho_{738}, \rho_{750}, \rho_{762}, \rho_{946}$	Blue, Green, Yellow, Red, Red-edge, Near-infrared 2
	Lablab	$\rho_{690}, \rho_{698}, \rho_{706}, \rho_{722}, \rho_{726}, \rho_{734}, \rho_{750}, \rho_{826}, \rho_{918}, \rho_{930}, \rho_{946}, \rho_{950}, \rho_{954}, \rho_{958}$	Blue, Green, Yellow, Red, Red edge
	Maize	$\rho_{474}, \rho_{478}, \rho_{674}, \rho_{682}, \rho_{690}, \rho_{694}, \rho_{794}, \rho_{802}, \rho_{806}, \rho_{822}, \rho_{870}, \rho_{874}, \rho_{890}, \rho_{898}, \rho_{906}, \rho_{930}, \rho_{954}$	Blue, Green, Yellow, Red, Red edge, Near-infrared 2
LCC **($\mu g/cm^2$)**	Finger Millet	$\rho_{746}, \rho_{750}, \rho_{754}, \rho_{758}, \rho_{762}, \rho_{766}$	Blue, Green, Yellow, Red, Red edge, Near-infrared 1, Near-infrared 2
	Lablab	$\rho_{574}, \rho_{638}, \rho_{718}, \rho_{742}, \rho_{750}$	Blue, Green, Yellow, Red, Red edge
	Maize	$\rho_{682}, \rho_{690}, \rho_{698}, \rho_{702}$	Blue, Green, Yellow, Red, Red edge
CWC **(kg/m^2)**	Finger Millet	$\rho_{470}, \rho_{478}, \rho_{522}, \rho_{526}, \rho_{694}, \rho_{706}, \rho_{710}, \rho_{722}, \rho_{742}, \rho_{746}$	Blue, Green, Yellow, Red, Red edge, Near-infrared 2
	Lablab	$\rho_{502}, \rho_{606}, \rho_{614}, \rho_{618}, \rho_{630}, \rho_{666}, \rho_{678}, \rho_{682}, \rho_{742}, \rho_{802}, \rho_{834}$	Blue, Green, Yellow, Red, Red edge
	Maize	$\rho_{866}, \rho_{878}, \rho_{886}, \rho_{918}, \rho_{966}, \rho_{970}$	Blue, Green, Yellow, Red, Red edge

3.4. Crop Vegetation Parameter Estimation with Random Forest Regression

3.4.1. Key Wavebands

Important WBs for crop vegetation parameter estimation were identified using Boruta feature selection algorithms. Table 5 summarises the identified WBs from each RS datasets (CUB or WV3) for each crop vegetation parameter.

3.4.2. Model Performance

RFR models were built to estimate crop-wise vegetation parameters using the identified best WBs. Irrespective of the RS datasets and crop type, the RFR models for LAI estimation yielded less than 16.1% nRMSE$_{cv}$ and over 0.70 R$^2_{cv}$ (Table 6).

Table 6. Summary of the crop parameter estimation model results from random forest regression (RFR) using selected wavebands. Bold values indicate the lowest nRMSEcv values among the two remote sensing datasets for each crop type. (LAI: leaf area index, LCC: leaf chlorophyll content, CWC: canopy water content, CUB: Cubert, WV3: WorldView3, R$^2_{cv}$: coefficient of determination from cross-validation, and nRMSE$_{cv}$: normalised root means squares error from cross-validation).

Parameter	Crop	RS Data	RFR Model with Selected Wavebands		
			No. of Wavebands	R$^2_{cv}$	nRMSE$_{cv}$ (%)
LAI (m^2/m^2)	Finger millet	CUB	13	0.74	**16.1**
		WV3	6	0.70	17.1
	Lablab	CUB	14	0.84	12.9
		WV3	5	0.87	**12.0**
	Maize	CUB	18	0.79	13.9
		WV3	6	0.80	**13.9**
LCC (µg/cm^2)	Finger millet	CUB	6	0.45	22.1
		WV3	7	0.51	**20.8**
	Lablab	CUB	5	0.23	**25.8**
		WV3	5	0.13	27.4
	Maize	CUB	4	0.16	**24.9**
		WV3	5	0.01	31.5
CWC (kg/m^2)	Finger millet	CUB	10	0.43	**19.9**
		WV3	6	0.23	22.9
	Lablab	CUB	11	0.51	**16.9**
		WV3	5	0.42	18.2
	Maize	CUB	4	0.24	21.4
		WV3	5	0.26	**21.4**

The LAI estimation for lablab resulted in the lowest error among the three crop types (nRMSE$_{cv}$ =12.9% and 12.0%, respectively, from CUB and WV3 data). The LAI estimation models for finger millet showed better performance for CUB data (nRMSE$_{cv}$ = 16.1%) compare to WV3 data (nRMSE$_{cv}$ = 17.1%). In contrast, CUB data and WV3 data had similar accuracy for maize LAI estimation (nRMSE$_{cv}$ = 13.9%).

LCC estimation based on CUB data was more accurate than WV3 data for maize and lablab. For finger millet, the opposite was found (Table 6). The nRMSE$_{cv}$ for LCC estimation with RFR was above 20.5% for all crops, regardless of the RS datatype. The nRMSE$_{cv}$ values for LCC estimation from CUB data were 22.1%, 25.8% and 24.9%, and from WV3 data were 20.8%, 27.4 and 31.5 %, respectively, for finger millet, lablab, and maize. Based on the nRMSE$_{cv}$, the RFR models were less accurate than the LR models for LCC estimation irrespective of the RS data type and crop type.

The R$^2_{cv}$ was less than 0.5 for CWC estimation for all three crops (Table 6). For finger millet (nRMSE$_{cv}$ = 19.9%) and lablab (nRMSE$_{cv}$ = 16.9%), CWC estimation with CUB data performed better than models with WV3 data (nRMSE$_{cv}$ = 22.9% and 18.2%, respectively). Both RS datasets showed similar performance for maize CWC estimation ($nRMSE_{cv}$ = 21.4%).

3.5. Best Models and Distribution of Residuals

The best models from two RS datasets (CUB vs. WV3) and two modelling methods (LR vs. RFR) for each crop vegetation parameter were identified based on $nRMSE_{cv}$. Observed vs predicted values for crop-wise vegetation parameters from the best models are plotted in Figure 4.

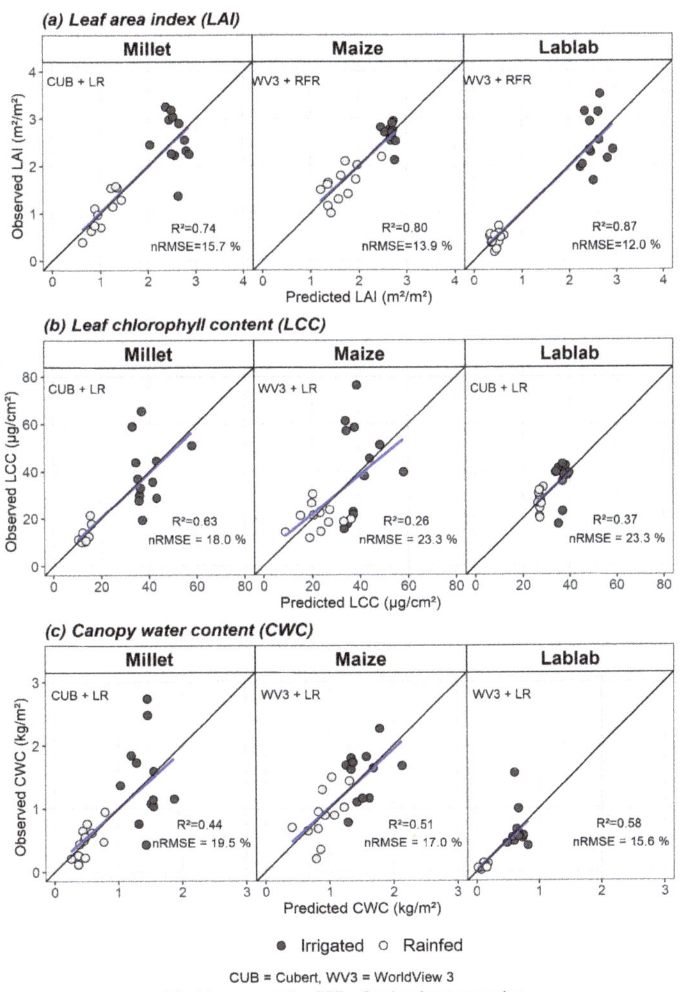

Figure 4. Observed vs predicted values of the best performing models for (**a**) leaf area index (LAI), (**b**) leaf chlorophyll content (LCC), and (**c**) canopy water content (CWC). The remote sensing data type (CUB or WV3) and modelling method (LR or RFR) for the best models are indicated as "RS data type + modelling method" (e.g., CUB + LR). The blue line is the fitted regression line between predicted and observed values, and the black line is the 1:1 line.

The normalised residual distribution values against irrigation and fertiliser treatments are shown in Figure A3. The normalised residuals of LAI, LCC, and CWC were not significantly affected ($p > 0.05$) by irrigation for any of the crops. In comparison, only the residuals from finger millet LAI and CWC prediction were significantly affected by fertiliser, with residuals decreasing from low to medium to high N fertiliser treatments.

4. Discussion

The main objective of this study was to evaluate two different spectral RS datasets (multispectral WV3 and hyperspectral CUB) for estimating three crop vegetation parameters (LAI, LCC, and CWC) of three major tropical crop types (finger millet, maize, and lablab). Considering the modelling method, out of the best nine (three vegetation parameters × three crop types) LR models based on VIs, CUB data provided six of the best models, while WV3 data provided three of the best models (Table 4). In contrast, out of the best nine RFR models with selected WBs, five of the best models were based on CUB data, whereas the other four relied on WV3 data. Overall, these results did not show a definite pattern between the RS datasets and the vegetation parameter estimation model's accuracy. Similarly, [10] reported that maize LAI estimation accuracy did not significantly differ between data with two different spectral resolutions and two different modelling methods (LR vs machine learning regression). In contrast, [54] detailed that narrow band VIs derived from hyperspectral data models yielded 20% higher R^2 values than multispectral data models for wheat and barley LAI estimation.

4.1. Finger Millet Vegetation Parameter Estimation

According to the authors' knowledge, only a few studies have utilised RS data to estimate crop vegetation parameters of finger millet and lablab [29,55]. Finger millet is a small-grained cereal (C4 type) with similar crop characteristics as pearl millet, sorghum, and foxtail millet [56]. This study revealed that the hyperspectral CUB data clearly showed the substantial potential to estimate finger millet vegetation parameters irrespective of the modelling method. For finger millet LAI estimation, $NDVI_{800\_670}$ from CUB data showed the minimum error, which confirmed that NDVI has a closer relationship with LAI at lower LAI values (less than $3.2 \text{ m}^2/\text{m}^2$) [8]. Similar to these results, NDVI showed the best estimation accuracy for sorghum LAI than other VIs (i.e., greenNDVI, EVI, and MTVI2) [57].

DATT4 is a VI for leaf chlorophyll a and chlorophyll a+b content estimation [36] and, when derived from CUB data, showed the strongest correlation with finger millet LCC (Table 4). However, DATT4 from WV3 was the least correlated VI (Figure A1). The central wavelengths of the WV3 bands do not match with the exact wavelengths of the DATT4's formula, which may have reduced the sensitivity of the index. In contrast, Two NDVIs ($NDVI_{800\_670}$ and $NDVI_{750\_550}$) from CUB and WV3 data also showed a strong correlation with finger millet LCC (Figure A1). However, sorghum's LCC showed the highest correlation with hyperspectral data NDVI [58] and indicated a lower correlation with multispectral data NDVI [59].

Models with VIs showed better finger millet CWC estimation results for both RS datasets. $NDVI_{750\_550}$ was the best correlated VI from both datasets, which predicts CWC indirectly [25] and contained green and near-infrared bands. CWC estimation with VI derived from green and near-infrared bands ($CI_{green} = (\rho_{750}/\rho_{550}) - 1$) also showed the best results among other VIs that predict CWC indirectly (i.e., NDVI, NDVIrededge, and CIrededge) [24]. When it comes to RFR modelling with selected WBs, WBs above 750 nm were not selected for finger millet CWC estimation. Nevertheless, some of the identified vital WBs were comparable with important WBs for finger millet fresh biomass estimation using multi-temporal terrestrial CUB data (e.g., 694 nm) [29].

4.2. Lablab Vegetation Parameter Estimation

Lablab is a legume crop similar to pea, beans, and lentils [60]. The lablab LAI values showed a strong correlation with NDVI values, but the LAI estimation error with NDVI was higher than the error from RFR models with selected WBs. The higher LAI values

(>3.0) from lablab may impede accurately estimating LAI with NDVI due to the saturation effect, which also demonstrated by [39] with pea LAI values. In comparison to lablab LAI estimation, LR models with VI showed improved results for lablab LCC estimation. $NDVI_{750\_550}$, which contains the green band with the near-infrared band instead of the red band, was the most highly correlated VI with lablab LCC. $NDVI_{750\_550}$ is also known as 'Green NDVI', and according to [38], shows a strong relationship with Chlorophyll a.

NDVI and REIP, respectively, from CUB and WV3 data, delivered the lowest error for lablab CWC estimation. Even though these VI do not directly relate to the leaf water content, they could determine CWC because they are linked to crop biomass [25]. Furthermore, the identified best WBs from CUB data for lablab CWC estimation (Table 5 and Figure A2) were similar to the critical WBs for lablab fresh biomass estimation [29].

4.3. Maize Crop Vegetation Parameter Estimation

As opposed to finger millet and lablab, maize has been frequently explored with RS data for its vegetation parameter estimation. LR modelling with hyperspectral (CUB) data to calculate NDVI showed a lower error than NDVI from multispectral (WV3) data for maize LAI estimation. [10] also revealed the same pattern for maize LAI estimation using VI from hyperspectral (field spectrometer) and multispectral (Sentinel-2) data. RFR models with essential WBs showed similar relative errors for maize LAI estimation using both RS datasets. Likewise, maize LAI estimation models from hyperspectral data and multispectral data also demonstrated similar cross-validation error ($nRMSE_{cv}$ = 14.9%) with a support vector machine algorithm [10].

VI derived from green, red-edge, and near-infrared bands were usually better for LCC estimation [61,62] Logically, VI containing those bands (i.e., $NDIV_{800\_670}$, DATT4) were strongly correlated with maize LCC values. However, RFR models with WV3 data had > 31% relative error, although the centre wavelength of the red band from WV3 data is 660.1 nm, which is the region absorbed by leaf chlorophyll a [63]. In comparison, RFR models with CUB data obtained slightly lower error, but all the essential WBs were between 682–702 nm (red-edge region) (Table 5 and Figure A2). This contrasts with results from another study using the same hyperspectral sensor (CUB) data, which reported the usefulness of WBs from blue, red, red-edge, and near-infrared regions for maize LCC estimation [64].

Indirectly linked VIs could estimate maize CWC in this study, while WI, which is a directly sensitive VI for CWC, showed the weakest relationship with CWC for all crops. This could be because crop parameters were highly correlated, and the variation of CWC somehow directly linked with the crop LAI and biomass values [25]. Nevertheless, water absorption at 970 nm due to O-H bonds in liquid canopy water [65] was one of the key WBs for maize CWC estimation by CUB data only (Table 5 and Figure A2).

4.4. Overall Discussion

This study could not conclude which RS data (spaceborne multispectral or UAV-borne hyperspectral) is better for the evaluated crop parameters for three crop types. Nevertheless, it is worth to mention the pros and cons of the two RS systems in terms of practical aspects of general crop monitoring. The spaceborne multispectral WV3 data hugely affected by cloud coverage in tropical regions, especially in the rainy season. Proper atmospheric corrections are needed to obtain accurate surface reflectance data from WV3 images to relate spectral values with crop vegetation parameters, which might not be easy to achieve. Additionally, the WV3 data cannot be acquired whenever it is needed because of its revisit frequency of one to five days, depending on the latitude. However, applying WV3 data to estimate crop parameter in the entire crop field can be efficiently performed because of the large spatial coverage of each satellite scene.

On the other hand, the UAV-borne CUB data can be collected whenever the data is needed, and there is no effect on the data due to cloud cover (when a proper radiometric correction is applied). However, coverage of a larger field needs to done using several

UAV flight sessions, which could be a disadvantage over the WV3 data. Additionally, UAV-borne data is also challenging to collect in extreme weather conditions such as rain and wind, typical of the tropical region's monsoon seasons.

This study's third sub-objective explored how the crop parameter estimation accuracy was affected by the crop's water and fertiliser treatments. The collected field data showed a significant positive effect due to irrigation in all three crops. However, finger millet (inflorescence emergence) and maize (development of fruit) were in similar phenological stages in both water treatments, while lablab showed two different phenological stages for irrigated and rainfed crops. (Table 1). The results clearly showed that the prediction accuracy of crop vegetation parameters did not significantly affect irrigation, and only finger millet's LAI and CWC prediction error had a significant difference due to fertiliser treatments (Figure A3). Confirming these findings, [29] also reported no significant difference for biomass prediction error between two water treatments and fertiliser treatments for the same three crops with three-year data using in-situ hyperspectral data with machine learning methods.

This study utilised only a few ($n = 24$) samples for model building for vegetation parameter estimation. For this reason, separate models for the irrigation treatments were not employed, even though the data showed a significant difference between treatments. Therefore, the CV was applied to build unbiased models, which facilitated evaluating models with a limited number of data points from both treatments. However, the number of sample points for both training ($n = 22$) and validation ($n = 2$) in the CV was not enough to capture the dataset's total variability. For example, when the model was trained with a unique range of dataset and the validation data points were out of the range from the trained model, then the model tends to under or overestimates the prediction value. It is necessary to have more data points to increase the model sensitivity to the dataset's total variability. However, having many sample points is always challenging for RS-based crop parameter estimation for many reasons, including human and physical resource availability.

The two RS datasets used in this study were sensitive from the visible to the near-infrared region. According to published studies, usage of the spectral region until the shortwave infrared (2500 nm) could increase crop parameter estimation potential [12,24]. The two RS datasets utilised in this study could accurately estimate three crop vegetation parameters from three crop types with different agriculture treatments. Hence these results could be utilised as a starting point to an in-depth examination of how to use RS data without shortwave infrared spectral data for modelling LAI, LCC, and specifically CWC. Additionally, these research findings could be employed to monitor monsoon crops using the currently available spaceborne and UAV-borne high spatial resolution remote sensors with similar spectral sensitivity (e.g., Parrot Sequoia, Micasense RedEdge, and microsatellite constellations such as Planet).

5. Conclusions

This study focused on uncovering how two different spectral resolution RS data can be utilised for estimating crop vegetation parameters from three crops (finger millet, maize, and lablab) prominently grown in Southern India. This study evaluated two different very high spatial resolution (>1.5 m) RS spectral datasets (UAV-borne hyperspectral Cubert–CUB, spaceborne multispectral WorldView3–WV3) for estimating LAI, LCC, and CWC for the three target crops. Two distinct modelling methods, namely linear regression with best-correlated vegetation index and random forest regression with important wavebands, were also evaluated. According to the results, irrespective of the RS datatype, crop type, and modelling method, the average relative estimation error was less than 16%, 25%, and 22%, respectively, for LAI, LCC, and CWC estimation. However, there was no clear evidence to identify the best RS dataset or the best modelling method to estimate the examined crop parameters. Nevertheless, there was a trend that hyperspectral (CUB) data was better for estimation of vegetation parameters of finger millet while multispectral (WV3) data was better for both lablab and maize vegetation parameter estimation. Overall, vegetation indices derived from the combination of either green, red, red-edge, and near-infrared wavebands showed clear potential from either multi or hyperspectral data for an accurate estimation of the investigated vegetation parameters regardless of the crop type.

Author Contributions: T.A., M.W. and J.W. conceptualised the study's idea; J.W. and S.D. conducted the fieldwork; J.W. processed the data and analysed the results; All authors have read and agreed to the published version of the manuscript.

Funding: The authors gratefully acknowledge the financial support provided by the German Research Foundation, DFG, through the grant number WA 2135/4-1 and by the Indian Department of Biotechnology, DBT, through grant number BT/IN/German/DFG/14/BVCR/2016 as part(s) of the Indo-German consortium of DFG Research Unit FOR2432/1 and DBT (The Rural-Urban Interface of Bangalore: A Space of Transitions in Agriculture, Economics, and Society).

Institutional Review Board Statement: Not applicable.

Informed Consent Statement: Not applicable.

Data Availability Statement: Not applicable.

Acknowledgments: The authors would like to thank the former B.K. Ramachandrappa and Nagaraju, for monitoring and setting up the experiments in the initial years. The authors are grateful to Rajanna and Dhananjaya for supporting the field data collection and Ashoka Trust for Research in Ecology and the Environment (ATREE), India, for providing field assistants and support.

Conflicts of Interest: The authors declare no conflict of interest. The funders had no role in the study's design; in the collection, analyses, or interpretation of data; in the writing of the manuscript; and in the decision to publish.

Appendix A

Table A1. Summary of the crop parameter data (LAI: leaf area index, LCC = leaf chlorophyll content, CWC: canopy water content, SD: standard deviation, CV: coefficient of variation).

Crop	Water	Min	Mean	SD	Max	CV
		LAI (m^2/m^2)				
Finger millet	Irrigated	1.4	2.6	0.5	3.2	19.2%
	Rainfed	0.4	1.0	0.4	1.6	40.0%
Lablab	Irrigated	1.7	2.5	0.6	3.5	24.0%
	Rainfed	0.2	0.5	0.2	0.7	40.0%
Maize	Irrigated	2.1	2.7	0.2	3.0	7.4%
	Rainfed	1.0	1.6	0.4	2.2	25.0%
		LCC ($\mu g/cm^2$)				
Finger millet	Irrigated	19.3	39.7	13.6	65.6	34.3%
	Rainfed	10.2	12.8	3.4	21.4	26.6%
Lablab	Irrigated	17.5	36.3	7.8	43.0	21.5%
	Rainfed	20.1	27.5	4.3	33.3	15.6%
Maize	Irrigated	15.7	42.1	19.6	76.4	46.6%
	Rainfed	11.9	20.3	5.3	30.6	26.1%

Table A1. *Cont.*

Crop	Water	Min	Mean	SD	Max	CV
			CWC (kg/m^2)			
Finger	Irrigated	0.4	1.4	0.7	2.7	46.5%
millet	Rainfed	0.1	0.5	0.2	1.0	52.1%
Lablab	Irrigated	0.4	0.7	0.3	1.6	48.5%
	Rainfed	0.03	0.08	0.04	0.1	50.0%
Maize	Irrigated	0.8	1.5	0.4	2.3	26.6%
	Rainfed	0.2	0.9	0.4	1.5	45.5%

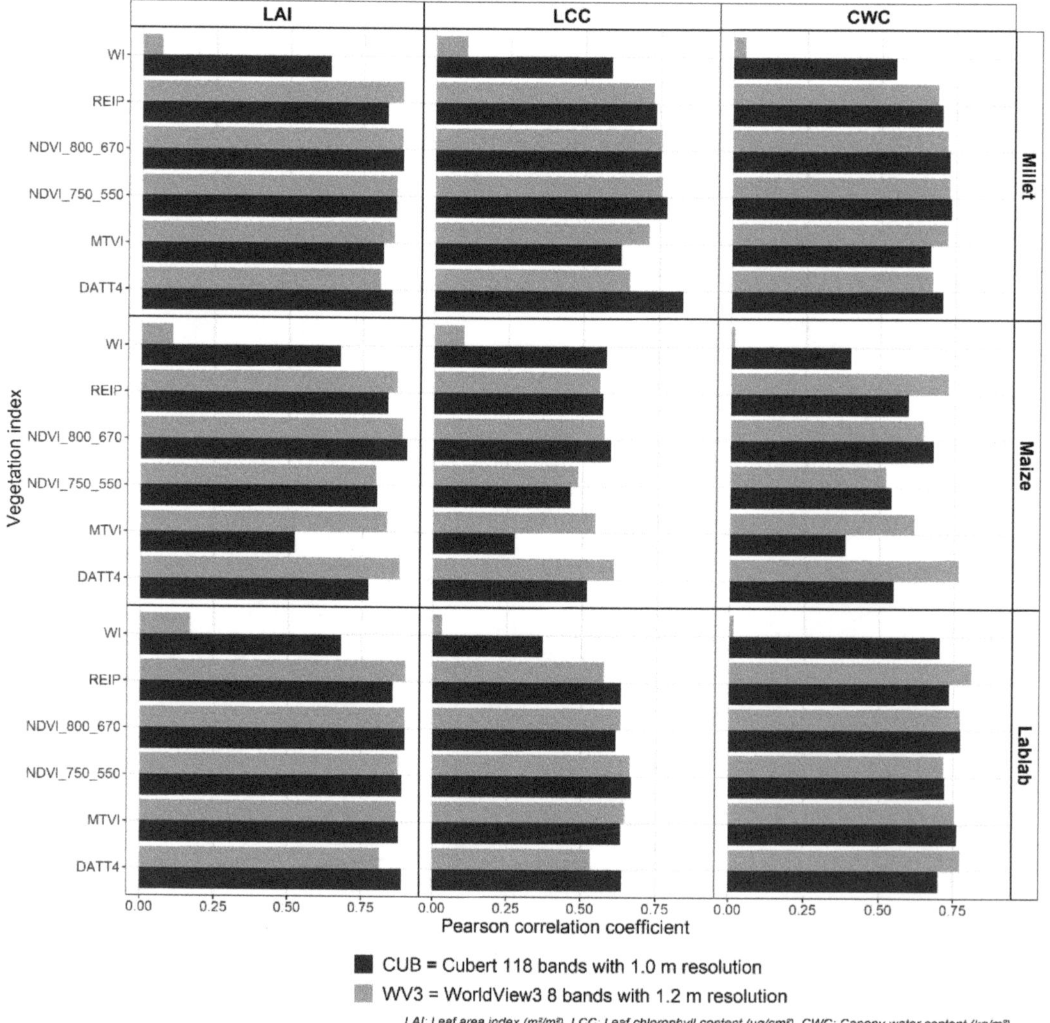

Figure A1. Correlation between vegetation indexes from two remote sensing data Cubert (black) and WorldView3 (grey) and crop vegetation parameters leaf area index (LAI), leaf chlorophyll content (LCC), and crop water content (CWC) for finger millet, maize, and lablab.

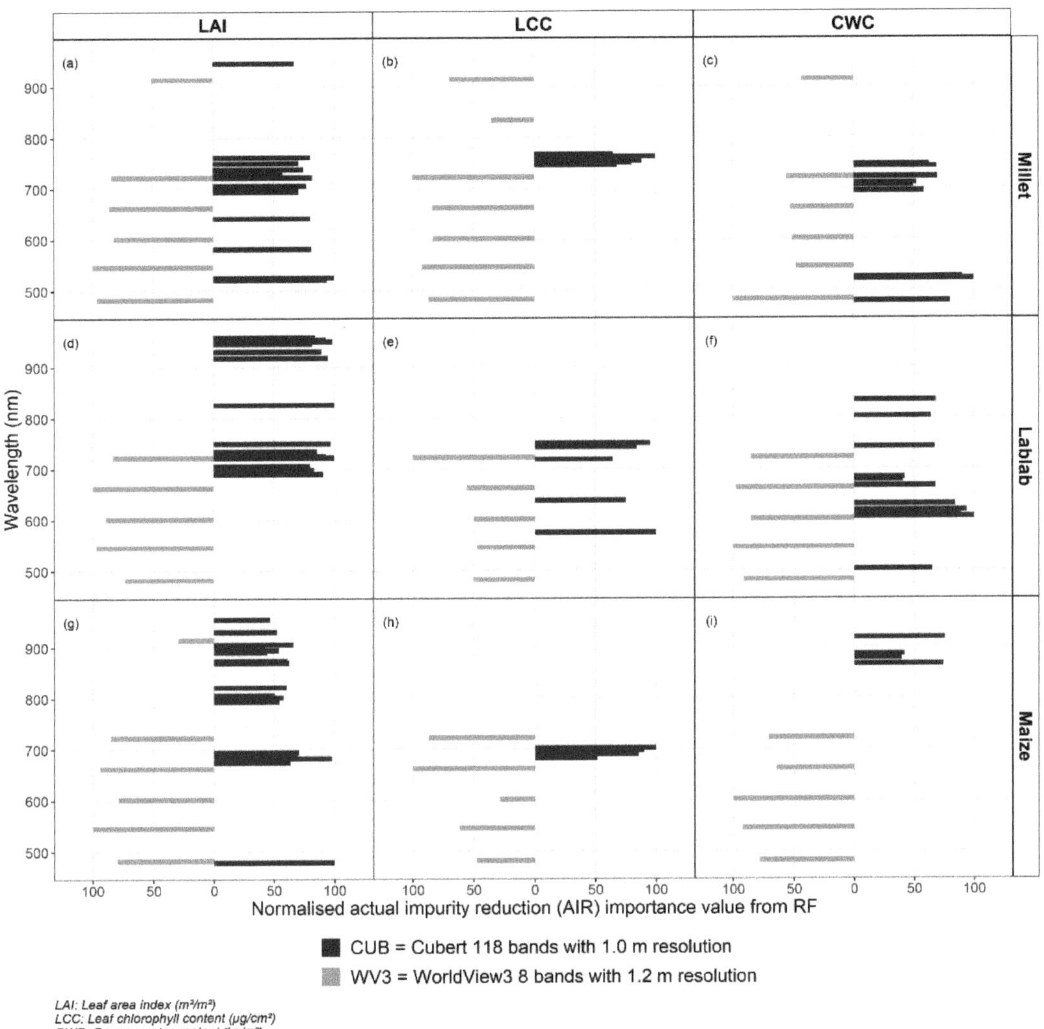

Figure A2. Distribution of actual impurity reduction value-based important wavebands for two remote sensing datasets (Cubert–black and WorldView3–grey) for leaf area index (LAI) estimation (**a,d,g**), leaf chlorophyll content (LCC) estimation (**b,e,h**), canopy water content (CWC) estimation (**c,f,i**) for finger millet (**a,b,c**), lablab (**d,e,f**), and maize (**g,h,i**) crops.

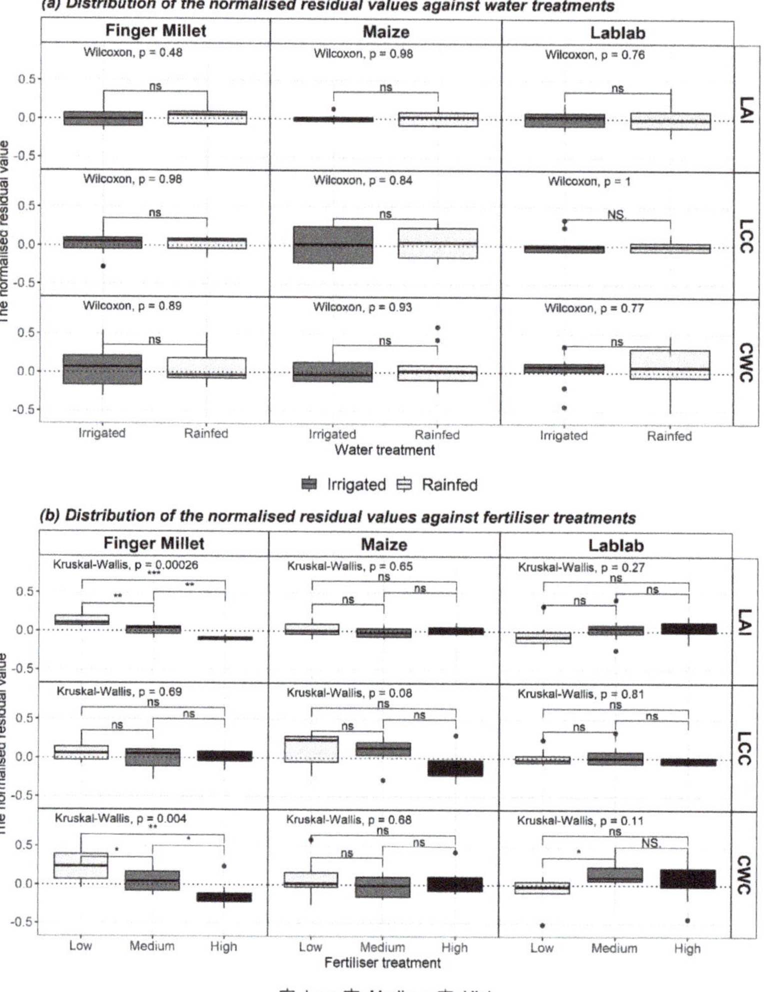

Figure A3. Distribution of the normalised residuals values against (**a**) water treatments and (**b**) fertiliser treatments from the best models for leaf area index (LAI) estimation, leaf chlorophyll content (LCC) estimation, and canopy water content (CWC) estimation for finger millet, lablab, and maize. The dashed line at y = 0 represents zero normalised residual value. (ns or NS: not significant, *: $p < 0.05$, **: $p < 0.01$, ***: $p < 0.001$).

References

1. D'Amour, C.B.; Reitsma, F.; Baiocchi, G.; Barthel, S.; Güneralp, B.; Erb, K.H.; Haberl, H.; Creutzig, F.; Seto, K.C. Future urban land expansion and implications for global croplands. *Proc. Natl. Acad. Sci. USA* **2017**, *114*, 8939–8944. [CrossRef]
2. Kübler, D.; Lefèvre, C. Megacity governance and the state. *Urban Res. Pract.* **2018**, *11*, 378–395. [CrossRef]
3. Patil, S.; Dhanya, B.; Vanjari, R.S.; Purushothaman, S. Urbanisation and new agroecologies. *Econ. Polit. Wkly.* **2018**, *LIII*, 71–77.
4. Directorate of Economics and Statistics. *Report on Area, Production, Productivity and Prices of Agriculture Crops in Karnataka, 2009–2010*; DES: Bengaluru, India, 2012.
5. Food and Agriculture Organization of the United Nations. *The Future of Food and Agriculture: Trends and Challenges*; FAO: Rome, Italy, 2017.
6. Mulla, D.J. Twenty five years of remote sensing in precision agriculture: Key advances and remaining knowledge gaps. *Biosyst. Eng.* **2013**, *114*, 358–371. [CrossRef]

7. Sishodia, R.P.; Ray, R.L.; Singh, S.K. Applications of remote sensing in precision agriculture: A review. *Remote Sens.* **2020**, *12*, 3136. [CrossRef]
8. Jones, H.G.; Vaughan, R.A. *Remote Sensing of Vegetation—Principles, Techniques, and Applications*; Oxford University Press: New York, NY, USA, 2010; ISBN 978-0-19-920779-4.
9. Thenkabail, P.S.; Lyon, J.G.; Huete, A. *Hyperspectral Remote Sensing of Vegetation*; Prasad, S.T., John, G.L., Alfredo, H., Eds.; CRC Press: Boca Raton, FL, USA, 2011; ISBN 1439845387.
10. Mananze, S.; Pôças, I.; Cunha, M. Retrieval of maize leaf area index using hyperspectral and multispectral data. *Remote Sens.* **2018**, *10*, 1942. [CrossRef]
11. Verrelst, J.; Malenovský, Z.; Van der Tol, C.; Camps-Valls, G.; Gastellu-Etchegorry, J.P.; Lewis, P.; North, P.; Moreno, J. Quantifying Vegetation Biophysical Variables from Imaging Spectroscopy Data: A Review on Retrieval Methods. *Surv. Geophys.* **2019**, *40*, 589–629. [CrossRef]
12. Rivera, J.P.; Verrelst, J.; Delegido, J.; Veroustraete, F.; Moreno, J. On the semi-automatic retrieval of biophysical parameters based on spectral index optimization. *Remote Sens.* **2014**, *6*, 4927–4951. [CrossRef]
13. Verrelst, J.; Camps-Valls, G.; Muñoz-Marí, J.; Rivera, J.P.; Veroustraete, F.; Clevers, J.G.P.W.; Moreno, J. Optical remote sensing and the retrieval of terrestrial vegetation bio-geophysical properties—A review. *ISPRS J. Photogramm. Remote Sens.* **2015**, *108*, 273–290. [CrossRef]
14. Caicedo, J.P.R. Optimized and Automated Estimation of Vegetation Properties: Opportunities for Sentinel-2. Ph.D. Thesis, Universitat De València, Valencia, Spain, 2014.
15. Chen, J.M.; Black, T.A. Defining leaf area index for non-flat leaves. *Plant. Cell Environ.* **1992**, *15*, 421–429. [CrossRef]
16. Xu, J.; Quackenbush, L.J.; Volk, T.A.; Im, J. Forest and crop leaf area index estimation using remote sensing: Research trends and future directions. *Remote Sens.* **2020**, *12*, 2934. [CrossRef]
17. Houborg, R.; McCabe, M.F.; Cescatti, A.; Gitelson, A.A. Leaf chlorophyll constraint on model simulated gross primary productivity in agricultural systems. *Int. J. Appl. Earth Obs. Geoinf.* **2015**, *43*, 160–176. [CrossRef]
18. Croft, H.; Chen, J.M.; Wang, R.; Mo, G.; Luo, S.; Luo, X.; He, L.; Gonsamo, A.; Arabian, J.; Zhang, Y.; et al. The global distribution of leaf chlorophyll content. *Remote Sens. Environ.* **2020**, *236*. [CrossRef]
19. Caicedo, J.P.R.; Verrelst, J.; Munoz-Mari, J.; Moreno, J.; Camps-Valls, G. Toward a semiautomatic machine learning retrieval of biophysical parameters. *IEEE J. Sel. Top. Appl. Earth Obs. Remote Sens.* **2014**, *7*, 1249–1259. [CrossRef]
20. Liang, L.; Qin, Z.; Zhao, S.; Di, L.; Zhang, C.; Deng, M.; Lin, H.; Zhang, L.; Wang, L.; Liu, Z. Estimating crop chlorophyll content with hyperspectral vegetation indices and the hybrid inversion method. *Int. J. Remote Sens.* **2016**, *37*, 2923–2949. [CrossRef]
21. Penuelas, J.; Filella, I.; Biel, C.; Serrano, L.; Save, R. The reflectance at the 950–970 nm region as an indicator of plant water status. *Int. J. Remote Sens.* **1993**, *14*, 1887–1905. [CrossRef]
22. Clevers, J.G.P.W.; Kooistra, L.; Schaepman, M.E. Estimating canopy water content using hyperspectral remote sensing data. *Int. J. Appl. Earth Obs. Geoinf.* **2010**, *12*, 119–125. [CrossRef]
23. Penuelas, J.; Pinol, J.; Ogaya, R.; Filella, I. Estimation of plant water concentration by the reflectance Water Index WI (R900/R970). *Int. J. Remote Sens.* **1997**, *18*, 2869–2875. [CrossRef]
24. Zhang, F.; Zhou, G. Estimation of vegetation water content using hyperspectral vegetation indices: A comparison of crop water indicators in response to water stress treatments for summer maize. *BMC Ecol.* **2019**, *19*, 1–12. [CrossRef]
25. Zhang, F.; Zhou, G. Estimation of canopy water content by means of hyperspectral indices based on drought stress gradient experiments of maize in the north plain China. *Remote Sens.* **2015**, *7*, 15203–15223. [CrossRef]
26. Weiss, M.; Jacob, F.; Duveiller, G. Remote sensing for agricultural applications: A meta-review. *Remote Sens. Environ.* **2020**, *236*, 111402. [CrossRef]
27. Tsouros, D.C.; Bibi, S.; Sarigiannidis, P.G. A Review on UAV-Based Applications for Precision Agriculture. *Information* **2019**, *10*, 349. [CrossRef]
28. Lu, B.; He, Y.; Dao, P.D. Comparing the performance of multispectral and hyperspectral images for estimating vegetation properties. *IEEE J. Sel. Top. Appl. Earth Obs. Remote Sens.* **2019**, *12*, 1784–1797. [CrossRef]
29. Dayananda, S.; Astor, T.; Wijesingha, J.; Chickadibburahalli Thimappa, S.; Dimba Chowdappa, H.; Mudalagiriyappa; Nidamanuri, R.R.; Nautiyal, S.; Wachendorf, M. Multi-Temporal Monsoon Crop Biomass Estimation Using Hyperspectral Imaging. *Remote Sens.* **2019**, *11*, 1771. [CrossRef]
30. Danner, M.; Locherer, M.; Hank, T.; Richter, K. *Measuring Leaf Area Index (LAI) with the LI-Cor LAI 2200C or LAI-2200*; EnMAP Field Guide Technical Report; GFZ Data Services: Potsdam, Germany, 2015.
31. Cerovic, Z.G.; Masdoumier, G.; Ghozlen, N.B.; Latouche, G. A new optical leaf-clip meter for simultaneous non-destructive assessment of leaf chlorophyll and epidermal flavonoids. *Physiol. Plant.* **2012**, *146*, 251–260. [CrossRef]
32. Kuester, M. *Radiometric Use of WorldView-3 Imagery*; Digital Globe: Longmont, CO, USA, 2016.
33. Digital Globe. *WorldView-3*; Digital Globe: Longmont, CO, USA, 2014.
34. Davaadorj, A. Evaluating Atmospheric Correction Methods Using Worldview-3 Image. Master's Thesis, University of Twente, Enschede, The Netherlands, 2019.
35. Liu, K.; Zhou, Q.B.; Wu, W.B.; Xia, T.; Tang, H.J. Estimating the crop leaf area index using hyperspectral remote sensing. *J. Integr. Agric.* **2016**, *15*, 475–491. [CrossRef]

36. Datt, B. Remote Sensing of Chlorophyll a, Chlorophyll b, Chlorophyll a+b, and Total Carotenoid Content in Eucalyptus Leaves. *Remote Sens. Environ.* **1998**, *66*, 111–121. [CrossRef]
37. Rouse, J.W.; Haas, R.H.; Schell, J.A.; Deering, D.W. Monitoring the vernal advancement and retrogradation (green wave effect) of natural vegetation. *Prog. Rep. RSC 1978-1* **1973**, *112*.
38. Gitelson, A.A.; Kaufman, Y.J.; Merzlyak, M.N. Use of a green channel in remote sensing of global vegetation from EOS- MODIS. *Remote Sens. Environ.* **1996**, *58*, 289–298. [CrossRef]
39. Haboudane, D.; Miller, J.R.; Pattey, E.; Zarco-Tejada, P.J.; Strachan, I.B. Hyperspectral vegetation indices and novel algorithms for predicting green LAI of crop canopies: Modeling and validation in the context of precision agriculture. *Remote Sens. Environ.* **2004**, *90*, 337–352. [CrossRef]
40. Clevers, J.G.P.W. Imaging Spectrometry in Agriculture—Plant Vitality And Yield Indicators. In *Imaging Spectrometry—A Tool for Environmental Observations*; Hill, J., Mégier, J., Eds.; Springer: Dordrecht, The Netherlands, 1994; pp. 193–219; ISBN 978-0-585-33173-7.
41. Aasen, H.; Burkart, A.; Bolten, A.; Bareth, G. Generating 3D hyperspectral information with lightweight UAV snapshot cameras for vegetation monitoring: From camera calibration to quality assurance. *ISPRS J. Photogramm. Remote Sens.* **2015**, *108*, 245–259. [CrossRef]
42. Cubert GmbH. *Cubert S185*; Cubert GmbH: Ulm, Germany, 2016.
43. Wijesingha, J.; Astor, T.; Schulze-Brüninghoff, D.; Wengert, M.; Wachendorf, M. Predicting forage quality of grasslands using UAV-borne imaging spectroscopy. *Remote Sens.* **2020**, *12*, 126. [CrossRef]
44. Yang, G.; Li, C.; Wang, Y.; Yuan, H.; Feng, H.; Xu, B.; Yang, X. The DOM generation and precise radiometric calibration of a UAV-mounted miniature snapshot hyperspectral imager. *Remote Sens.* **2017**, *9*, 642. [CrossRef]
45. Breiman, L. Random forests. *Mach. Learn.* **2001**, *45*, 5–32. [CrossRef]
46. Soukhavong, D. (Laurae) Ensembles of tree-based models: Why correlated features do not trip them—And why NA matters. Available online: https://medium.com/data-design/ensembles-of-tree-based-models-why-correlated-features-do-not-trip-them-and-why-na-matters-7658f4752e1b (accessed on 19 November 2020).
47. Kursa, M.B.; Rudnicki, W.R. Feature selection with the boruta package. *J. Stat. Softw.* **2010**, *36*, 1–13. [CrossRef]
48. Probst, P.; Wright, M.N.; Boulesteix, A.L. Hyperparameters and tuning strategies for random forest. *Wiley Interdiscip. Rev. Data Min. Knowl. Discov.* **2019**, *9*, 1–19. [CrossRef]
49. Nembrini, S.; König, I.R.; Wright, M.N. The revival of the Gini importance? *Bioinformatics* **2018**, *34*, 3711–3718. [CrossRef]
50. Lang, M.; Binder, M.; Richter, J.; Schratz, P.; Pfisterer, F.; Coors, S.; Au, Q.; Casalicchio, G.; Kotthoff, L.; Bischl, B. mlr3: A modern object-oriented machine learning framework in R. *J. Open Source Softw.* **2019**, *4*, 1903. [CrossRef]
51. R Core Team. *R: A Language and Environment for Statistical Computing*; R Foundation for Statistical Computing: Vienna, Austria, 2020.
52. Wright, M.N.; Ziegler, A. Ranger: A fast implementation of random forests for high dimensional data in C++ and R. *J. Stat. Softw.* **2017**, *77*. [CrossRef]
53. Kvalseth, T.O. Cautionary note about R2. *Am. Stat.* **1985**, *39*, 279–285.
54. Afrasiabian, Y.; Noory, H.; Mokhtari, A.; Nikoo, M.R.; Pourshakouri, F.; Haghighatmehr, P. Effects of spatial, temporal, and spectral resolutions on the estimation of wheat and barley leaf area index using multi- and hyper-spectral data (case study: Karaj, Iran). *Precis. Agric.* **2020**. [CrossRef]
55. Lambert, M.J.; Traoré, P.C.S.; Blaes, X.; Baret, P.; Defourny, P. Estimating smallholder crops production at village level from Sentinel-2 time series in Mali's cotton belt. *Remote Sens. Environ.* **2018**, *216*, 647–657. [CrossRef]
56. Taylor, J.R.N.; Kruger, J. Millets. *Encycl. Food Heal.* **2015**, 748–757. [CrossRef]
57. Shafian, S.; Rajan, N.; Schnell, R.; Bagavathiannan, M.; Valasek, J.; Shi, Y.; Olsenholler, J. Unmanned aerial systems-based remote sensing for monitoring sorghum growth and development. *PLoS ONE* **2018**, *13*, e0196605. [CrossRef] [PubMed]
58. Bhadra, S.; Sagan, V.; Maimaitijiang, M.; Maimaitiyiming, M.; Newcomb, M.; Shakoor, N.; Mockler, T.C. Quantifying leaf chlorophyll concentration of sorghum from hyperspectral data using derivative calculus and machine learning. *Remote Sens.* **2020**, *12*, 2082. [CrossRef]
59. Li, J.; Shi, Y.; Veeranampalayam-Sivakumar, A.N.; Schachtman, D.P. Elucidating sorghum biomass, nitrogen and chlorophyll contents with spectral and morphological traits derived from unmanned aircraft system. *Front. Plant Sci.* **2018**, *9*, 1–12. [CrossRef]
60. Allen, L.H. Legumes. *Encycl. Hum. Nutr.* **2012**, *3–4*, 74–79.
61. Haboudane, D.; Miller, J.R.; Tremblay, N.; Pattey, E.; Vigneault, P. Estimation of leaf area index using ground spectral measurements over agriculture crops: Prediction capability assessment of optical indices. *Int. Arch. Photogramm. Remote Sens. Spat. Inf. Sci. ISPRS Arch.* **2004**, *35*, 108–113.
62. Schlemmera, M.; Gitelson, A.; Schepersa, J.; Fergusona, R.; Peng, Y.; Shanahana, J.; Rundquist, D. Remote estimation of nitrogen and chlorophyll contents in maize at leaf and canopy levels. *Int. J. Appl. Earth Obs. Geoinf.* **2013**, *25*, 47–54. [CrossRef]
63. Lichtenthaler, H.K. Chlorophylls and Carotenoids: Pigments of Photosynthetic Biomembranes. *Methods Enzymol.* **1987**, *148*, 350–382.
64. Zhu, W.; Sun, Z.; Yang, T.; Li, J.; Peng, J.; Zhu, K.; Li, S.; Gong, H.; Lyu, Y.; Li, B.; et al. Estimating leaf chlorophyll content of crops via optimal unmanned aerial vehicle hyperspectral data at multi-scales. *Comput. Electron. Agric.* **2020**, *178*, 105786. [CrossRef]
65. Dawson, T.P.; Curran, P.J. Technical note A new technique for interpolating the reflectance red edge position. *Int. J. Remote Sens.* **1998**, *19*, 2133–2139. [CrossRef]

sensors

Article

Above-Ground Biomass Estimation in Oats Using UAV Remote Sensing and Machine Learning

Prakriti Sharma [1], Larry Leigh [2], Jiyul Chang [1], Maitiniyazi Maimaitijiang [3] and Melanie Caffé [1,*]

1   Department of Agronomy, Horticulture and Plant Science, South Dakota State University, Brookings, SD 57007, USA; prakriti.sharma@sdstate.edu (P.S.); jiyul.chang@sdstate.edu (J.C.)
2   Image Processing Lab., Department of Electrical Engineering and Computer Science, South Dakota State University, Brookings, SD 57007, USA; larry.leigh@sdstate.edu
3   Department of Geography & Geospatial Sciences, South Dakota State University, Brookings, SD 57007, USA; maitiniyazi.maimaitijiang@sdstate.edu
*   Correspondence: melanie.caffe@sdstate.edu

Abstract: Current strategies for phenotyping above-ground biomass in field breeding nurseries demand significant investment in both time and labor. Unmanned aerial vehicles (UAV) can be used to derive vegetation indices (VIs) with high throughput and could provide an efficient way to predict forage yield with high accuracy. The main objective of the study is to investigate the potential of UAV-based multispectral data and machine learning approaches in the estimation of oat biomass. UAV equipped with a multispectral sensor was flown over three experimental oat fields in Volga, South Shore, and Beresford, South Dakota, USA, throughout the pre- and post-heading growth phases of oats in 2019. A variety of vegetation indices (VIs) derived from UAV-based multispectral imagery were employed to build oat biomass estimation models using four machine-learning algorithms: partial least squares (PLS), support vector machine (SVM), Artificial neural network (ANN), and random forest (RF). The results showed that several VIs derived from the UAV collected images were significantly positively correlated with dry biomass for Volga and Beresford (r = 0.2–0.65), however, in South Shore, VIs were either not significantly or weakly correlated with biomass. For Beresford, approximately 70% of the variance was explained by PLS, RF, and SVM validation models using data collected during the post-heading phase. Likewise for Volga, validation models had lower coefficient of determination (R^2 = 0.20–0.25) and higher error (RMSE = 700–800 kg/ha) than training models (R^2 = 0.50–0.60; RMSE = 500–690 kg/ha). In South Shore, validation models were only able to explain approx. 15–20% of the variation in biomass, which is possibly due to the insignificant correlation values between VIs and biomass. Overall, this study indicates that airborne remote sensing with machine learning has potential for above-ground biomass estimation in oat breeding nurseries. The main limitation was inconsistent accuracy in model prediction across locations. Multiple-year spectral data, along with the inclusion of textural features like crop surface model (CSM) derived height and volumetric indicators, should be considered in future studies while estimating biophysical parameters like biomass.

Keywords: high throughput phenotyping; remote sensing; machine learning; UAV/drone; biomass estimation; oats

Citation: Sharma, P.; Leigh, L.; Chang, J.; Maimaitijiang, M.; Caffé, M. Above-Ground Biomass Estimation in Oats Using UAV Remote Sensing and Machine Learning. *Sensors* **2022**, *22*, 601. https://doi.org/10.3390/s22020601

Academic Editor: Sindhuja Sankaran

Received: 10 December 2021
Accepted: 9 January 2022
Published: 13 January 2022

Publisher's Note: MDPI stays neutral with regard to jurisdictional claims in published maps and institutional affiliations.

1. Introduction

Oat (*Avena sativa* L.) is a cool-season, multipurpose grain crop which ranks sixth among the most produced cereal in the world [1]. According to USDA-National Agricultural Statistics Service small grains 2020 summary statistics, out of 1.2 million hectares of oats farmed in the United States, approximately 406,000 hectares were harvested for grain, accounting for less than half of the entire planted area [2]. The crop has traditionally been collected for fodder, forage, straw, hay, silage, and chaff production in addition to grain production [1]. Oat forage is preferred over other annual forage crops because of its high palatability and dry matter content [3,4]. In accordance with previous findings, oat forage dry matter production ranged from 4000 kg per hectare in water-stressed conditions [5] to 8000 kg per hectare for the humid north-central US [6].

Breeding for improved forage yield necessitates an accurate estimation of the performance of genotypes for biomass production across the target environment [7,8]. Visual scoring, sample clipping, and mowing of individual breeding plots are some of the approaches utilized for the phenotypic assessment of forage productivity. Although visual scoring is non-destructive and ratings on individually spaced plants or rows can be correlated to dry matter yield, they are still time-consuming and vulnerable to subjectivity [9]. The clipping of small samples for the measurement of biomass is often constrained by greater sampling error resulting from soil variability and other factors. Full plot harvest provides a means to collect a representative sample, but it is destructive and time-consuming. With limited resources, full plot harvest restricts the number of seasons and places that can be sampled, the number of experimental lines that can be evaluated, and thus the genetic gain that can be obtained [10–12]. To maximize genetic gain for dry matter yield, high-throughput, cost-effective, resilient, and precise in-field forage phenotyping techniques are required [13]. Remote sensing platforms such as low altitude unmanned aerial vehicles (UAV) are becoming a common tool to increase the throughput of phenotypic data collection in plant breeding nurseries [14–16]. UAV are capable of rapid assessment of phenotypes in varietal trials with high spatial and temporal resolutions [17], and per consequent, can increase selection intensity, improve selection accuracy, and provide valuable selection decision support [18]. Such platforms can be equipped with different types of sensors such as RGB sensor (red (R), green (G), and blue (B)) and a multispectral sensor including near-infrared spectral bands (wavelength ranging between 400 and 1000 nm). These are commonly used for phenotyping various agronomic traits, including biomass [19–22], yield, disease resistance, crop/soil water status, and ground cover [23–27].

A variety of spectral features, also known as vegetative indices (VIs), have been used for biomass estimation, which also offers to quantitatively evaluate the richness, greenness, and vitality of vegetation in field experiments [28]. Several studies have utilized VIs for biomass monitoring in various crop species, including maize (*Zea mays* L.) [22,29,30], barley (*Hordeum vulgare*) [15], rice (*Oryza sativa*) [31,32], wheat (*Triticum* spp.) [19,20], and other small grain crops [33]. One of the most used indices is the normalized difference vegetation index (NDVI) [34,35], which responds to variation in chlorophyll absorption in red spectra and multi-scattering in NIR spectra, causing high reflectance [36]. The NDVI has been used for the prediction of biomass and percentage of ground cover in winter forage crops [37]. An NDVI value less than 0 indicates no vegetation covering, whereas a value larger than 0.1 indicates vegetation coverage [38] as the index is directly proportional to vegetation density, the higher the NDVI score, the greater the vegetation covering. However, the use of multiple indices is recommended for biomass prediction as different types of VIs are subject to different sensitivity depending on the amount of biomass and the stage of the crop. The NDVI, GNDVI (Green Normalized Differential Vegetation Index), SAVI (Soil-Adjusted Vegetation Index) and G-R (Green-Red Vegetation Index) are more accurate for estimating the biomass at early crop stages [37], while they get saturated at later stages [36,39] and TVI (Triangular Vegetation Index) is useful for predicting canopy biomass at later stages [40].

Accurate detection and mapping of crop canopy through remote sensing is challenging because of background effects like soil, shadow, and non-target canopies with high morphological similarities. An object-based classification method, particularly machine learning-based supervised and unsupervised pixel classification, has been widely used for canopy identification. Gašparović et al. [41] implemented automatic/manual and object-based/pixel-based classification algorithms for oats (*Avena sativa* L.) mapping using UAV-based red, green, and blue (RGB) imagery. Random forest supervised classification followed K-means unsupervised classification to differentiate oats from background soil and weed effects [41]. Likewise, Devia et al. [42] utilized the K-mean clustering algorithm for pixel classification for the identification of rice plants over soil and grasses.

Statistical models have been implemented to relate spectral information with biophysical attributes of crops [43,44]. Traditional modeling approaches are limited by statistical assumptions failing to address outlier data, nonlinearity, heteroscedasticity, and multicollinearity issues [45]. Recently, machine learning algorithms have been widely employed for the exploration and analysis of big data sets to identify meaningful correlations, patterns, and rules among data, which are frequently found to outperform traditional regression analysis [46]. The relationship between spatial and temporal changes of various predictor factors determines biomass estimation. Machine learning techniques could be highly relevant for biomass estimation as it has excellent capacity to treat multidimensional data via incorporating several predictor features [47]. Expected biomass being a continuous variable, machine learning methods such as support vector machine (SVM) [24], partial least square (PLS) [48,49], random forest (RF) [50], and artificial neural network (ANN) [51] have been used for biomass estimation. Training data is often required for supervised machine learning algorithms, however, obtaining a large dataset is often challenging because of the difficulty in manually harvesting large numbers of plots and the limited crop growing season [28]. In order to get reliable and unbiased estimates of model performance in these cases, validation techniques such as leave one out for cross-validation and k-fold cross-validation have been used in previous studies [22,52].

There are a limited number of studies that have used UAV-based canopy spectral information and machine learning to predict the biomass in oats. Various studies related to above-ground biomass estimation in cereal crops have seen lower estimation accuracy after the heading stage, which could be due to higher biomass amount or other inflorescence/stem interference overleaf canopy after heading [25,53]. Few studies have explored the impacts of canopy spectral information from different growth phases on biomass estimation for oats. Thus, the objectives of this study are; to (i) evaluate the potential of UAV multispectral imagery-derived VIs in estimation of above ground biomass in oats, (ii) evaluate the performance of UAV imagery collected at pre- and post-heading phases for oat biomass estimation, and (iii) compare the performance of different machine-learning algorithms for estimating above ground biomass of oats.

2. Materials and Methods

2.1. Field Experiments

Thirty-five oat genotypes adapted to the Northern Great Plains were cultivated in 2019 at three locations in South Dakota (Figure 1): Volga (44.321994, −96.924565), South Shore (45.105087, −96.927985), and Beresford (43.080859, −96.776148). The experimental design followed a randomized complete block design (RCBD) with three replications. Each plot (experimental unit) was approximately 2.78 m^2. Oats were planted at a density of approximately 300 seeds per square meter and at a depth of approximately 0.038 m. Beresford, Volga, and South Shore were planted on 26 April, 14 May, and 7 May, respectively, and were harvested on 11 July, 18 July, and 19 July, respectively. Agronomic practices such as fertilization and weed management were carried out in accordance with regional practices. Based on the information extracted from the Agacis website (https://agacis.rcc-acis.org, accessed on 1 July 2021), the average temperature during the growing season (May to July) was 16.4 °C in South Shore, 18.8 °C in Beresford and 17.2 °C in Volga. In 2019, precipitations during the growing season (May to July) totaled 11.93 cm in South Shore, 9.90 cm in Volga, and 11.93 cm in Beresford.

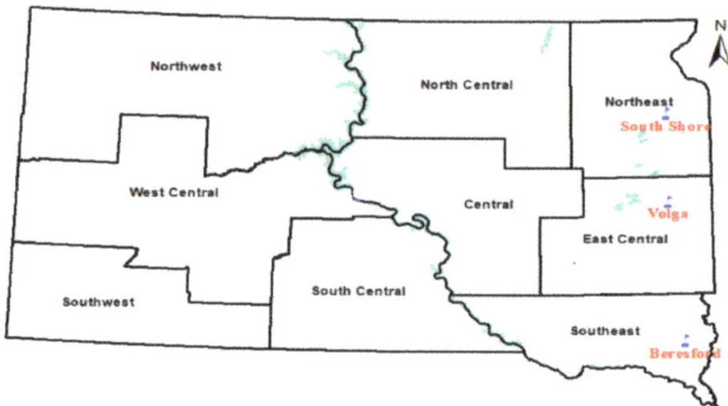

Figure 1. Three different experimental locations (South Shore, Beresford, and Volga) in South Dakota.

2.2. Ground Data Collection

Several phenotypic traits, such as heading time and crown rust severity, which can directly or indirectly affect forage yield, were collected for this study. Crown rust severity was scored as the percentage of leaf area covered by pustules over the entire plot. When plants were between late milk and early dough, oats were harvested for forage. The plants were cut close to the soil surface (approximately 7.6 cm) with a *Jari* mower or a forage harvester (Figure 2a), depending on the location. The above-ground biomass of each plot (fresh weight) was recorded immediately after harvest. For each plot, a sub-sample was collected and subjected to air-dried oven set at 70 degrees Celsius until the weight was constant (approximately a week). Dry matter content was calculated and used to measure dry matter yield for each plot; the details of dry biomass calculation are as follows:

$$\text{Dry mater content } (\%) = \frac{\text{Subsample dry weight}}{\text{Subsample fresh weight}} * 100\% \tag{1}$$

$$\text{Dry biomass } = \frac{\text{Fresh biomass} * \text{dry matter content}}{100} \tag{2}$$

(a)

Figure 2. *Cont.*

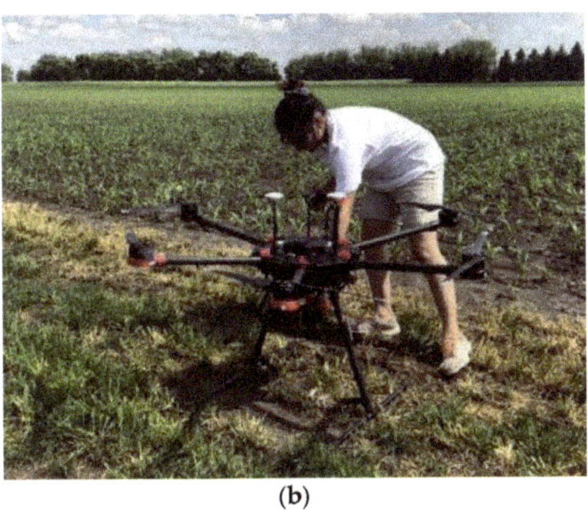

(b)

Figure 2. Harvesting of forage for biomass yield in Beresford (**a**); preparation for drone flight (**b**).

2.3. Sensor and Aerial Platform

The UAV deployed is a DJI (Dà-Jiāng Innovations) Matrice 600 hexcopter (SZ DJI Technology Co., Ltd., Shenzhen 518057, China) (Figure 2b). Multispectral images were collected with a MicaSense RedEdge-MX camera (MicaSense, Inc., Seattle, WA, USA). Micasense RedEdge-MX has a 3.2-megapixel resolution, and five bands with central wavelengths of 457 nm (blue), 560 nm (green), 668 nm (red), 717 nm (red-edge), and 840 nm (near-NIR). The spectral range covered by the green, red, red-edge, and NIR bands were 545–555 nm, 640–660 nm, 710–720 nm, and 840–860 nm, respectively. For UAV waypoint navigation and flights, an autopilot system was applied using Drone Deploy (Drone Deploy, San Francisco, CA, USA) software over the fields. Drone Deploy software was used for autonomous takeoff, flight, and landing purposes, and for capturing consistent data over time. Each of the flights was performed at an altitude of 25 m and with a front and side overlap of 80%. The flights were performed in either sunny or overcast conditions with wind gusts less than 12 miles per hour. Aerial images were collected on multiple days: Beresford (14 June, 1 July, 8 July, and 12 July), Volga (13 June, 25 June, 4 July, and 11 July), and South Shore (16 June, 25 June, 6 July, 11 July, and 18 July). The UAV flights were conducted between 10 a.m. to 12 p.m. to ensure constant daylight operation.

2.4. UAV Data Processing

2.4.1. Image Preprocessing

The processing of raw images captured by UAV was conducted by using Pix4DMappersoftware (Pix4D Inc., San Francisco, CA, USA) to generate orthomosaic images in tiff format (Figure 3). The orthomosaic images were generated with a spatial resolution of 0.7 cm. Following the orthomosaic, 10 ground control points (GCPs) were employed across the field area to geo-reference the imageries from various flights. The GCP coordinates were measured with a Magellan GPS device (Magellan Navigation Inc., San Dimas, CA, USA). Four white tarps were evenly spaced around each corner of each field for radiometric correction. The reflectance value of the tarps was determined using a CROPSCAN MSR16R (CROPSCAN Inc., 1932 Viola Heights Lane NE Rochester, MN 55906, USA). Four white tarps were used in the development of the linear relationship between DN (digital number) and surface reflectance. The average DN of white tarps from drone imageries from all the flights was used to develop an equation for each band. A linear regression-based calibration [54] was used where slope and intercept from the equation was later used to convert DN values from each band to reflectance as described. The DN values were converted to reflectance using the following equation:

$$SR_{ij} = Slope \times DN_{ij} + Intercept \tag{3}$$

where DN_{ij} is the digital number for ith band at jth flight period, and SR_{ij} is the surface reflectance for ith band at jth flight period.

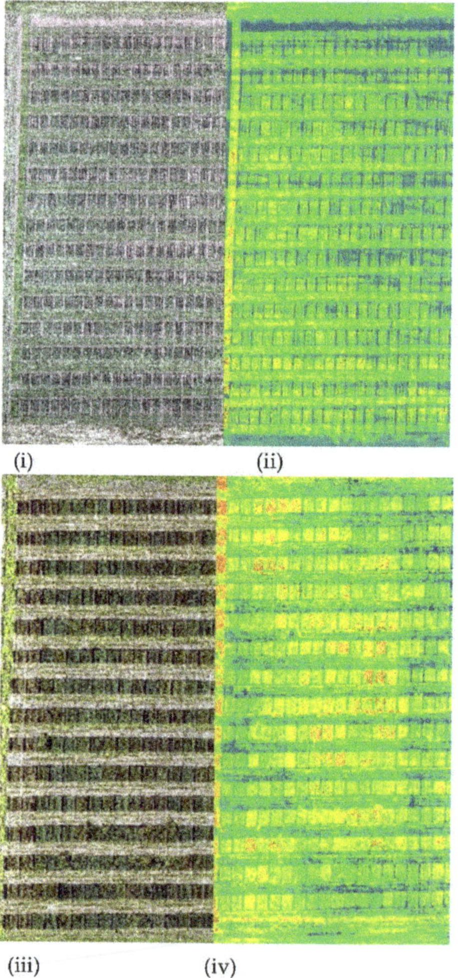

Figure 3. Orthomosaic RGB raster and corresponding NDVI map obtained during first flight (**i,ii**) and last flight (**iii,iv**).

2.4.2. Spectral Vegetation Indices Extraction

Two methodologies were used to derive vegetation indices. The first one (hereafter referred to as "average reflectance over ROI") was based on averaging the spectral reflectance for all pixels within the region of interest (ROI). However, the spectral information derived from average reflectance over ROI included shadows, background soil, and panicles (for imagery collected after heading), which could affect the overall VIs values. Spectral indices are sensitive to green living vegetation, therefore, only pixels with high NIR reflectance values within ROI were selected in the second methodology (hereafter referred to as "pixel classification").

Average Reflectance over Region of Interest

The orthomosaic images were processed using ArcGIS software (Version 10.7. Redlands, CA, USA) to extract the spectral indices. They were first converted to float from raster format. Then, using the raster calculator tool in the software, a variety of VIs were generated (Table 1). The shape file polygons were created using the same software and used for the identification of each sampling plot as an experimental unit. Finally, the zonal statistics tool was used to derive plot-level mean VIs from each experimental unit.

Table 1. List of spectral vegetation indices calculated.

Vegetative Index	Source	Mathematical Formula
Normalized Differential Vegetation Index (NDVI)	Rouse et al. (1974) [55]	$(NIR - R)/(NIR + R)$
Green Normalized Differential Vegetation Index (GNDVI)	Moges et al. (2004) [56]	$(NIR - G)/(NIR + G)$
Triangular Vegetation index (TVI)	Broge and Leblanc (2000) [57]	$0.5 \times (120 \times (NIR - G) - 200 \times (R - G))$
Red edge Triangular Vegetation Index (RTVI)	Chen (2010) [58]	$100 \times (NIR - RE) - 10 \times (NIR - G)$
Normalized Red-Green Difference Index (NGRDI)	Tucker (1979) [59]	$(G - R)/(G + R)$
Visual Atmospheric Resistance Index (VARI)	Gitelson et al. (2002) [60]	$(G - R)/(G + R - B)$
Excess Green Minus Red (ExGR)	Camargo and Neto (2014) [61]	$EXG - (1.4R - G)$

NIR, Near Infra-Red; R, Red; G, Green; and RE, Red Edge.

Pixel Classification Using K-Mean Clustering Algorithm

Pixel classification was used based on the K-mean algorithm using MATLAB. The processing software imported stacked mosaic images to create 6 cluster classes. This differentiation of clusters was based on the color feature of the image. Based on higher NIR reflectance, cluster types with green pixels were identified. A binary vegetation image was created after masking non-canopy type cluster classes. Then DN values for that cluster were extracted for all bands (NIR, red edge, red, green, and blue) and converted to surface reflectance using a calibration method. The same VIs was computed as previously described (Table 1).

2.5. Statistical Analysis

2.5.1. Data Pre-Processing

Multispectral imagery from each flight was aggregated, resulting in a comprehensive dataset for all three locations. For accessing spectral properties in accordance with the specific growth phase of oats and its relationship with biomass yield, the dataset was divided into two subsets, i.e., pre-heading and post-heading stages. This division was based on the heading date noted for each genotype in different field conditions. The spectral information collected prior to panicle emergence was separated as the pre-heading dataset, and the spectral information collected after panicle emergence in most genotypes was separated as the post-heading dataset. More explanation could be obtained from histograms plotted for each location (Figure 4) representing the distribution of heading occurrence in different genotypes measured after days of planting. The vertical dotted line represents spectral data collection through UAV. For Beresford and South Shore, spectral data from the first two flights were averaged and considered as pre-heading sample data. Likewise, remaining later flights were averaged and considered as post-heading data. While in Volga, the first three flights were averaged for the pre-heading data frame and the last single flight was considered as the post-heading data frame.

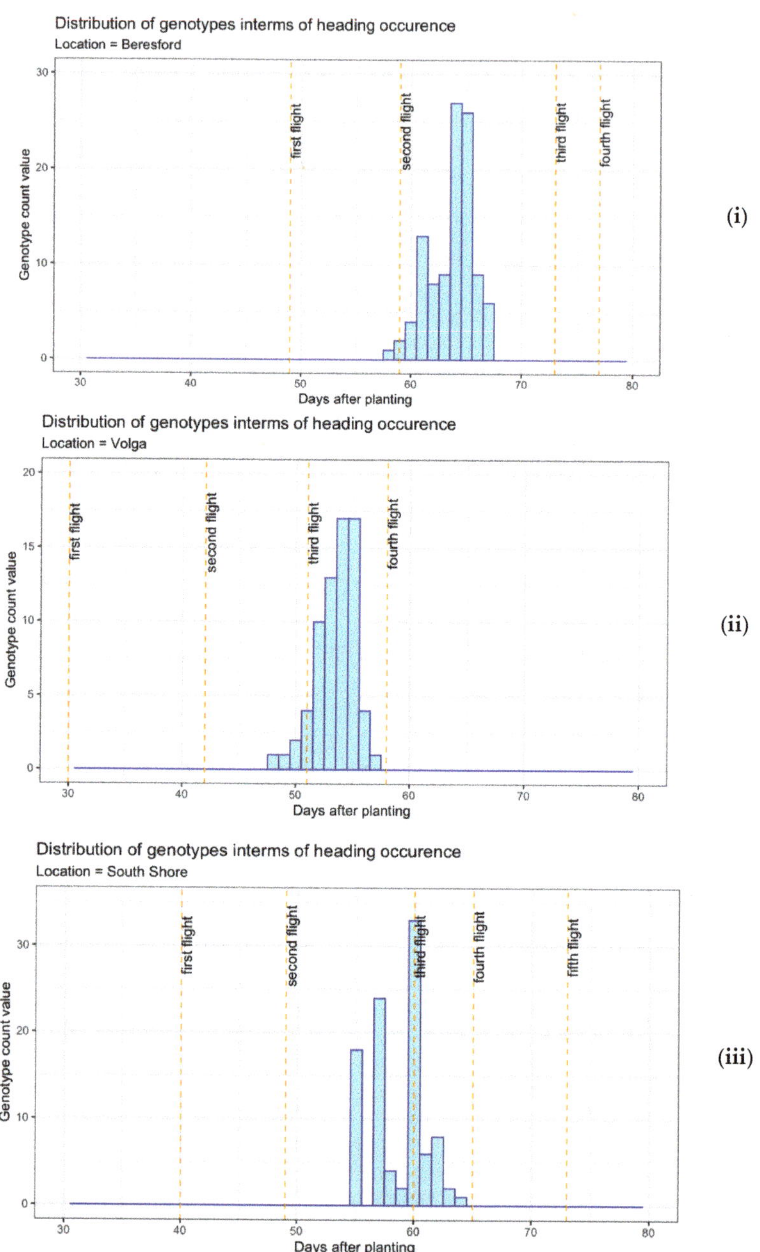

Figure 4. Distribution of heading dates for 35 oats genotypes at Beresford (**i**), Volga (**ii**), and South Shore (**iii**).

2.5.2. Correlation Analysis between VIs and Biomass

The package "hmisc" in R (version 3.5.1, R Development Core Team, 2018) [62] was used to calculate the correlation matrix, including VIs and biomass. The function "rcorr" was used to generate a matrix of Pearson's rank correlation coefficients for all possible pairs of columns of the matrix.

2.5.3. Broad Sense Heritability Estimate

Broad sense heritability estimate refers to the proportion of phenotypic variance in a trait that is attributed to the genetic variance in a population. Based on the linear mixed model approach, "Minimum norm quadratic unbiased estimation (MINQUE)" was used for estimating variance components and random effects. The jackknife as resampling technique was implemented to generalize statistical test using R package "minque" [63].

2.5.4. Modeling

The spectral data retrieved from image processing were combined with ground truth dry biomass to create the final dataset for modeling. The dataset included many variables as each VI was considered over different time frames. Hence, various linear and non-linear regression-based machine learning techniques were evaluated, and their performance was compared. The "caret" package (Version 6.0-88) in R (version 3.5.1, R Development Core Team, 2018) was used for implementing all different model algorithms [64]. In this study, four machine learning algorithms, i.e., PLS (partial least square regression), SVM (support vector machine), RF (random forest), and ANN (artificial neural network), were used to predict biomass.

The PLS approach is known for its convenience in highly correlated predictors by dimension reduction techniques as in principle component analysis [65]. The SVM algorithm aims to find a hyperplane in an n-dimensional space that distinctly classifies the data points. These hyperplanes are known as the decision boundary and are used to predict continuous output [66]. In our study, SVM was implemented using a linear variant, "svmLinear" method that was chosen from the caret package in R for this purpose. The RF algorithm principle works on a combination of tree predictors, such that each tree is dependent on the values of a random vector that is sampled independently having similar distribution for rest of trees in forest [67]. The ANN adopts the computing environment by repeated adjustment using neuron weights and thresholds. The network training completes its task once the output error of the network reaches its expected value [68].

For all four modeling approaches, tuning parameters were set (Table 2). For example, in the PLS method, the model was subjected to tuning for finding the optimal number of principal components ("ncomp") to be incorporated. While in the case of SVM, parameter C, known as "Cost", was used as a tuning parameter, allowing different iterations of C to maximize model accuracy. The cost-penalty parameter relates tolerance to error, which means that when C gets large, the model gets flexible, and it leads to overfitting. In other cases, with a small value of C, the model is rigid and subjected to underfitting. For the RF analysis, the number of trees defaulted to 500, while to obtain the best predictor combination for split candidate, the "mtry" parameter was tuned with its corresponding cross-validation error. For the ANN analysis, size and decay were hyper-parameters used to tune, where size is the number of units in the hidden layer and decay acts as a regularization parameter to avoid over-fitting. To change the candidate values of the tuning parameters, the "tuneLength" or "tuneGrid" arguments were used in the train function.

Table 2. Types of models implemented with their tuning parameters.

Model	Source	Strategy	Tuning Parameter
PLS	Abdi (2003) [69]	Linear regression	ncomp (#component)
SVM	Vapnik (1995) [70]	Linear regression	Cost (C)
RF	Livingston (2005) [67]	Tree-based regression	mtry
ANN	Zou (2008) [68]	Non-linear regression	Size and decay

For Beresford and South Shore, seventy percent of the data for each location was used for training the model and the rest was used as a validation set for evaluating the model performance. In Volga, only the first two replications of the field trial were used in our data analysis because heavy precipitation after planting caused delayed emergence in the third replication. Because of the smaller number of datapoints, the set was split 50:50 for training and validation. Random-number seeds were applied before training each model such that every model had the same data partition and had stable result output. For PLS, SVM, and ANN models, data were transformed using the "preProcess" function, which forced all predictors to be centered and scaled. In addition, "trainControl" was used to specify the type of resampling methods to estimate performance of model.

For resampling methods, k-fold cross-validation (CV) was performed on the training data set. The CV approach divides data into folds, estimating the error rate of machine learning-based classifications on iteration and outputs the final model with the least error rate [71]. In this study, repeated k-fold CV was implemented using 10 folds with three replications. The default metric used for accuracy assessment in each model was the root mean square error (RMSE). The comparison analysis was performed for both the training set (cross-validation) and the test set data using RMSE and coefficient of determination (R^2). Those parameters were calculated as

$$R^2 = \frac{\sum_{i=0}^{n}\left(X_i - \overline{X}\right)^2\left(Y_i - \overline{Y}\right)^2}{n\sum_{i=0}^{n}\left(X_i - \overline{X}\right)^2 \sum_{i=0}^{n}\left(Y_i - \overline{Y}\right)^2}$$

$$RMSE = \sqrt{\frac{1}{n}\sum_{i=1}^{n}\left(Y_i - X_i\right)^2}$$

where X_i and Y_i were estimated biomass and measured biomass, respectively, and $\overline{X}$, $\overline{Y}$ were the average estimated biomass and measured biomass, respectively, and n was the number of samples.

The predictor or variable importance for each model was derived using the generic function "varImp" using the caret package. For the PLS model, the variable importance was calculated based on weighted sums of the absolute regression coefficients. While in RF model, variable importance was derived from mean square error, computed out-of-bag data for each tree, then recomputed again after permuting each predictor variable. For ANN and SVM, there was no model-specific way for calculating variable importance; hence, the importance of each predictor was evaluated individually by using the "filter" approach [64]. The overall workflow for machine learning modeling using UAV remote-sensing data for above-ground biomass estimation is explained in Figure 5.

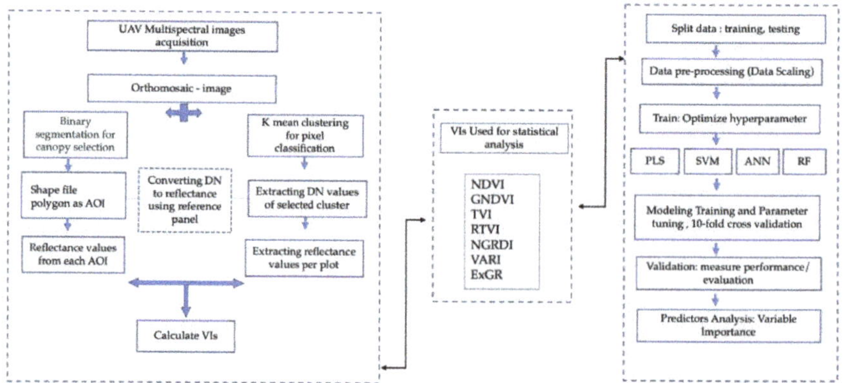

Figure 5. Workflow diagram representing methodology for UAV data processing and modeling for biomass estimation.

3. Results

3.1. Ground-Based Dry Biomass Measurements

The highest dry biomass was produced at South Shore, with an average of 13,674.4 kg/ha. The lowest dry biomass was produced in Volga, with an average of 9191.0 kg/ha (Figure 6i). Wet conditions favored the development of crown rust in all three locations. Crown rust severity was least severe in South Shore, where it averaged 25%, but 50% at the other two locations (Figure 6ii). There was a negative correlation between fresh biomass and crown rust severity at Beresford ($r = -0.59$ **) and Volga ($r = -0.4$ **), and this shows that biomass was negatively affected by the presence of crown rust infection on leaves at those two locations (Table 3). The correlation between biomass and crown rust severity was, however, not significant in South Shore. The average height for each plot was also correlated to dry biomass yield. Plant height had a significant positive correlation with dry biomass in Beresford ($r = 0.38$) and South Shore ($r = 0.24$).

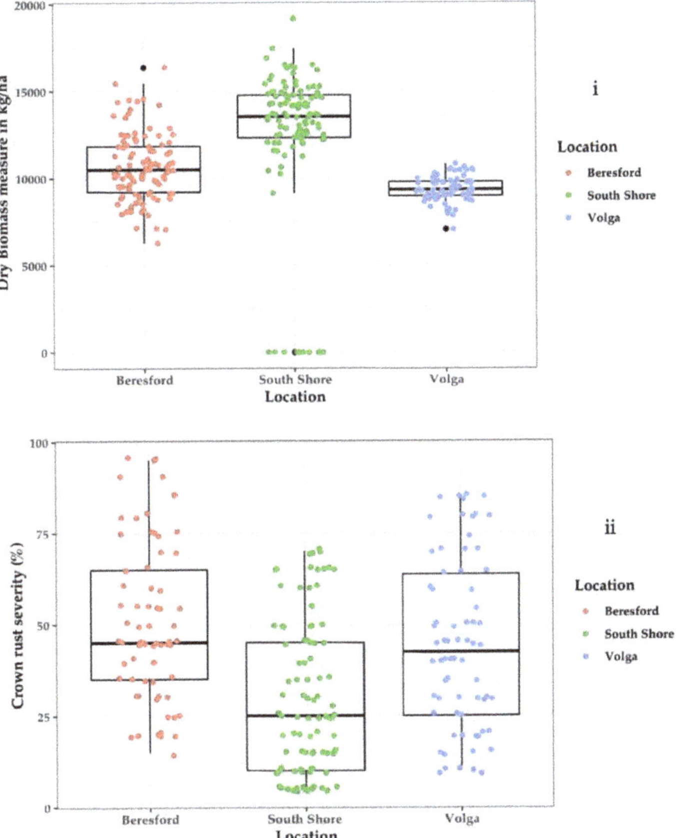

Figure 6. Boxplot representation of dry biomass yield (**i**) and crown rust severity (**ii**) for thirty-five oat genotypes evaluated at three South Dakota locations.

Table 3. Pearson correlation coefficient (r) of dry biomass with plant height and crown rust severity.

Location	Plant Height	Crown Rust Severity
Beresford	0.38 **	−0.59 **
Volga	0.15	−0.4 **
South Shore	0.24 **	0.01

** are significant at 95% CI.

3.2. Broad Sense Heritability Estimates for Vegetative Indices

Broad-sense heritability (H^2) estimates were calculated for dry biomass yield and VIs. The broad-sense heritability for dry biomass yield was 0.55 for Beresford and 0.24 for Volga. In South Shore, however, the heritability was 0.01, which shows that variation in dry biomass yield was primarily due to other factors than the genotype. Among the VIs considered, VARI and NDVI were found to consistently have higher heritability across growth phases and locations. The broad-sense heritability estimates were lower for VIs derived from pre-heading flights (NDVI: H^2 = 0.46 and VARI: H^2 = 0.47 for Beresford; NDVI: H^2 = 0.46 and VARI: H^2 = 0.45 for Volga; and NDVI: H^2 = 0.55 and VARI: H^2 = 0.64 for South Shore) than for VIs derived from post-heading flights for all locations (NDVI: H^2 = 0.53 and VARI: H^2 = 0.5 for Beresford; NDVI: H^2 = 0.63 and VARI: H^2 = 0.7 for Volga; and NDVI: H^2 = 0.55 and VARI: H^2 = 0.63 for South Shore) (Figure 7).

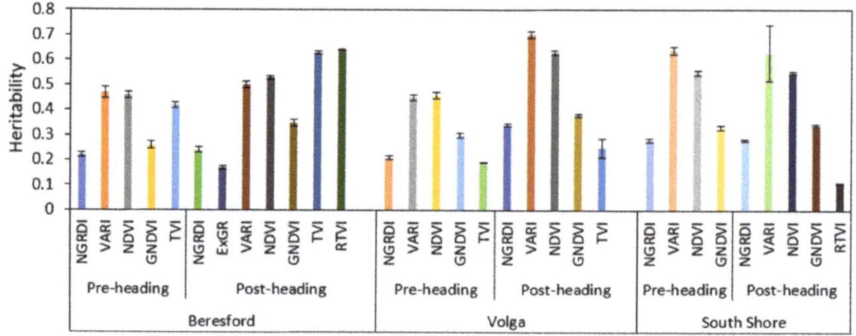

Figure 7. Bar plot representation of broad-sense heritability estimates for vegetative indices collected from pre-heading and post-heading phases across all three locations. (Only the VIs with significant heritability estimate at 95% CI are presented in the figure).

3.3. Comparison of Vegetation Indices Derived through "Average Reflectance over ROI" and "Pixel Classification" Methods

3.3.1. Relationship between Dry Biomass Yield and Vegetation Indexes Derived through Average Reflectance over ROI Method

Pearson correlation coefficients (r) were calculated between dry biomass and VIs obtained through average reflectance over the ROI method (Table 4). In Beresford, the highest correlations between VIs and dry biomass yield (0.45 to 0.6) were obtained for later flights (post-heading). For Volga, the strength of correlations between VIs and dry biomass yield was similar for both post- and pre-heading flights. Among the VIs, NDVI and RTVI were most highly correlated with dry biomass yield for both pre-heading (r = 0.43 and 0.57, respectively) and post-heading flights (r = 0.42 and 0.41, respectively). In South Shore, few VIs (TVI, ExGR, VARI) had significant correlations with dry biomass yield for flights before heading. For post-heading flights, only GNDVI was significantly positively correlated with biomass (r = 0.23).

Table 4. Pearson correlation coefficients (r) between dry biomass yield and VIs from pre- and post-heading flights.

Location	Stage	NGRDI	ExGR	VARI	NDVI	GNDVI	TVI	RTVI
Beresford	pre-heading	0.24 **	0.3 **	0.24 **	0.32 ***	0.26 **	0.27 **	0.3 **
	post-heading	0.6 ***	0.55 ***	0.55 ***	0.57 ***	0.54 ***	0.45 ***	0.54 ***
Volga	pre-heading	0.35 **	0.25 *	0.33 **	0.43 **	0.38 **	0.47 ***	0.57 ***
	post-heading	0.38 **	0.3 *	0.39 **	0.42 ***	0.35 **	0.38 **	0.41 ***
South Shore	pre-heading	0.17	0.3 *	0.28 *	0.08	0.3	0.24 *	0.1
	post-heading	0.23	−0.11	0.04	0.1	0.23 **	0.1	0.2

p value significance: * = $p \leq 0.05$, ** = $p \leq 0.01$, *** = $p \leq 0.001$.

3.3.2. Relationships between Dry Biomass Yield and Vegetation Indexes Derived through Pixel Classification

For VIs derived from the pixel classification method, post-heading flights were more strongly correlated ($r = 0.4$–0.7) with dry biomass yield than those derived from pre-heading flights ($r = 0.3$–0.5) in Beresford (Table 5). Similar results were obtained for Volga. For South Shore, however, dry biomass was not significantly correlated with any of the VIs except TVI ($r = 0.23$) for pre-heading flights (Table 5). The use of pixel classification resulted in higher correlations between VIs and dry biomass for both pre-heading and post-heading flights in Beresford.

Table 5. Pearson correlation coefficients (r) of dry biomass yield with VIs from pre- and post-heading flights.

Location	Stage	NGRDI	ExGR	VARI	NDVI	GNDVI	TVI	RTVI
Beresford	pre-heading	0.42 **	0.3 **	0.44 **	0.56 **	0.35 **	0.36 **	0.4 **
	post-heading	0.53 ***	0.61 ***	0.47 **	0.72 ***	0.52 **	0.40 **	0.44 **
Volga	pre-heading	0.28 *	0.20 *	0.30 **	0.33 **	0.32 **	0.36 *	0.45 **
	post-heading	0.44 **	0.38 *	0.42 **	0.54 **	0.45 **	0.42 **	0.46 **
South Shore	pre-heading	0.17	0.3	0.19	0.20	0.3	0.23 *	0.2
	post-heading	0.1	0.2	0.1	0.12	0.1	0.12	0.10

p value significance: * = $p \leq 0.05$, ** = $p \leq 0.01$, *** = $p \leq 0.001$.

For Beresford, the correlation between dry biomass and NDVI was $r = 0.57$ for the average reflectance over the ROI method and $r = 0.72$ after pixel classification. For Volga, correlation coefficients between dry matter yield and VIs derived from pre-heading flights were quite similar for both methods (average reflectance over ROI and pixel classification) irrespective of VIs. For the post-heading phase, VIs derived from the pixel classification method had significantly greater correlation values ($r = 0.38$–0.54) with dry matter yield as compared to average reflectance over the ROI method. For the post-heading stage in Volga, the correlation between dry biomass and NDVI was $r = 0.42$ for the average reflectance over the ROI method and $r = 0.54$ in the pixel classification method. No substantial differences were observed between the two methods for South Shore. In both cases only some VIs was significantly correlated to biomass during pre-heading, i.e., ExGR ($r = 0.3$), VARI ($r = 0.28$) and TVI ($r = 0.24$) in average reflectance over ROI method and TVI ($r = 0.23$) in the pixel classification method.

3.4. Analysis of Oat Biomass Estimation

3.4.1. Biomass Prediction from Spectral Information Collected Pre- and Post-Heading

Biomass estimation models were built with 7 VIs derived from flights during pre-heading and post-heading phases using machine learning regression methods. To assess each model's performance, the RMSE and R^2 for the testing data set were compared for each model (Table 6). For UAV data collected prior to heading, the RF model was the best model for Beresford (RMSE = 1726.3 and $R^2 = 0.3$) and South Shore (RMSE = 1659.1 and

$R^2 = 0.2$), but the SVM model was best for Volga (RMSE = 695 and $R^2 = 0.4$). For UAV data collected post-heading, the PLS model performed best for Beresford (RMSE = 1098.6 and $R^2 = 0.7$) and Volga (RMSE = 717.4 and $R^2 = 0.3$), and the SVM model worked best for South Shore (RMSE = 1681.5 and $R^2 = 0.1$). For Beresford, most models had a good fit; data points were distributed close to the fitted line as compared to the other two locations (Figure 8). We found no single model that performed best in all three sites, no matter if it was based on pre-heading or post-heading flights. The interval in the dot plot (Figure 9) shows the difference in performance, with wider intervals indicative of greater variation in performance. The overlapping confidence interval for RMSE values for the different models (Figure 9) represents the performance gap which could be due to the small sample size used for modeling.

For Beresford, models' validation using testing dataset indicates higher R^2 for models developed based on data from post-heading flights as compared to models based on data from pre-heading flights. For Volga and South Shore, however, the model's performance was very similar whether pre- or post-heading data was used for model development.

A. Beresford

i.   *Pre-heading*

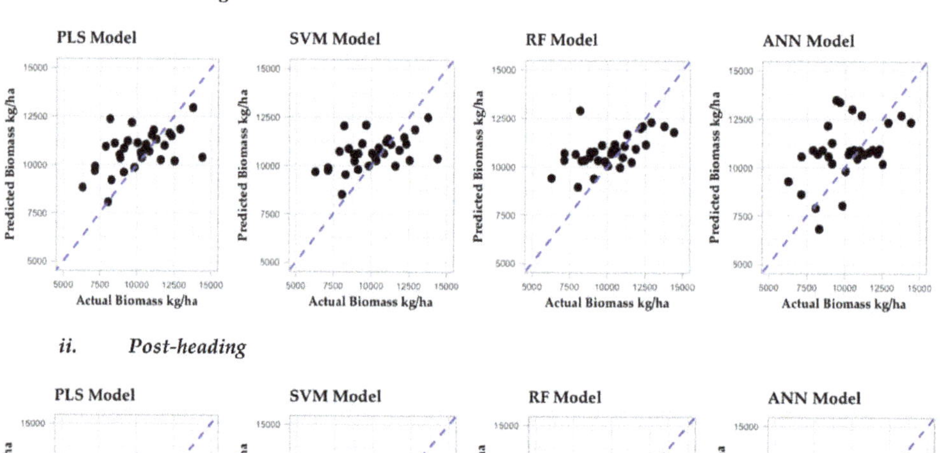

ii.   *Post-heading*

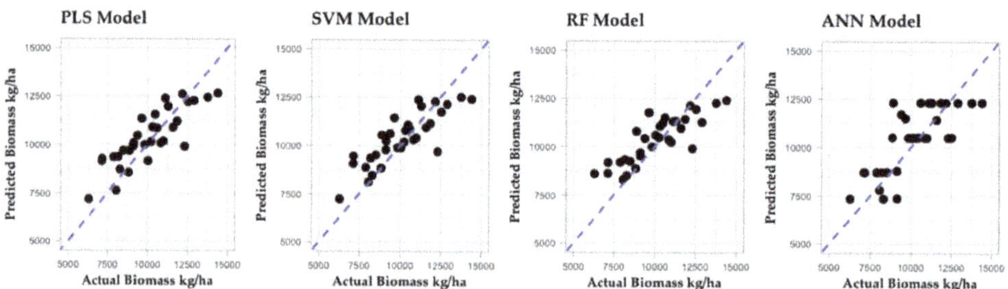

Figure 8. *Cont.*

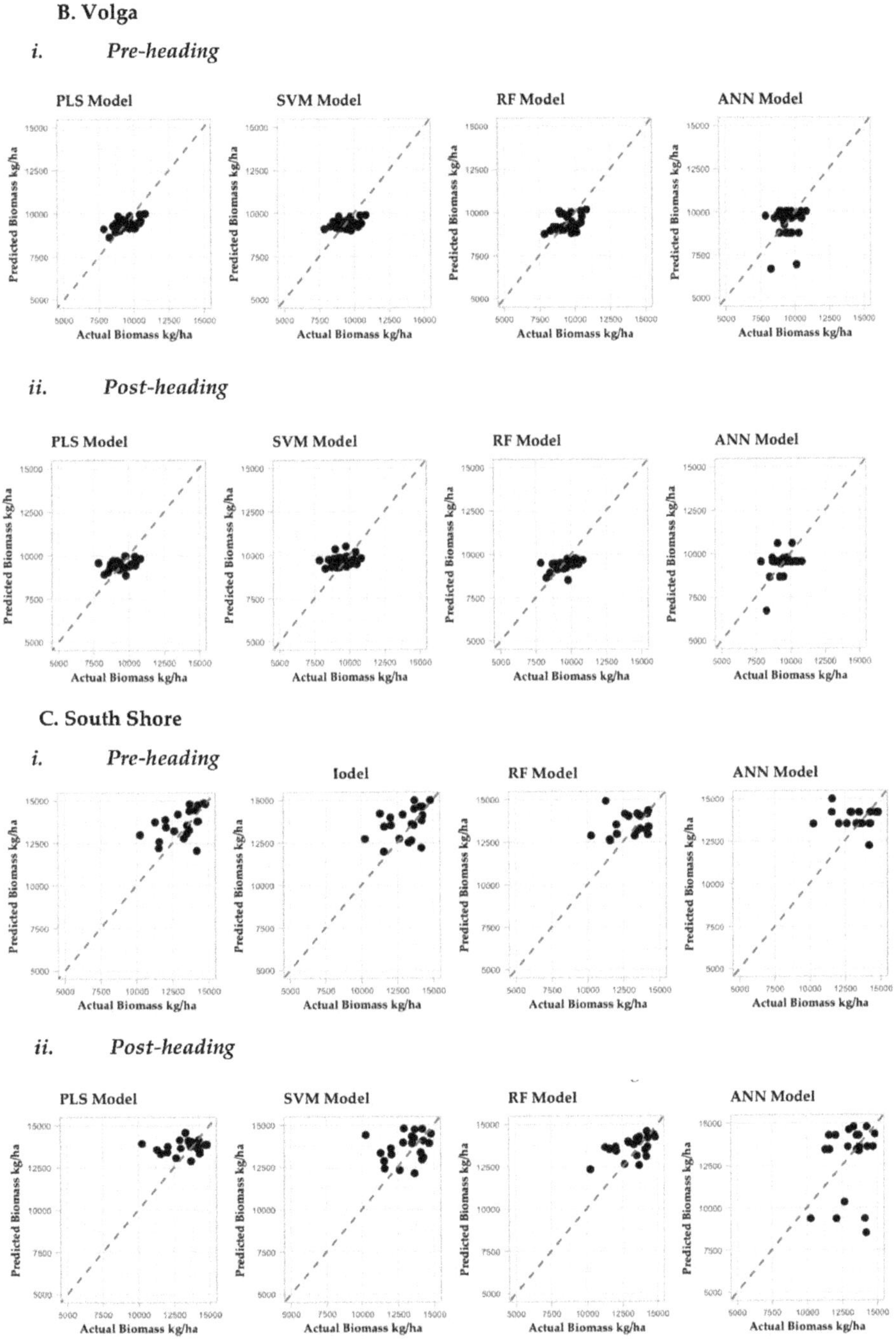

B. Volga

i. *Pre-heading*

ii. *Post-heading*

C. South Shore

i. *Pre-heading*

ii. *Post-heading*

Figure 8. Plots of Predicted Vs Actual biomass yield for 35 oat genotypes grown in Beresford (**A**), Volga (**B**), and South Shore (**C**) for pre-heading (**i**) and post-heading (**ii**) phase. The horizontal axis represents the predicted biomass yield obtained from the model, and the vertical axis represents the biomass measured manually at ground level.

Table 6. Performance of prediction models for dry matter yield in oats based on VIs derived from imagery collected pre- and post-heading (A and B) using RGB and multispectral sensors.

A. Pre-Heading						
Training Data	**Beresford**		**Volga**		**South Shore**	
	RMSE	R2	RMSE	R2	RMSE	R2
PLS	1546.98	0.30	538.08	0.55	1502.14	0.29
RF	1682.61	0.28	538.08	0.56	1546.98	0.30
SVM	1636.66	0.33	605.34	0.51	1479.72	0.28
ANN	1860.86	0.20	582.92	0.52	1703.92	0.29
Test data						
PLS	1771.18	0.26	695.02	0.3	1703.92	0.15
RF	1726.34	0.30	717.44	0.22	1659.08	0.20
SVM	1793.6	0.22	695.02	0.36	1793.6	0.10
ANN	1860.86	0.24	695.02	0.32	2264.42	0.10

B. Post-Heading						
Training Data	**Beresford**		**Volga**		**South Shore**	
	RMSE	R2	RMSE	R2	RMSE	R2
PLS	1233.10	0.60	605.34	0.61	1659.08	0.18
RF	1345.20	0.54	538.08	0.56	1748.76	0.30
SVM	1233.10	0.59	560.50	0.56	1726.34	0.13
ANN	1300.36	0.56	695.02	0.52	1771.18	0.25
Test data						
PLS	1098.58	0.70	717.44	0.27	1703.92	0.15
RF	1188.26	0.70	739.86	0.24	1771.18	0.10
SVM	1121.00	0.71	784.70	0.20	1681.50	0.14
ANN	1143.42	0.68	739.86	0.16	1771.18	0.18

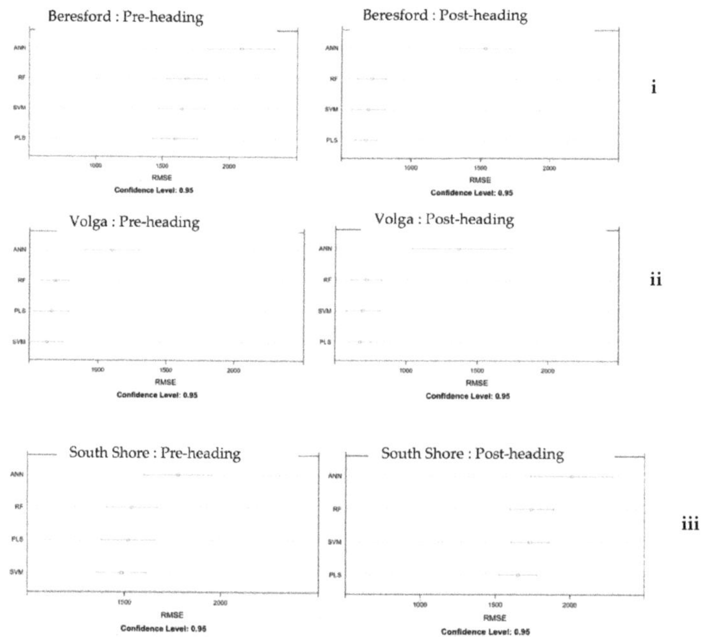

Figure 9. Dot plots from "caret" package show model comparisons using the resampling technique for Beresford (**i**), Volga (**ii**), and South Shore (**iii**). Each plot shows the mean estimated RMSE value for all four algorithms. Error bars are 95% confidence intervals on the metrics for each algorithm.

3.4.2. Assessing Variable Importance in Various Models

All four regression methods considered for model development were implemented with seven predictor variables (VIs), but the relative importance of each predictor varied depending on the algorithm, location, and time of spectral information collection (i.e., pre-heading or post-heading). For Beresford, GNDVI and ExGR had high importance for both pre- and post-heading across the models (Figure 10a). For Volga, RTVI had the greatest importance among the VIs (Figure 10b). For South Shore, results were variable across models (i.e., GNDVI in SVM and PLS, ExGR in ANN and RTVI in RF) (Figure 10c).

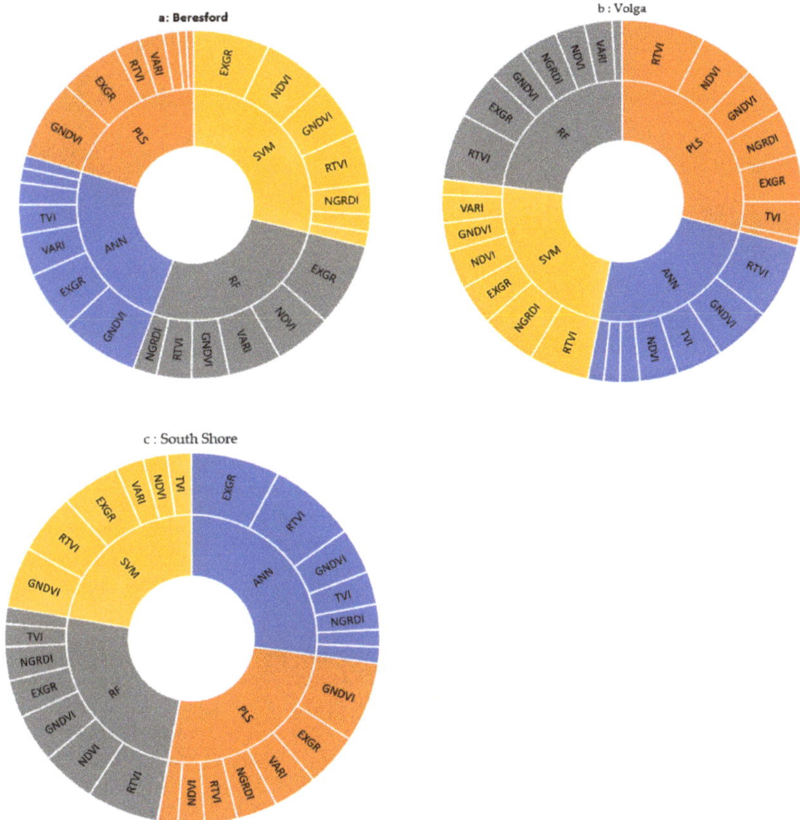

Figure 10. Importance scores for predictor variables at Beresford (**a**), Volga (**b**), and South Shore (**c**) aggregating data from pre- and post-heading flights.

Variable importance was also accessed for pre- and post-heading by aggregating information for all locations and models. For models based on pre-heading data, ExGR, GNDVI, and RTVI had a greater value of importance in comparison to another VIs (Figure 11). The same three predictor variables also had higher importance in models developed using data from post-heading flights (Figure 11). This suggests that both RGB based (ExGR) and NIR based (GNDVI and RTVI) indices were influential for biomass prediction.

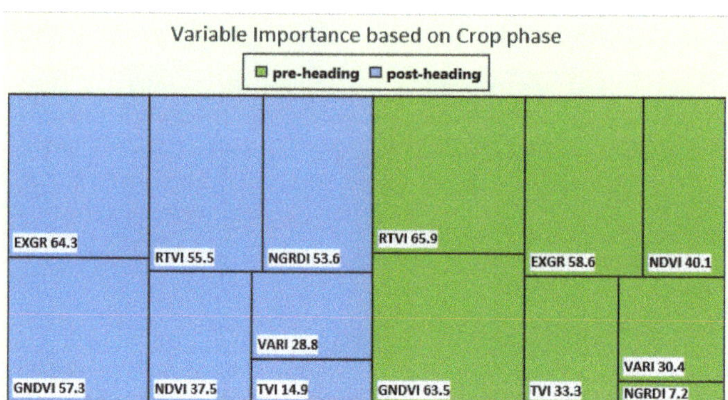

Figure 11. Importance scores for predictor variables. The importance scores of predictors are summarized considering all locations and model types.

4. Discussion

4.1. Vegetative Indices on Predicting Biomass

Significant correlations between VIs and dry biomass yield were observed in Beresford and Volga. In South Shore, however, very few VIs were significantly correlated to dry biomass. This means that spectral information from aerial multispectral sensors may not be fully efficient for biomass monitoring in certain cases. The principle of VIs is based on photosynthetically active material, which could lead to error for the prediction of total biomass [72,73]. The indicators of plant performance in remote sensing are color, structure, and shapes of leaves. This is determined by properties like chlorophyll content and leaf morphological and surface structures, which are dependent on the genotypes and on environmental stresses and plant nutrition status. In our case, the higher moisture and lower temperature in South Shore likely resulted in the higher biomass production along with a low correlation of biomass yield with a disease like crown rust which led to minimal spectral differences amongst genotypic plots. Another possible reason for the indices not being able to predict biomass could be optical saturation. VIs saturation has been reported previously in different studies. Prabhakara et al. [37] reported that VIs was not able to detect the amount of biomass when there was high vegetation for barley and rye. In their study, NDVI, GNDVI, and G-R saturated after reaching a value of approximately 0.8 and were only related to biomass under ~1500 kg/ha, beyond which an increase in biomass did not increase vegetative index value. In our study, although every location had average biomass measured above 1500 kg/ha, in South Shore, VIs reached the highest value (average NDVI value of 0.63) during the second flight (before heading) and gradually declined in later flights. Whereas, for Beresford and Volga, the average value of VIs consistently increased over time and reached to peak for the last flight before forage harvest.

In addition, during the 2019 growing season, precipitations were frequent at South Shore, where the soil was saturated with water, and dew was frequent. The average soil moisture over the growing season in South Shore was relatively 37.5% higher than in Beresford (29.2%) [74]. The presence of dew on the canopies at the time of flight could have affected the spectral reflectance quality and resulted in inaccurate vegetation indices. Pinter et al. [75], in their study on the effect of dew on canopy reflectance, found that moderate to high dew levels enhanced reflectance in visible wavelengths by 40–60% in wheat cultivars. The wetness on leaves has been observed to affect the canopy reflectance in a variety of plants, particularly in visible wavelengths [76,77].

The thirty-five oat genotypes used in this study had different maturity. The interval for heading occurrence varied depending on the location. The heading stage for all 105 plots occurred within nine days in Beresford, within six days in Volga, and within nine days in South Shore. Plots also had different maturity stages on the day of forage harvest. There is evidence that the vegetation indices are affected not only by environmental conditions but also by the growth stage of the crop [78]. Future studies should include soil moisture status, weather information, crop stage of each genotype, and other environmental factors to investigate the possible cause for failure of VIs to predict biomass.

Several studies reported using plant height derived from the crop surface model (CSM) in combination with VIs for the accurate prediction of biomass for crops like barley [79] and winter wheat [80]. Using a volume metric to estimate crop biomass within a plot (combination of spectral and structural information) has significantly improved above-ground biomass in corn [22]. Overall, these studies, along with our findings, suggest that a combination of spectral and structural information from an aerial sensor may be necessary to predict biophysical parameters like biomass more precisely.

4.2. Broad-Sense Heritability Estimates for VIs

For all three locations, NDVI and VARI had higher broad-sense heritability than dry biomass yield. Another study reported a strong genetic correlation between winter wheat grain yield and spectral reflectance and found Multispectral/RGB-based VIs with heritability (H^2 = 0.6–0.8), greater than for yield (H^2 = 0.4–0.7) [81]. With these criteria, spectral data can be used for indirect selection in plant breeding operations to increase genetic gains [18]. However, in this study, biomass and VIs were not significantly correlated in all locations. Evaluating the performance of UAV as a breeding tool for phenotyping should be evaluated over multiple locations and years before determining if VIs can be used as an indirect selection tool for oat biomass.

4.3. Comparison of Methodologies for VIs Computation

Several studies [82,83] have used pixel classification to enhance the accuracy of UAV-based data to differentiate canopy and non-canopy areas. Booth et al. [82] used the single pixel sample point method to differentiate shrub and grass species from other background pixels. Patrignani et al. [83] used Canopeo (automatic color threshold classification in MATLAB, which classified pixels to the canopy and non-canopy categories in various crops (turf, corn, sorghum, etc.). In our study, NDVI correlation to biomass improved with the pixel classification method in almost all cases (except for Volga for pre-heading flights). Nevertheless, it is essential to note that improvement seen with average reflectance over the ROI method was not consistent for every VIs. The lack of significant correlations between VIs and biomass remained unchanged for most cases in South Shore even when the pixel classification method was applied.

When considering different planophile and erectophile species, Myneni and Williams [84] reported that NDVI was unaffected by pixel heterogeneity for estimating canopy vigor based on biomass and color. Pixel heterogeneity, in our case, was comprised of panicle structure and other background effects (shadow). But resolving problems through the selection of pure canopy pixels was successful for one location (Beresford), but it did not quite improve the relationship with ground truth biomass in all cases.

4.4. Evaluation of Prediction Models for Biomass

The proportion of variance in dry biomass yield explained by the models developed in this study ranged from 70% in Beresford to 0.1% in South Shore. Similar to our results, Wengert et al. [23] used VIs (RGB and multispectral) along with texture and plant height as the predictor variable with the RF algorithm to predict above-ground biomass in barley with a R^2 of 0.62. Lu et al. [19], using VIs only as predictor variables, found that RF had a higher R^2 (0.69) than SVM and other linear-based models for predicting biomass in wheat. For Beresford, model performance marginally fluctuated between model development and model validation. Validation R^2 for Volga, however, drastically decreased for all types of models. One of the possible reason could be the lower range of dry biomass yield among plots at that location. The low performance metrics for the models developed for

South Shore are expected considering the insignificant correlations between VIs and dry biomass yield.

Comparatively, all machine learning approaches yielded similar performances, except ANN. The sample size in this study was very small, while a high number of training data points is required to build optimal neural network models. Small datasets are subject to overfitting [80,85–87].

RTVI, GNDVI, and ExGR consistently ranked as highly important variables. GNDVI was also reported to be a highly ranked variable for above-ground biomass prediction of a legume–grass mixture using UAV-borne spectral information [21]. Several studies [88,89] have reported that red-edge VIs were not as important as NIR-based VIs for model prediction. In our study, a red-edge-based VI (RTVI) was ranked as an important predictor.

5. Conclusions

The purpose of the study was to estimate oat biomass using VIs derived from high resolution UAV imagery. Differences in growing conditions between the three locations resulted in significant variations in oat biomass production. The VIs derived from multi-spectral imagery was found to be positively correlated to above-ground biomass for two of the locations. In the third location, however, very few UAV-derived VIs were significantly correlated with biomass yield. Two different methodologies for VI extraction were compared, i.e., the pixel classification method and average reflectance over ROI method. While the use of pixel classification appears useful to increase the strength of the correlation between VIs and biomass as observed in Beresford, this was not consistent across locations.

Four machine learning algorithms for estimating dry biomass yield were developed using VIs from UAV imagery. Approximately 70% of the variance was explained by RF, SVM, and PLS models for biomass prediction at one location. Additional sampling points with multi-year trials should be considered to improve prediction models because advanced machine learning algorithms, such as deep learning, often requires larger number of data points and long training periods to improve model accuracy.

The same crop in different environments exhibited distinct physical properties, hence, a single algorithm may not suffice the need for precise biomass monitoring. Multi-sensor data fusion, multi-index combination, the inclusion of a range of characteristics not directly linked to crop biomass monitoring, and the use of sophisticated algorithms are all viable options for enhancing the accuracy of oat biomass predictions [90].

Author Contributions: Conceptualization, M.C.; methodology, M.C., P.S. and J.C.; formal analysis, P.S., M.C., L.L. and J.C.; investigation., P.S., M.C. and J.C.; data curation, M.C. and P.S.; writing—original draft preparation, P.S.; writing—review and editing, P.S., M.C., J.C. and M.M.; visualization, P.S.; supervision, M.C.; funding acquisition, M.C. All authors have read and agreed to the published version of the manuscript.

Funding: This work is supported by Hatch (project accession no. 0215292) from the USDA National Institute of Food and Agriculture. This research was also supported by the South Dakota Crop Improvement Association and the South Dakota Agricultural Experiment Station.

Data Availability Statement: The data presented in this study are available on request from the corresponding author.

Acknowledgments: The authors wish to acknowledge the technical support of Nicholas Hall and Paul Okello. Thanks to the South Dakota State University Agronomy, Horticulture, and Plant Science Department for providing access to the drone and some of the software used in this study.

Conflicts of Interest: The authors declare no conflict of interest. The funders had no role in the design of the study; in the collection, analyses, or interpretation of data; in the writing of the manuscript, or in the decision to publish the results.

References

1. Suttie, J.M.; Reynolds, S.G. *Fodder Oats: A World Overview*; Food & Agriculture Organization: Rome, Italy, 2004.
2. *United States Department of Agriculture—National Agricultural Statistics Service QuickStats*; USDA-NASS: Washington, DC, USA, 2020.
3. Kim, K.S.; Tinker, N.A.; Newell, M.A. Improvement of oat as a winter forage crop in the Southern United States. *Crop Sci.* **2014**, *54*, 1336–1346. [CrossRef]
4. McCartney, D.; Fraser, J.; Ohama, A. Annual cool season crops for grazing by beef cattle. A Canadian review. *Can. J. Anim. Sci.* **2008**, *88*, 517–533. [CrossRef]
5. Carr, P.M.; Horsley, R.D.; Poland, W.W. Barley, oat, and cereal–pea mixtures as dryland forages in the northern Great Plains. *Agron. J.* **2004**, *96*, 677–684. [CrossRef]
6. Contreras-Govea, F.E.; Albrecht, K.A. Forage production and nutritive value of oat in autumn and early summer. *Crop Sci.* **2006**, *46*, 2382–2386. [CrossRef]
7. Lee, J.M.; Matthew, C.; Thom, E.R.; Chapman, D.F. Perennial ryegrass breeding in New Zealand: A dairy industry perspective. *Crop Pasture Sci.* **2012**, *63*, 107–127. [CrossRef]
8. Lin, Z.; Cogan, N.O.; Pembleton, L.W.; Spangenberg, G.C.; Forster, J.W.; Hayes, B.J.; Daetwyler, H.D. Genetic gain and inbreeding from genomic selection in a simulated commercial breeding program for perennial ryegrass. *Plant Genome* **2016**, *9*, plantgenome2015-06. [CrossRef] [PubMed]
9. Riday, H. Correlations between visual biomass scores and forage yield in space planted red clover (*Trifolium pratense* L.) breeding nurseries. *Euphytica* **2009**, *170*, 339–345. [CrossRef]
10. Pembleton, L.W.; Inch, C.; Baillie, R.C.; Drayton, M.C.; Thakur, P.; Ogaji, Y.O.; Spangenberg, G.C.; Forster, J.W.; Daetwyler, H.D.; Cogan, N.O. Exploitation of data from breeding programs supports rapid implementation of genomic selection for key agronomic traits in perennial ryegrass. *Theor. Appl. Genet.* **2018**, *131*, 1891–1902. [CrossRef]
11. Watanabe, K.; Guo, W.; Arai, K.; Takanashi, H.; Kajiya-Kanegae, H.; Kobayashi, M.; Yano, K.; Tokunaga, T.; Fujiwara, T.; Tsutsumi, N. High-throughput phenotyping of sorghum plant height using an unmanned aerial vehicle and its application to genomic prediction modeling. *Front. Plant Sci.* **2017**, *8*, 421. [CrossRef]
12. Minervini, M.; Scharr, H.; Tsaftaris, S.A. Image analysis: The new bottleneck in plant phenotyping [applications corner]. *IEEE Signal Processing Mag.* **2015**, *32*, 126–131. [CrossRef]
13. Barrett, B.; Faville, M.; Nichols, S.; Simpson, W.; Bryan, G.; Conner, A. Breaking through the feed barrier: Options for improving forage genetics. *Anim. Prod. Sci.* **2015**, *55*, 883–892. [CrossRef]
14. Ballesteros, R.; Ortega, J.F.; Hernandez, D.; Moreno, M.A. Onion biomass monitoring using UAV-based RGB imaging. *Precis. Agric.* **2018**, *19*, 840–857. [CrossRef]
15. Bendig, J.; Bolten, A.; Bennertz, S.; Broscheit, J.; Eichfuss, S.; Bareth, G. Estimating biomass of barley using crop surface models (CSMs) derived from UAV-based RGB imaging. *Remote Sens.* **2014**, *6*, 10395–10412. [CrossRef]
16. Díaz-Varela, R.; de la Rosa, R.; León, L.; Zarco-Tejada, P. High-resolution airborne UAV imagery to assess olive tree crown parameters using 3D photo reconstruction: Application in breeding trials. *Remote Sens.* **2015**, *7*, 4213–4232. [CrossRef]
17. Shi, Y.; Thomasson, J.A.; Murray, S.C.; Pugh, N.A.; Rooney, W.L.; Shafian, S.; Rajan, N.; Rouze, G.; Morgan, C.L.; Neely, H.L. Unmanned aerial vehicles for high-throughput phenotyping and agronomic research. *PLoS ONE* **2016**, *11*, e0159781. [CrossRef]
18. Araus, J.L.; Kefauver, S.C.; Zaman-Allah, M.; Olsen, M.S.; Cairns, J.E. Translating high-throughput phenotyping into genetic gain. *Trends Plant Sci.* **2018**, *23*, 451–466. [CrossRef] [PubMed]
19. Lu, N.; Zhou, J.; Han, Z.; Li, D.; Cao, Q.; Yao, X.; Tian, Y.; Zhu, Y.; Cao, W.; Cheng, T. Improved estimation of aboveground biomass in wheat from RGB imagery and point cloud data acquired with a low-cost unmanned aerial vehicle system. *Plant Methods* **2019**, *15*, 17. [CrossRef] [PubMed]
20. Yue, J.; Yang, G.; Tian, Q.; Feng, H.; Xu, K.; Zhou, C. Estimate of winter-wheat above-ground biomass based on UAV ultrahigh-ground-resolution image textures and vegetation indices. *ISPRS J. Photogramm. Remote Sens.* **2019**, *150*, 226–244. [CrossRef]
21. Grüner, E.; Wachendorf, M.; Astor, T. The potential of UAV-borne spectral and textural information for predicting aboveground biomass and N fixation in legume-grass mixtures. *PLoS ONE* **2020**, *15*, e0234703. [CrossRef]
22. Han, L.; Yang, G.; Dai, H.; Xu, B.; Yang, H.; Feng, H.; Li, Z.; Yang, X. Modeling maize above-ground biomass based on machine learning approaches using UAV remote-sensing data. *Plant Methods* **2019**, *15*, 10. [CrossRef]
23. Wengert, M.; Piepho, H.-P.; Astor, T.; Graß, R.; Wijesingha, J.; Wachendorf, M. Assessing spatial variability of barley whole crop biomass yield and leaf area index in silvoarable agroforestry systems using UAV-borne remote sensing. *Remote Sens.* **2021**, *13*, 2751. [CrossRef]
24. Duan, B.; Liu, Y.; Gong, Y.; Peng, Y.; Wu, X.; Zhu, R.; Fang, S. Remote estimation of rice LAI based on Fourier spectrum texture from UAV image. *Plant Methods* **2019**, *15*, 124. [CrossRef]
25. Zhou, X.; Zheng, H.; Xu, X.; He, J.; Ge, X.; Yao, X.; Cheng, T.; Zhu, Y.; Cao, W.; Tian, Y. Predicting grain yield in rice using multi-temporal vegetation indices from UAV-based multispectral and digital imagery. *ISPRS J. Photogramm. Remote Sens.* **2017**, *130*, 246–255. [CrossRef]
26. Bhandari, M.; Ibrahim, A.M.; Xue, Q.; Jung, J.; Chang, A.; Rudd, J.C.; Maeda, M.; Rajan, N.; Neely, H.; Landivar, J. Assessing winter wheat foliage disease severity using aerial imagery acquired from small Unmanned Aerial Vehicle (UAV). *Comput. Electron. Agric.* **2020**, *176*, 105665. [CrossRef]

27. Acharya, B.S.; Bhandari, M.; Bandini, F.; Pizarro, A.; Perks, M.; Joshi, D.R.; Wang, S.; Dogwiler, T.; Ray, R.L.; Kharel, G. Unmanned aerial vehicles in hydrology and water management–applications, challenges and perspectives. *Water Resour. Res.* **2021**, *57*, e2021WR029925. [CrossRef]

28. Wang, T.; Liu, Y.; Wang, M.; Fan, Q.; Tian, H.; Qiao, X.; Li, Y. Applications of UAS in Crop Biomass Monitoring: A Review. *Front. Plant Sci.* **2021**, *12*, 595. [CrossRef]

29. Li, W.; Niu, Z.; Chen, H.; Li, D.; Wu, M.; Zhao, W. Remote Estimation of Canopy Height and Aboveground Biomass of Maize Using High-Resolution Stereo Images from a Low-Cost Unmanned Aerial Vehicle System. *Ecol. Indic.* **2016**, *67*, 637–648. [CrossRef]

30. Li, W.; Niu, Z.; Huang, N.; Wang, C.; Gao, S.; Wu, C. Airborne LiDAR technique for estimating biomass components of maize: A case study in Zhangye City, Northwest China. *Ecol. Indic.* **2015**, *57*, 486–496. [CrossRef]

31. Cen, H.; Wan, L.; Zhu, J.; Li, Y.; Li, X.; Zhu, Y.; Weng, H.; Wu, W.; Yin, W.; Xu, C. Dynamic monitoring of biomass of rice under different nitrogen treatments using a lightweight UAV with dual image-frame snapshot cameras. *Plant Methods* **2019**, *15*, 32. [CrossRef]

32. Swain, K.C.; Zaman, Q.U. Rice crop monitoring with unmanned helicopter remote sensing images. In *Remote Sensing of Biomass-Principles and Applications*; IntechOpen: London, UK, 2012.

33. Royo, C.; Villegas, D. Field measurements of canopy spectra for biomass assessment of small-grain cereals. In *Biomass-Detection, Production and Usage*; IntechOpen: London, UK, 2011.

34. Calvao, T.; Palmeirim, J. Mapping Mediterranean scrub with satellite imagery: Biomass estimation and spectral behaviour. *Int. J. Remote Sens.* **2004**, *25*, 3113–3126. [CrossRef]

35. Tucker, C.J.; Slayback, D.A.; Pinzon, J.E.; Los, S.O.; Myneni, R.B.; Taylor, M.G. Higher northern latitude normalized difference vegetation index and growing season trends from 1982 to 1999. *Int. J. Biometeorol.* **2001**, *45*, 184–190. [CrossRef] [PubMed]

36. Mutanga, O.; Skidmore, A.K. Narrow band vegetation indices overcome the saturation problem in biomass estimation. *Int. J. Remote Sens.* **2004**, *25*, 3999–4014. [CrossRef]

37. Prabhakara, K.; Hively, W.D.; McCarty, G.W. Evaluating the relationship between biomass, percent groundcover and remote sensing indices across six winter cover crop fields in Maryland, United States. *Int. J. Appl. Earth Obs. Geoinf.* **2015**, *39*, 88–102. [CrossRef]

38. Zhe, L.; Die, H.; Deng-zhong, Z.; Da-xiang, X. Research Advance of Broadband Vegetation Index UsingRemotely Sensed Images. *J. Yangtze River Sci. Res. Inst.* **2015**, *32*, 125.

39. Thenkabail, P.S.; Smith, R.B.; De Pauw, E. Hyperspectral vegetation indices and their relationships with agricultural crop characteristics. *Remote Sens. Environ.* **2000**, *71*, 158–182. [CrossRef]

40. Chen, J.; Gu, S.; Shen, M.; Tang, Y.; Matsushita, B. Estimating aboveground biomass of grassland having a high canopy cover: An exploratory analysis of in situ hyperspectral data. *Int. J. Remote Sens.* **2009**, *30*, 6497–6517. [CrossRef]

41. Gašparović, M.; Zrinjski, M.; Barković, Đ.; Radočaj, D. An automatic method for weed mapping in oat fields based on UAV imagery. *Comput. Electron. Agric.* **2020**, *173*, 105385. [CrossRef]

42. Devia, C.A.; Rojas, J.P.; Petro, E.; Martinez, C.; Mondragon, I.F.; Patiño, D.; Rebolledo, M.C.; Colorado, J. High-throughput biomass estimation in rice crops using UAV multispectral imagery. *J. Intell. Robot. Syst.* **2019**, *96*, 573–589. [CrossRef]

43. Chang, K.-W.; Shen, Y.; Lo, J.-C. Predicting rice yield using canopy reflectance measured at booting stage. *Agron. J.* **2005**, *97*, 872–878. [CrossRef]

44. Teal, R.; Tubana, B.; Girma, K.; Freeman, K.; Arnall, D.; Walsh, O.; Raun, W. In-season prediction of corn grain yield potential using normalized difference vegetation index. *Agron. J.* **2006**, *98*, 1488–1494. [CrossRef]

45. Lewis-Beck, C.; Lewis-Beck, M. *Applied Regression: An Introduction*; Sage Publications: Thousand Oaks, CA, USA, 2015; Volume 22.

46. Du, P.; Bai, X.; Tan, K.; Xue, Z.; Samat, A.; Xia, J.; Li, E.; Su, H.; Liu, W. Advances of four machine learning methods for spatial data handling: A review. *J. Geovis. Spat. Anal.* **2020**, *4*, 13. [CrossRef]

47. Sharifi, A. Estimation of biophysical parameters in wheat crops in Golestan province using ultra-high resolution images. *Remote Sens. Lett.* **2018**, *9*, 559–568. [CrossRef]

48. Liu, Y.; Liu, S.; Li, J.; Guo, X.; Wang, S.; Lu, J. Estimating biomass of winter oilseed rape using vegetation indices and texture metrics derived from UAV multispectral images. *Comput. Electron. Agric.* **2019**, *166*, 105026. [CrossRef]

49. Borra-Serrano, I.; De Swaef, T.; Muylle, H.; Nuyttens, D.; Vangeyte, J.; Mertens, K.; Saeys, W.; Somers, B.; Roldán-Ruiz, I.; Lootens, P. Canopy height measurements and non-destructive biomass estimation of Lolium perenne swards using UAV imagery. *Grass Forage Sci.* **2019**, *74*, 356–369. [CrossRef]

50. Viljanen, N.; Honkavaara, E.; Näsi, R.; Hakala, T.; Niemeläinen, O.; Kaivosoja, J. A novel machine learning method for estimating biomass of grass swards using a photogrammetric canopy height model, images and vegetation indices captured by a drone. *Agriculture* **2018**, *8*, 70. [CrossRef]

51. Yang, S.; Feng, Q.; Liang, T.; Liu, B.; Zhang, W.; Xie, H. Modeling grassland above-ground biomass based on artificial neural network and remote sensing in the Three-River Headwaters Region. *Remote Sens. Environ.* **2018**, *204*, 448–455. [CrossRef]

52. Jiang, Z.; Huete, A.R.; Didan, K.; Miura, T. Development of a two-band enhanced vegetation index without a blue band. *Remote Sens. Environ.* **2008**, *112*, 3833–3845. [CrossRef]

53. Cheng, T.; Song, R.; Li, D.; Zhou, K.; Zheng, H.; Yao, X.; Tian, Y.; Cao, W.; Zhu, Y. Spectroscopic estimation of biomass in canopy components of paddy rice using dry matter and chlorophyll indices. *Remote Sens.* **2017**, *9*, 319. [CrossRef]

54. Guo, Y.; Senthilnath, J.; Wu, W.; Zhang, X.; Zeng, Z.; Huang, H. Radiometric calibration for multispectral camera of different imaging conditions mounted on a UAV platform. *Sustainability* **2019**, *11*, 978. [CrossRef]

55. Rouse, J.W., Jr.; Haas, R.; Schell, J.; Deering, D. Monitoring vegetation systems in the Great Plains with ERTS. *NASA Spec. Publ.* **1974**, *351*, 309.

56. Moges, S.; Raun, W.; Mullen, R.; Freeman, K.; Johnson, G.; Solie, J. Evaluation of green, red, and near infrared bands for predicting winter wheat biomass, nitrogen uptake, and final grain yield. *J. Plant Nutr.* **2005**, *27*, 1431–1441. [CrossRef]

57. Broge, N.H.; Leblanc, E. Comparing prediction power and stability of broadband and hyperspectral vegetation indices for estimation of green leaf area index and canopy chlorophyll density. *Remote Sens. Environ.* **2001**, *76*, 156–172. [CrossRef]

58. Chen, P.-F.; Nicolas, T.; Wang, J.-H.; Philippe, V.; Huang, W.-J.; Li, B.-G. New index for crop canopy fresh biomass estimation. *Spectrosc. Spectr. Anal.* **2010**, *30*, 512–517.

59. Tucker, C.J. Red and photographic infrared linear combinations for monitoring vegetation. *Remote Sens. Environ.* **1979**, *8*, 127–150. [CrossRef]

60. Gitelson, A.A.; Kaufman, Y.J.; Stark, R.; Rundquist, D. Novel algorithms for remote estimation of vegetation fraction. *Remote Sens. Environ.* **2002**, *80*, 76–87. [CrossRef]

61. Neto, J.C.; Meyer, G.E.; Jones, D.D. Individual leaf extractions from young canopy images using Gustafson–Kessel clustering and a genetic algorithm. *Comput. Electron. Agric.* **2006**, *51*, 66–85. [CrossRef]

62. Harrell, F.E., Jr.; Harrell, M.F.E., Jr. Package 'hmisc'. *CRAN2018* **2019**, *2019*, 235–236.

63. Wu, J.; Wu, M.J. Package 'Minque'. 2019. Available online: www.Cran.Rproject.Org/Web/Packages/Minque/Minque.Pdf (accessed on 25 May 2020).

64. Kuhn, M. Building predictive models in R using the caret package. *J. Stat. Softw.* **2008**, *28*, 1–26. [CrossRef]

65. Godoy, J.L.; Vega, J.R.; Marchetti, J.L. Relationships between PCA and PLS-regression. *Chemom. Intell. Lab. Syst.* **2014**, *130*, 182–191. [CrossRef]

66. Cherkassky, V.; Ma, Y. Practical selection of SVM parameters and noise estimation for SVM regression. *Neural Netw.* **2004**, *17*, 113–126. [CrossRef]

67. Livingston, F. Implementation of Breiman's random forest machine learning algorithm. *ECE591Q Mach. Learn. J. Pap.* **2005**, 1–13. Available online: http://datajobstest.com/data-science-repo/Random-Forest-[Frederick-Livingston].pdf (accessed on 21 May 2020).

68. Zou, J.; Han, Y.; So, S.-S. Overview of artificial neural networks. *Artif. Neural Netw.* **2008**, *458*, 14–22.

69. Abdi, H. Partial least square regression (PLS regression). *Encycl. Res. Methods Soc. Sci.* **2003**, *6*, 792–795.

70. Vapnik, V.; Guyon, I.; Hastie, T. Support vector machines. *Mach. Learn* **1995**, *20*, 273–297.

71. Yang, K.; Wang, H.; Dai, G.; Hu, S.; Zhang, Y.; Xu, J. Determining the repeat number of cross-validation. In Proceedings of the 2011 4th International Conference on Biomedical Engineering and Informatics (BMEI), Shanghai, China, 15–17 October 2011; pp. 1706–1710.

72. Bratsch, S.; Epstein, H.; Buchhorn, M.; Walker, D.; Landes, H. Relationships between hyperspectral data and components of vegetation biomass in Low Arctic tundra communities at Ivotuk, Alaska. *Environ. Res. Lett.* **2017**, *12*, 025003. [CrossRef]

73. Räsänen, A.; Virtanen, T. Data and resolution requirements in mapping vegetation in spatially heterogeneous landscapes. *Remote Sens. Environ.* **2019**, *230*, 111207. [CrossRef]

74. Mesonet, S.D. *South Dakota State University*; SD Mesonet Archive: Brookings, SD, USA, 2020.

75. Pinter Jr, P.J. Effect of dew on canopy reflectance and temperature. *Remote Sens. Environ.* **1986**, *19*, 187–205. [CrossRef]

76. Madeira, A.; Gillespie, T.; Duke, C. Effect of wetness on turfgrass canopy reflectance. *Agric. For. Meteorol.* **2001**, *107*, 117–130. [CrossRef]

77. Guan, J.; Nutter, F., Jr. Factors that affect the quality and quantity of sunlight reflected from alfalfa canopies. *Plant Dis.* **2001**, *85*, 865–874. [CrossRef] [PubMed]

78. Aparicio, N.; Villegas, D.; Casadesus, J.; Araus, J.L.; Royo, C. Spectral vegetation indices as nondestructive tools for determining durum wheat yield. *Agron. J.* **2000**, *92*, 83–91. [CrossRef]

79. Tilly, N.; Aasen, H.; Bareth, G. Fusion of plant height and vegetation indices for the estimation of barley biomass. *Remote Sens.* **2015**, *7*, 11449–11480. [CrossRef]

80. Yue, J.; Yang, G.; Li, C.; Li, Z.; Wang, Y.; Feng, H.; Xu, B. Estimation of winter wheat above-ground biomass using unmanned aerial vehicle-based snapshot hyperspectral sensor and crop height improved models. *Remote Sens.* **2017**, *9*, 708. [CrossRef]

81. Babar, M.; Van Ginkel, M.; Reynolds, M.; Prasad, B.; Klatt, A. Heritability, correlated response, and indirect selection involving spectral reflectance indices and grain yield in wheat. *Aust. J. Agric. Res.* **2007**, *58*, 432–442. [CrossRef]

82. Booth, D.T.; Cox, S.E.; Berryman, R.D. Point sampling digital imagery with 'SamplePoint'. *Environ. Monit. Assess.* **2006**, *123*, 97–108. [CrossRef] [PubMed]

83. Patrignani, A.; Ochsner, T.E. Canopeo: A powerful new tool for measuring fractional green canopy cover. *Agron. J.* **2015**, *107*, 2312–2320. [CrossRef]

84. Myneni, R.; Williams, D. On the relationship between FAPAR and NDVI. *Remote Sens. Environ.* **1994**, *49*, 200–211. [CrossRef]

85. Yao, X.; Huang, Y.; Shang, G.; Zhou, C.; Cheng, T.; Tian, Y.; Cao, W.; Zhu, Y. Evaluation of six algorithms to monitor wheat leaf nitrogen concentration. *Remote Sens.* **2015**, *7*, 14939–14966. [CrossRef]

86. Yuan, H.; Yang, G.; Li, C.; Wang, Y.; Liu, J.; Yu, H.; Feng, H.; Xu, B.; Zhao, X.; Yang, X. Retrieving soybean leaf area index from unmanned aerial vehicle hyperspectral remote sensing: Analysis of RF, ANN, and SVM regression models. *Remote Sens.* **2017**, *9*, 309. [CrossRef]

87. Zheng, H.; Li, W.; Jiang, J.; Liu, Y.; Cheng, T.; Tian, Y.; Zhu, Y.; Cao, W.; Zhang, Y.; Yao, X. A comparative assessment of different modeling algorithms for estimating leaf nitrogen content in winter wheat using multispectral images from an unmanned aerial vehicle. *Remote Sens.* **2018**, *10*, 2026. [CrossRef]

88. Kross, A.; McNairn, H.; Lapen, D.; Sunohara, M.; Champagne, C. Assessment of RapidEye vegetation indices for estimation of leaf area index and biomass in corn and soybean crops. *Int. J. Appl. Earth Obs. Geoinf.* **2015**, *34*, 235–248. [CrossRef]

89. Naito, H.; Ogawa, S.; Valencia, M.O.; Mohri, H.; Urano, Y.; Hosoi, F.; Shimizu, Y.; Chavez, A.L.; Ishitani, M.; Selvaraj, M.G. Estimating rice yield related traits and quantitative trait loci analysis under different nitrogen treatments using a simple tower-based field phenotyping system with modified single-lens reflex cameras. *ISPRS J. Photogramm. Remote Sens.* **2017**, *125*, 50–62. [CrossRef]

90. Maimaitijiang, M.; Sagan, V.; Sidike, P.; Hartling, S.; Esposito, F.; Fritschi, F.B. Soybean yield prediction from UAV using multimodal data fusion and deep learning. *Remote Sens. Environ.* **2020**, *237*, 111599. [CrossRef]

Article

Accurate Wheat Lodging Extraction from Multi-Channel UAV Images Using a Lightweight Network Model

Baohua Yang [1,2,3,*], Yue Zhu [1] and Shuaijun Zhou [1]

1   School of Information and Computer, Anhui Agricultural University, Hefei 230036, China;
    zhuyue@stu.ahau.edu.cn (Y.Z.); zhoushuaijun@stu.ahau.edu.cn (S.Z.)
2   Anhui Provincial Engineering Laboratory for Beidou Precision Agriculture Information,
    Anhui Agricultural University, Hefei 230036, China
3   Smart Agriculture Research Institute, Anhui Agricultural University, Hefei 230036, China
*   Correspondence: ybh@ahau.edu.cn

Abstract: The extraction of wheat lodging is of great significance to post-disaster agricultural production management, disaster assessment and insurance subsidies. At present, the recognition of lodging wheat in the actual complex field environment still has low accuracy and poor real-time performance. To overcome this gap, first, four-channel fusion images, including RGB and DSM (digital surface model), as well as RGB and ExG (excess green), were constructed based on the RGB image acquired from unmanned aerial vehicle (UAV). Second, a Mobile U-Net model that combined a lightweight neural network with a depthwise separable convolution and U-Net model was proposed. Finally, three data sets (RGB, RGB + DSM and RGB + ExG) were used to train, verify, test and evaluate the proposed model. The results of the experiment showed that the overall accuracy of lodging recognition based on RGB + DSM reached 88.99%, which is 11.8% higher than that of original RGB and 6.2% higher than that of RGB + ExG. In addition, our proposed model was superior to typical deep learning frameworks in terms of model parameters, processing speed and segmentation accuracy. The optimized Mobile U-Net model reached 9.49 million parameters, which was 27.3% and 33.3% faster than the FCN and U-Net models, respectively. Furthermore, for RGB + DSM wheat lodging extraction, the overall accuracy of Mobile U-Net was improved by 24.3% and 15.3% compared with FCN and U-Net, respectively. Therefore, the Mobile U-Net model using RGB + DSM could extract wheat lodging with higher accuracy, fewer parameters and stronger robustness.

Keywords: UAV; wheat lodging; deep learning; lightweight; digital surface model (DSM)

Citation: Yang, B.; Zhu, Y.; Zhou, S. Accurate Wheat Lodging Extraction from Multi-Channel UAV Images Using a Lightweight Network Model. *Sensors* **2021**, *21*, 6826. https://doi.org/10.3390/s21206826

Academic Editors: Jiyul Chang and Sigfredo Fuentes

Received: 12 September 2021
Accepted: 13 October 2021
Published: 14 October 2021

Publisher's Note: MDPI stays neutral with regard to jurisdictional claims in published maps and institutional affiliations.

1. Introduction

Wheat is the main food source in the world, the quality and yield of which are related to food security [1]. Lodging is a common agricultural natural disaster in wheat production, especially in the middle and late stages of wheat growth, and it is one of the important factors that limit the high yield of wheat [2]. On the one hand, lodging changes the individual development of wheat and, on the other hand, lodging changes the population structure of wheat. Previous studies have shown that lodging not only affects protein synthesis and nutrient transport, but also causes a sharp decline in photosynthetic rate and dry matter production capacity [3]. Therefore, it is of great significance for production management, prevention and control guidance, as well as disaster assessment for agricultural departments and agricultural insurance departments, to accurately and quickly obtain information, such as the location and area of wheat lodging.

The traditional method of obtaining lodging information is ground manual measurement, which is time-consuming and labor-intensive and its measurement results are subjectively affected. In addition, for large-scale lodging disasters, its low work efficiency often cannot meet actual needs [4]. In contrast, the rapid development based on remote sensing technology provides a practical means for large-scale and rapid monitoring of lodging information [5], such as near-ground remote sensing, satellite remote sensing and unmanned aerial vehicle (UAV) remote sensing monitoring. The low efficiency of near-ground remote sensing technology limits its further application on the farmland scale [6]. To achieve large-scale crop lodging monitoring, Yang et al. used the Radarsat-2 radar polarization index method to monitor wheat lodging [7]. Chauhan et al. used Sentinel 1 radar data and Sentinel 2 multispectral data to monitor the incidence of wheat lodging [8]. To make full use of the information provided by satellites, Chauhan et al. realized the classification of the degree of lodging of wheat by combining satellite data and the measured crop height on the ground [9]. However, for the limitation of time resolution, satellites cannot quickly obtain data to meet the needs of real-time identification. Therefore, it is necessary to develop a fast and reliable method for identifying wheat lodging. In recent years, UAV remote sensing has made up for the shortcomings of satellite remote sensing and near-ground remote sensing by virtue of its advantages of miniaturization, low cost, simple operation and high spatial and temporal resolution. UAV is the main tool for rapid and accurate acquisition of crop information in the application of agricultural quantitative remote sensing. UAV remote sensing is the current research hotspot and the future research trend. Previous studies have shown that remote sensing technology based on UAV can detect not only lodging in high-density crops, such as buckwheat [10], rice [11], barley [12], wheat [13] and jute [14], but low-density crop lodging information acquisition, such as corn [15], sunflower [16], cotton [17] and sugarcane [18], has also achieved good results. In addition, many scholars have also carried out analyses of crop lodging based on different features extracted by UAV, including spectral information [19], texture features [20], gray level co-occurrence matrix [21] and vegetation indices [22]. In any case, the above research papers showed the feasibility of extracting crop lodging based on digital images obtained from UAV. However, it is difficult to achieve accurate lodging detection tasks for traditional features. Therefore, it is expected that more robust features will be used to identify wheat lodging.

At present, UAV not only obtains digital images with three channels of R, G and B, but also can generate a variety of derivative models based on multiple aerial images, including digital orthophoto (DOM), digital elevation model (DEM) and digital surface model (DSM), which have been successfully used in the application of monitoring crop growth. Among them, DSM has received extensive attention because of its rich information and intuitive reflection of features such as canopy, location and height. Handique et al. used the difference of DSM to distinguish crops of different heights [23]. Feng et al. utilized DSM to successfully estimate crop yields [24]. In general, the DSM generated by UAV images can accurately represent the spatial variability of crops in different growth states. Yang et al. successfully realized the lodging detection of rice using DSM and texture features generated by UAV images [25]. In fact, fusion images based on RGB images contain multi-channel information, which can provide more heterogeneous features for lodging recognition. For example, some studies have focused on fusion image combining RGB and DSM to extract lodging, while other studies have developed a method of fusing RGB and the vegetation index to extract lodging [26]. At present, there is no universally accepted understanding of which information is better to fuse aerial images obtained by UAV. In addition, most of the research was still based on manually extracted features. Therefore, the extraction of crop lodging information still faces many challenges.

With the enhancement of computer processing power, the recognition of crop lodging based on deep learning has become a research hotspot in the field of agriculture. Many methods based on convolutional neural networks have been successfully applied to the research of lodging recognition. Yang et al. used EDANet to extract the lodging information of rice [27]. Zhao et al. utilized U-Net to extract the lodging area of rice [28]. Compared with traditional algorithms, the advantage of deep learning is that it can automatically extract effective features through a multi-layer neural network. In particular, the convolutional neural network model not only extracts the local detailed features of the image, but also extracts the high-level semantic features of the image. Research results showed the feasibility and superiority of extracting crop lodging information based on deep learning. However, the limitations of large amounts of calculation and high resource consumption still make the model complex, which makes it difficult to meet the needs of large-scale, real-time detection. In particular, it was not known whether the multi-channel image of fusion information could further improve the accuracy of lodging information extraction. Although Li et al. exploited deep learning methods to achieve lodging area segmentation based on multi-channel spectral information [29], so far, it is not clear how fusion-based multi-channel images could detect crop lodging based on lightweight neural network models.

Therefore, a method for extracting wheat lodging information based on a light-weight U-Net model with depthwise separable convolution is proposed in this study. Self-built data sets obtained from UAV were used to evaluate the performance of the model, including RGB of three channels, RGB + DSM of four channels and RGB + ExG of four channels. The purpose of this research study is to (1) train Mobile U-Net using self-built data sets and fine-tune model parameters to improve the robustness of the model, (2) verify the effectiveness of the multi-channel fusion image to improve the accuracy of wheat lodging extraction and (3) compare ours with other models to evaluate the performance of the proposed model.

2. Materials and Methods

2.1. Data Collection

The field experiment was conducted in the National Modern Agriculture Demonstration Zone (31°29′26″ N, 117°13′4″ E) located in Guohe Town, Lujiang County, Anhui Province, China. The area belongs to the subtropical monsoon climate, with four distinct seasons, obvious cold and heat and it is suitable for the cultivation of wheat. Thirty-six plots in the experimental area were selected as the study area, each plot covering the area of 144.3 square meters (78 × 1.85 m^2). The large row spacing was 0.3 m and the small row spacing was 0.1 m. The variety of wheat was 'Wanmai 55'. From 30 April to 26 May 2021, Lujiang County experienced severe convective weather such as severe storms and rains, with winds reaching up to 7–8 levels, and severe weather such as hail in some areas, leading to multiple lodging of wheat in the study area. The wheat in the experimental area was in the critical period of wheat growth. During this period, members of our team collected UAV images and ground information at different stages of wheat growth, including the flowering (7 May 2021), filling (17 May 2021) and maturity (27 May 2021) stages.

During the data collection process, a total of 298 UAV aerial images was obtained at a height of 30 m above the ground during the three growth stages of wheat, including flowering (98 images), filling (100 images) and maturity (100 images). The size of a single image was 4000 × 3000 pixels. The Pix4DMapper software (Pix4D, Prilly, Switzerland) was used to stitch the original images to obtain orthophotos of wheat fields in three periods. Then, the acquired aerial images were manually annotated, cropped and subjected to data augmentation.

Figure 1a shows the research location; Figure 1b is a partially enlarged display of the wheat field. It is easy to see that the lodging area was very large and the degree of lodging was very serious. Figure 1c shows a close-up map of lodging and healthy wheat in flowering stage; the image of the wheat field was acquired by UAV at a height of about 3 m above the ground and the shooting angle was about 65°. Figure 1d shows a close-up map of lodging and healthy wheat in filling stage, Figure 1e shows a close-up map of lodging and healthy wheat in maturity stage. We found that the height of lodging wheat is significantly lower than that of non-lodging wheat by at least 20 cm. Figure 1d,e was obtained using a mobile phone (nova5 pro, ISO: 50, focal length: 26 mm).

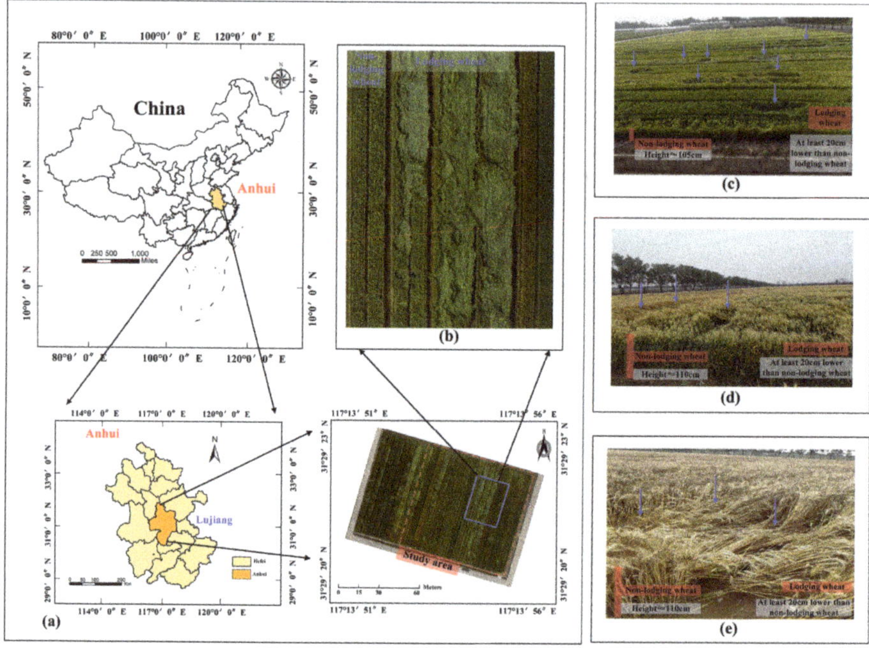

Figure 1. Location of UAV imaging area, study site and lodging samples: (**a**) study site; (**b**) partially enlarged display of the wheat field; the close-up maps of lodging and healthy wheat in (**c**) flowering, (**d**) filling and (**e**) maturity. The field indicated by the blue arrow is the lodging wheat.

2.2. Data Preprocessing

2.2.1. Image Annotation

Among them, the Labelme software (http://labelme.csail.mit.edu/Release3.0/, accessed on 10 May 2021) was used to manually mark; the non-lodging area of wheat was marked as wheat, the lodging area was marked as lodging, the other areas were marked as background. The label images were created and the annotated images were cropped into images with 256 × 256 pixels, as shown in Figure 2.

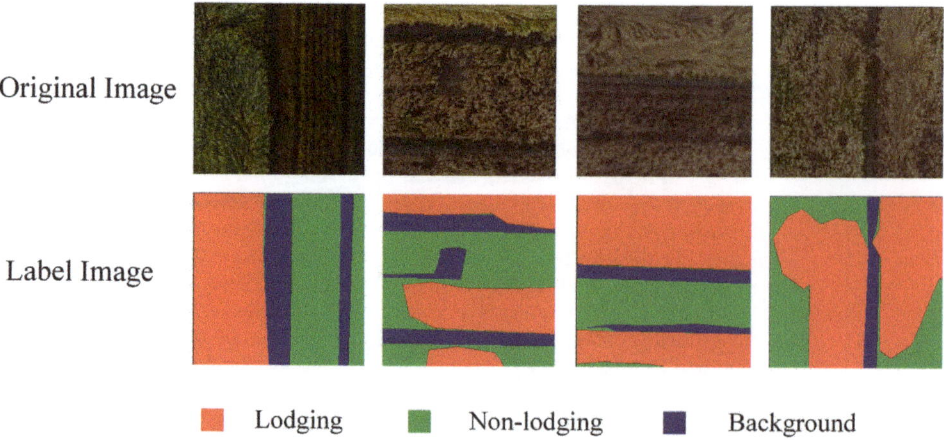

Figure 2. Example of original image and labeled image after cropping.

2.2.2. Image Fusion

To explore the influence of DSM and ExG on the recognition of the lodging effect based on the deep learning model, the RGB images collected by the UAV in this study were calculated to obtain the ExG index, the DSM was generated based on the dense point cloud and then the band was synthesized by the ENVI5.3 (Exelis Visual Information Solutions, USA) software. The ExG and DSM were added to the RGB image as the fourth band to obtain fusion images of RGB + ExG and RGB + DSM.

Among them, high-resolution, multi-view dense images were obtained from UAV and then Pix4Dmapper (Pix4D Company, Switzerland) software was used to adjust and match the images to generate dense point clouds; then, the triangulated irregular network (TIN) was constructed and, finally, a digital surface model (DSM) was obtained.

Excess green (ExG) can better distinguish vegetation and soil and it is often used for crop remote sensing monitoring [30]. To increase the extraction accuracy of wheat lodging information, the ENVI5.3 software was used to extract the gray values of the three bands of R, G and B from the RGB image obtained by UAV aerial photography and then the ExG index was calculated according to Equation (1).

$$\text{ExG} = \frac{2R - G - B}{R + G + B} \tag{1}$$

where G, B and R are the visible light green band, blue band and red band respectively.

2.2.3. Image Augmentation

To obtain more training samples, data augmentation was performed on training sample images and label images. A lossless transformation method was used, i.e., random horizontal or vertical flipping, random rotation at 90° and random x–y coordinate axis transposition. Therefore, data sets based on RGB and fusion images (four channels based on RGB + ExG and four-channel images based on RGB + DSM) were constructed, each including 1500 images. Different lodging detection models were trained based on three different data sets, training sets, validation sets and test sets, which included 1200, 150 and 150 images, respectively.

2.3. Model Construction and Evaluation Indicators

2.3.1. U-Net Model

U-Net is currently a popular deep learning model for semantic segmentation, which consists of a convolutional coding unit and a convolutional decoding unit [31]. Generally, the coding unit is mainly used to capture the context information in the image and the decoding unit is used to accurately locate the part that needs to be divided. Although the U-Net performance has been improved by improving the fully convolutional network (FCN), the standard U-Net neural network still needs to be further improved. To improve the detection accuracy, we proposed a wheat lodging recognition model combining Mo-bileNetV1 with depthwise separable convolution and U-Net to form a wheat lodging segmentation model.

2.3.2. Mobile U-Net Model

The Mobile U-Net model was composed of an encoder and a decoder. The ordinary convolution was replaced with a depthwise separable convolution to reduce the number of parameters and calculations of the entire network [32]. Among them, the pooling layer (Max pooling) and the convolutional layer were combined to construct a down-sampling unit, while the up-sampling layer and the convolutional layer were combined to construct an up-sampling unit. At the same time, depthwise separable convolution was used for feature extraction in the down-sampling unit, which enhances the feature extraction capability of the network model and reduces the computational cost. The addition of the convolutional layer could make up for the shortcomings of the Max Pooling layer and up-sampling layer that are not trainable, so it could reduce the loss of feature information during the sampling process and effectively improve the segmentation accuracy of the small boundary of the lodging edge of wheat, as shown in Figure 3.

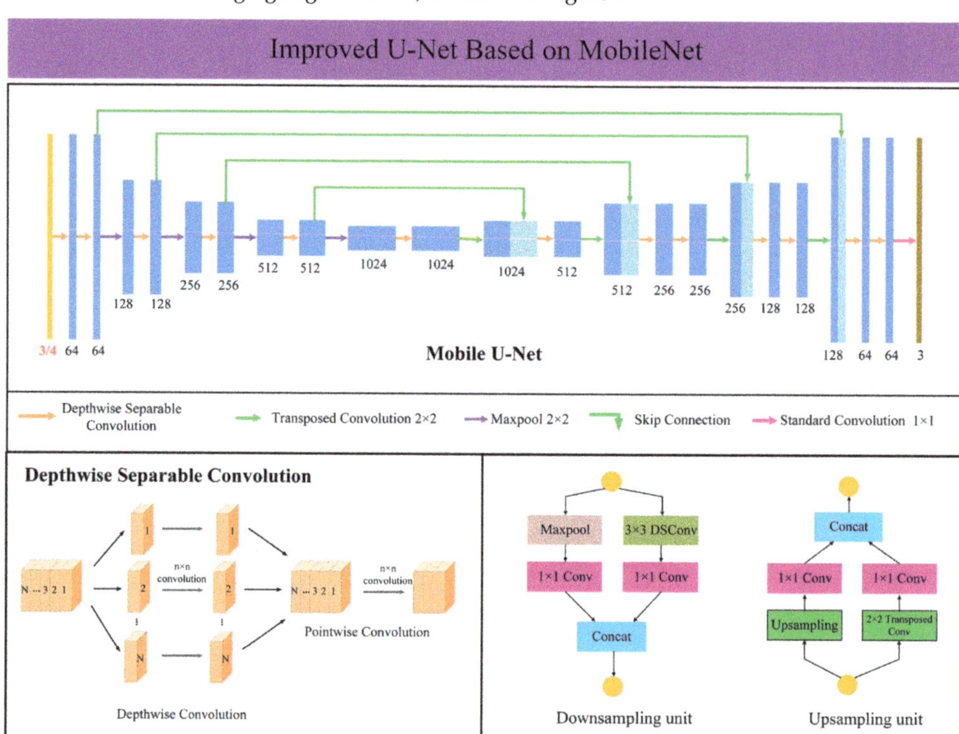

Figure 3. The structure of the Mobile U-Net model. (3/4 means that the parameter is set to 3 for RGB as input data and the parameter is set to 4 for four-channel image as input).

The input of the model was an image with a resolution of 256 × 256 pixels (3-channel image or 4-channel image) and the output was a single-channel segmented image. In the convolutional coding unit, a total of 4 up-samplings was performed and the first up-sampling unit included 2 repeated depthwise separable convolution modules and a Max pooling layer. The second, third and fourth up-sampling units had the same structure, including a depthwise separable convolution module and a Max pooling layer. After each pooling operation, the feature map size decreased and the number of channels doubled. The decoder performed down-sampling through transposed convolution and gradually restored image information. Corresponding to the encoder part, the decoder performed a total of 4 down-samplings. The first down-sampling unit included a depthwise separable convolution module and a transposed convolution module. The second, third and fourth down-sampling units also had the same structure, including two repeated depthwise separable convolution modules and one transposed convolution module, respectively. Each up-sampling expanded the feature map size and reduced the number of channels by half. Finally, a standard convolution module with a size of 1 × 1 was used to reduce the dimension and a normalized exponential function (SoftMax) was used to convert the value into a probability. The specific parameters are shown in the Table 1.

2.3.3. Wheat Lodging Segmentation Model

The technical process of this research study, shown in Figure 4, mainly included UAV digital image collection, data set construction, model training and verification, testing, model evaluation and optimization. Firstly, the DSM and ExG derived from the RGB image obtained by UAV were used to construct the RGB, RGB + ExG and RGB + DSM data sets. Secondly, the Mobile U-Net model proposed in this study was trained, verified and tested using different data sets. Furthermore, we compare the performance of Mobile U-Net with typical deep learning frameworks, such as FCN and U-Net. Finally, three data sets in different periods were used to predict the lodging area.

Table 1. Parameters of the Mobile U-Net model.

Layer Type	Size	Filter	Stride
Input	$256 \times 256 \times 3/4$		
Depthwise separable convolution	$256 \times 256 \times 64$	$3 \times 3, 1 \times 1$	2
Max pooling	$128 \times 128 \times 64$	2×2	1
Depthwise separable convolution	$128 \times 128 \times 128$	$3 \times 3, 1 \times 1$	2
Max pooling	$64 \times 64 \times 128$	2×2	1
Depthwise separable convolution	$64 \times 64 \times 256$	$3 \times 3, 1 \times 1$	2
Max pooling	$32 \times 32 \times 256$	2×2	1
Depthwise separable convolution	$32 \times 32 \times 512$	$3 \times 3, 1 \times 1$	2
Max pooling	$16 \times 16 \times 512$	2×2	1
Depthwise separable convolution	$16 \times 16 \times 1024$	$3 \times 3, 1 \times 1$	2
Transposed Convolution	$32 \times 32 \times 512$	3×3	1
Skip connection	$32 \times 32 \times 1024$		1
Depthwise separable convolution	$32 \times 32 \times 512$	$3 \times 3, 1 \times 1$	2
Transposed Convolution	$64 \times 64 \times 256$	3×3	1
Skip connection	$64 \times 64 \times 512$		1
Depthwise separable convolution	$64 \times 64 \times 256$	$3 \times 3, 1 \times 1$	2
Transposed Convolution	$128 \times 128 \times 128$	3×3	1
Skip connection	$128 \times 128 \times 256$		1
Depthwise separable convolution	$128 \times 128 \times 128$	$3 \times 3, 1 \times 1$	2
Transposed Convolution	$256 \times 256 \times 64$	3×3	1
Skip connection	$256 \times 256 \times 128$		1
Depthwise separable convolution	$256 \times 256 \times 64$	$3 \times 3, 1 \times 1$	2
Standard convolution	$256 \times 256 \times 3$	1×1	1

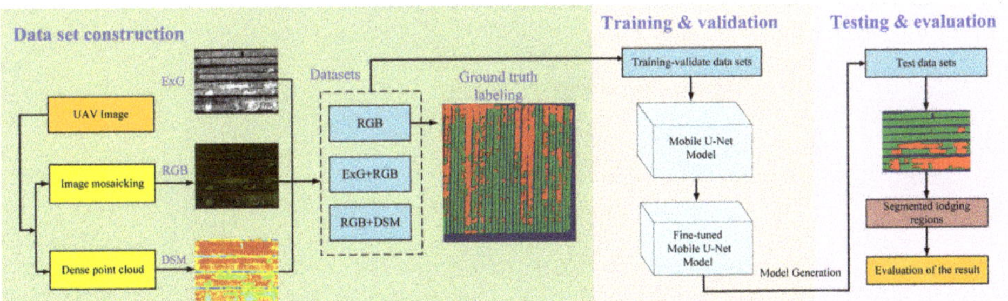

Figure 4. The technical flow chart of this study.

2.3.4. Evaluation Indicators

There were four indicators used to evaluate the performance of the model, including *precision*, *recall*, $F1 - score$ and mean Intersection over Union ($mIoU$). Among them, *precision* shows the proportion of samples that are predicted to be lodging wheat in the segmented image that are actually lodging wheat; *recall* shows to the proportion of samples that are predicted to be lodging wheat among all the samples that are actually lodging wheat; $F1 - score$ is the harmonic mean of accuracy and recall, reflecting the comprehensive performance of segmentation of lodging wheat in the wheat field; $mIoU$ is the ratio of overlap between the segmentation result of wheat lodging and ground truth. The values of the above evaluation indicators are all between 0 and 1 and the larger the value, the better the segmentation effect. In this study, *precision*, *recall*, $F1 - score$ and $mIoU$ are used as the evaluation indexes for evaluating the segmentation accuracy of lodging wheat and the calculation formulas are as follows:

$$precision = \frac{TP}{TP + FP} \tag{2}$$

$$recall = \frac{TP}{TP + FN} \tag{3}$$

$$F1 - score = 2 \times \frac{precision \times recall}{precision + recall} \tag{4}$$

$$mIoU = \frac{1}{k+1} \sum_{i=0}^{k} \frac{TP}{FN + FP + TP} \tag{5}$$

where TP refers to the correct segmentation of the wheat lodging area, which is the wheat lodging area; TN refers to the correct segmentation of the non-lodging area of wheat, which is a non-lodging area of wheat; FP refers to the correct segmentation of the wheat lodging area, which is a non-lodging area of wheat; FN refers to the correct segmentation of the non-lodging area of wheat, which is the wheat lodging area; k is the number of categories.

3. Results

3.1. DSM and ExG Images Derived from RGB

Pix4Dmapper was used to generate a high-precision DSM (digital surface model) and ExG (excess green) in the wheat research area with high-resolution digital images obtained from UAV in different growth periods, as shown in Figure 5. Among them, the first column represents the flowering period, the second column represents the filling period and the third column represents the maturity period, as shown in Figure 1a–c. The first row represents the RGB image of the study area, the second row represents the DSM extracted from the image of the study area and the third row represents the ExG extracted from the image of the study area.

It can be seen, from Figure 5 (a2, DSM of flowering period; b2, DSM of filling period; c2, DSM of maturity period), that the elevations of the digital surface models in different periods were still significantly different. Especially, in the same period, the elevation of the wheat field was also different, because the digital surface model covered the elevation of other surface information except the ground. In this study, DSM showed the ground elevation model of normal wheat and lodging wheat, which could most truly express the growth status of crops on the ground of wheat fields. Therefore, DSM was beneficial to distinguish between normal wheat and lodging wheat in the field.

In addition, to clarify the contribution of the ExG index in identifying lodging wheat, the digital numbers (DNs) of the R, G and B channels were extracted from the RGB images of the study area acquired in three different periods; then, ExG was calculated and the visualization of ExG is shown in Figure 5 (a3, ExG of flowering period; b3, ExG of filling period; c3, ExG of maturity period). It can be seen, from Figure 5, that ExG was different in different periods.

Figure 6a–d shows the specific values of the digital number of R, digital number of G, digital number of B and ExG of lodging and non-lodging wheat in different periods extracted from the set 30 regions of interest (ROI). It can be seen, from Figure 6a–d, that the distribution of R, G, B and ExG was different in the flowering, filling and maturity periods. Especially, Figure 6d shows that the ExG of non-lodging wheat was significantly lower than that of lodging wheat. The mean values of ExG were 0.193–0.307, 0.009–0.157 and 0.027–0.049 for non-lodging and 0.238–0.319, 0.053–0.227 and 0.032–0.07 for lodging at the flowering, filling and maturity stage, respectively. Among them, the average ExG values of lodging wheat fields were 0.281, 0.116 and 0.044 in the three periods, which were 10%, 39% and 12% higher than those of normal wheat fields. It can be seen that ExG had a positive effect on the identification of wheat lodging.

3.2. Model Parameter Setting and Training

The experimental environment of this research project was the Windows10 Professional 64-bit operating system and the deep learning framework was Keras 2.2.4, which was used to train the network model. Model training and verification environment were as follows: Intel(R) Core (TM) i7-8700 @3.20 GHz and 16 G NVIDIA GeForce RTX 2080. The images were stitched with Pix4Dmapper and were cropped with Python codes. The language of model development used was python.

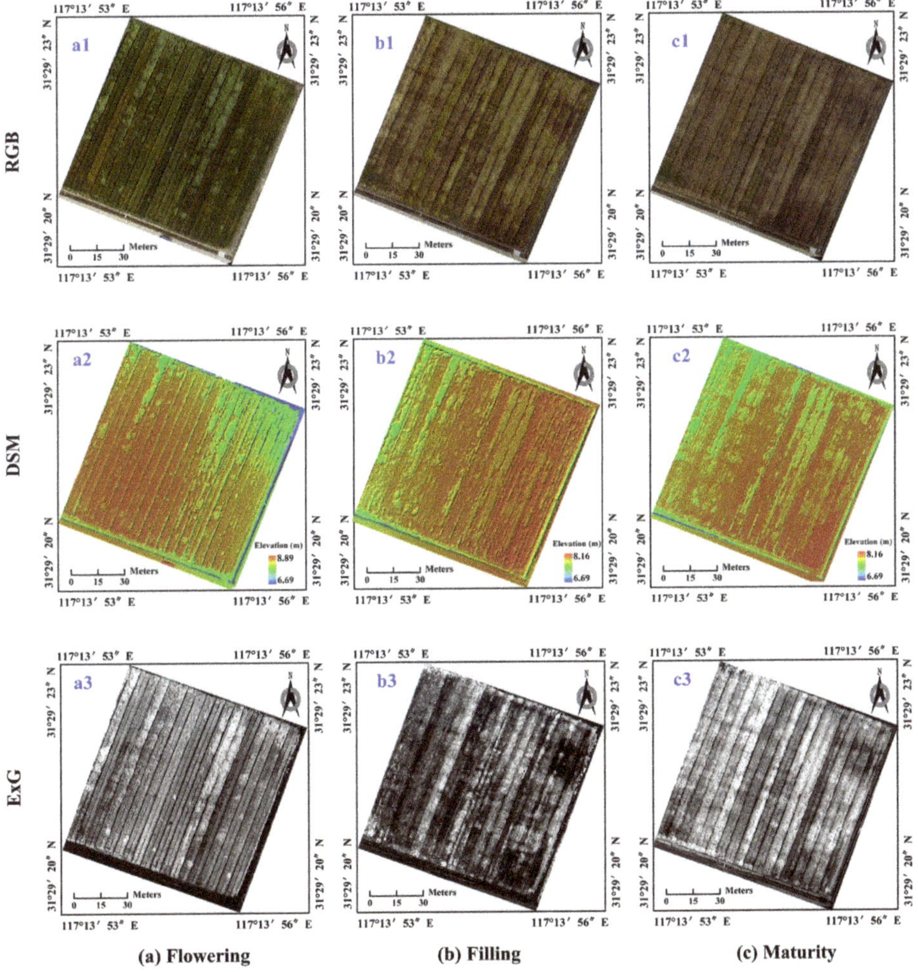

Figure 5. RGB, DSM and ExG of wheat fields in different growth periods: (**a**) flowering, (**b**) filling and (**c**) maturity.

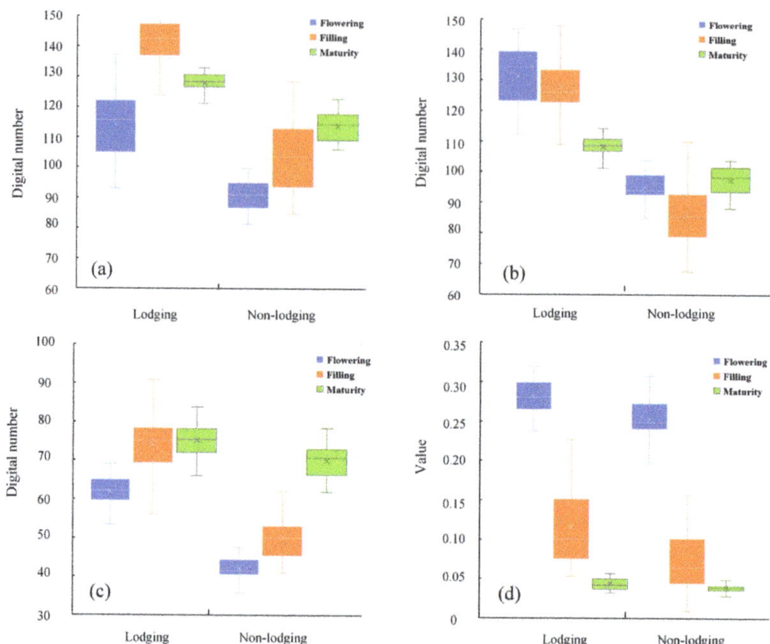

Figure 6. Comparison of the ExG values of lodging and non-lodging in different periods: (**a**) digital number of R; (**b**) digital number of G; (**c**) digital number of B; (**d**) value of ExG.

The model was trained using the Adam algorithm, the learning rate was 0.0001, the Batch size was 4 and the training iterations were 200 Epochs. After each Epoch training, not only the loss and accuracy were obtained by calculation, but the weights were also updated and saved. After the model was trained for 200 Epochs, the model with the highest accuracy was selected as the test model. Figure 7 shows the loss and accuracy curves of the training set and the validation set (RGB, RGB + ExG and RGB + DSM) of the Mobile U-Net model. It can be seen, from Figure 7, that that the error between the training set and the validation set decreased with the increase in the number of iterations and the error dropped below 0.1 when epoch = 65, then finally stabilized. On the one hand, this shows that the model can control the deviation. However, the close error of the training set and the verification set after stabilization indicated that the variance of the model was relatively low. In addition, the accuracy of the network increased as the number of iterations increased, until it stabilized. Therefore, when the training converged, the model with the highest accuracy was selected as the test model.

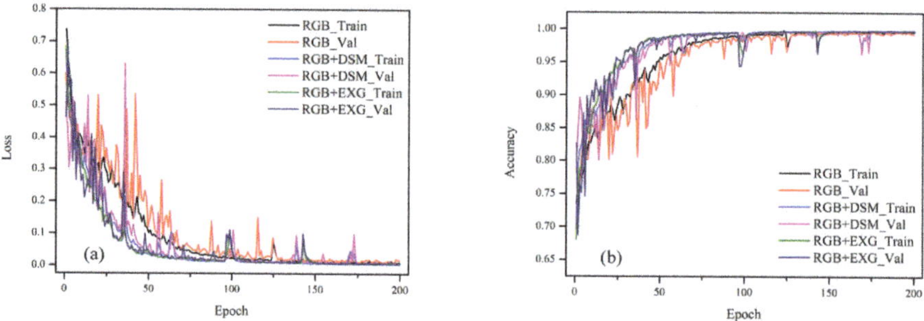

Figure 7. Loss and accuracy curves of the training set and validation set: (**a**) loss curve; (**b**) accuracy curve of training set and verification set. Train indicates training set; Val indicates validation set.

3.3. Results of Wheat Lodging Recognition with Different Data Sets

Table 2 showed the test results of data sets for different growth periods based on the Mobile U-Net model. Among them, the F1-score of the training set was 74.31–94.87% and the mIoU was 70.21–91.31%. The F1-score of the test set was 70.45–96.82% and the mIoU was 62.11–87.99%. Therefore, the Mobile U-Net model performed well in the extraction of wheat lodging. In particular, the F1-score of wheat lodging segmentation was 70.45–85.42% for RGB, 78.49–90.37% for RGB + ExG and 80.8–96.82% for RGB + DSM. The corresponding mIoU were 62.11–74.68%, 69.58–83.45% and 70.39–87.99%.

Table 2. Segmentation results using different data of three different periods.

Dataset		*F1-Score* (%)	*mIoU* (%)	*F1-Score* (%)	*mIoU* (%)
		Training Set		Test Set	
RGB	Flowering	74.31	70.21	70.45	62.11
	Filling	88.02	77.67	85.42	74.68
	Maturity	83.46	72.89	79.65	70.64
RGB + ExG	Flowering	81.32	76.53	78.49	69.58
	Filling	94.87	87.04	90.37	83.45
	Maturity	88.36	83.87	81.58	72.94
RGB + DSM	Flowering	89.69	85.94	80.8	70.39
	Filling	97.59	91.31	96.82	87.99
	Maturity	90.62	84.55	89.36	80.73

Figure 8 shows the lodging segmentation results of three different data sets in different periods, including RGB, RGB + ExG and RGB + DSM. It could be seen from Figure 8a that the lodging degree of wheat in the three different periods was quite different and the canopy structure was also different. Figure 8b represents the ground truth of wheat lodging. Figure 8c–e shows the results of wheat lodging recognition. Among them, the lodging recognition error rate with the RGB image was relatively high. There were many missed recognitions in lodging recognition using RGB + ExG. The result of lodging recognition using RGB + DSM was close to ground truth.

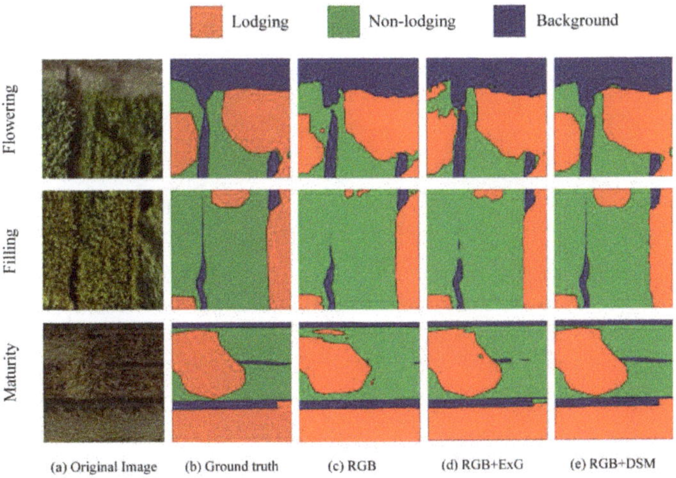

Figure 8. Test results of lodging recognition using different data.

4. Discussions

4.1. Compare the Identifying Results of Wheat Lodging Using Different Fusion Images

The visible light vegetation index could quantify the growth of vegetation under certain conditions, because it could reflect the difference between the reflection of vegetation under visible light and the soil background [33]. Some studies have used the vegetation index to successfully extract crop lodging information. For example, Wu et al. used NDVI to extract the lodging of rice [34]. Zhao et al. used a combination of three vegetation indices, including super green (ExG), super red (ExR) and the visible band difference vegetation index (VDVI), to successfully extract the lodging area of rice [28]. They only carried out the lodging extraction study based on the spectral characteristics of the vegetation, but did not carry out the comparison with the image fusion. In particular, the identification of lodging and non-lodging based only on the spectral characteristics of the wheat canopy could easily lead to misidentification and low recognition accuracy, because it was inevitable that the same objects had different spectra and the same spectrum reflected different objects. Therefore, it was necessary to study the different characteristics of the canopy in order to improve the accuracy of lodging detection.

In this study, three wheat field data sets of different growth periods were constructed, including RGB, RGB + ExG and RGB + DSM. Table 3 shows the comparison of the lodging recognition results based on data sets in different periods. Compared with RGB, F1-score and mIoU based on RGB + DSM increased by 12.8% and 11.8% in the flowering stage, increased by 11.8% and 15.1% in the filling stage and increased by 10.9% and 12.5% in the maturity period. In the corresponding period, F1-score and mIoU were 2.9% and 1.2%, 6.7% and 5.2%, and 8.7% and 9.6% higher than that of RGB + ExG, respectively. It is worth mentioning that there were significant differences in the elevations displayed in the DSM of the study area before and after the lodging of the wheat. Therefore, DSM fully expressed the difference in elevation between ground features in different periods, which was suitable for distinguishing lodging wheat from normal wheat.

Table 3. Comparison of lodging recognition results of different models.

Methods	Data	*F1-Score*	*mIoU*	Time-CPU (s/Image)	Parameter (Million)
FCN	RGB	59.45	56.87	0.53	17.08
	RGB + ExG	61.90	53.72	0.70	17.08
	RGB + DSM	67.33	55.89	0.73	17.08
U-Net	RGB	66.17	60.51	0.60	30.95
	RGB + ExG	69.06	59.78	0.80	30.95
	RGB + DSM	75.36	64.95	0.80	30.95
Mobile U-Net	RGB	78.51	69.14	0.33	9.49
	RGB + ExG	83.48	75.32	0.53	9.49
	RGB + DSM	88.99	80.7	0.53	9.49

4.2. Compare the Identifying Results of Wheat Lodging Based on Different Methods

To further verify the performance of our proposed method, the classic segmentation method U-Net model and FCN model under the deep learning framework were selected and compared with the model proposed in this paper on three identical test sets (150 images). The hardware environment for model testing was Intel(R) Core (TM) i7-1065G7 @1.30 GHz, 16 G. The results are shown in Table 3. It can be seen, from Table 3, that, compared with FCN and U-Net, the F1-Score of Mobile U-Net increased by 24.3% and 15.7% and mIoU increased by 17.7% and 12.5% for the RGB; 25.9% and 17.3% of F1-Score, 28.7% and 20.6% of mIoU for the RGB + ExG; 24.3% and 15.3% of F1-Score, 30.7% and 19.5% of mIoU for the RGB + DSM. Therefore, regardless of RGB, or the fused image RGB + ExG and RGB + DSM, the Mobile U-Net proposed in this study was superior to FCN and U-Net in wheat lodging recognition. In particular, F1-Score and mIoU based on Mobile U-Net using RGB + DSM was 88.99%, 80.7%, 11.8% and 14.3% higher than that of RGB and 6.2% and 6.7% higher than that of RGB + ExG.

Table 3 shows the comparison results of the average time for different models to process each image. After the model was tested using the test sets, the average time for Mobile U-Net to process each four-channel image with a size of 256 × 256 was 0.53 s using CPU (Intel(R) Core (TM) i7-1065G7, @1.30 GHz, 16 G). Both U-Net and FCN took longer to process the same types of images than the Mobile U-Net model. Regarding processing time per image, Mobile U-Net was 37.5% and 44.4% faster than U-Net and FCN for RGB, 23.8% and 33.3% faster for RGB + ExG, and 27.3% and 33.3% faster for RGB + DSM. In addition, regarding the parameters of the model, FCN was 17.08 million, U-Net was 30.95 million and Mobile U-Net was only 9.49 million, which was the model with the fewest parameters among the three models. Therefore, the model proposed in this study ensured that the accuracy was not reduced and improved the speed of image segmentation, aiming to achieve the goal of early warning of wheat lodging, reducing the impact of lodging, increasing production and income and benefiting farmers.

In fact, some studies have shown that semantic segmentation methods based on deep learning have strong advantages in lodging recognition. Yang [26] et al. used FCN (full neural network) to extract rice lodging based on RGB + ExG fusion information. Zhao et al. used UNet to extract lodging information [28]. Although the above-mentioned deep learning methods could effectively extract lodging features, too many parameters resulted in a low operating speed of the model. The possible reason is that the structure of the model they adopted was more complicated. For example, the standard U-Net neural network consists of 19 convolutional layers, the corresponding pooling layers and up-sampling layers. Therefore, it was necessary to improve the model, aiming to reduce the amount of calculation and improve the recognition effect.

To show the recognition of wheat lodging based on different models, only the recognition results using the RGB + DSM were provided here, as shown in Figure 9. It can be seen, from Figure 9, that there were many wrong recognitions based on the FCN. The lodging detection based on U-net was close to the result of our method and there were still some areas missing recognition. According to the analysis in Table 3, compared with FCN and U-net, the model we proposed not only maintained the premise of the same accuracy, but also improved the processing speed and reduced the parameters of the model, providing a technical basis for portable mobile devices that detect lodging in the field.

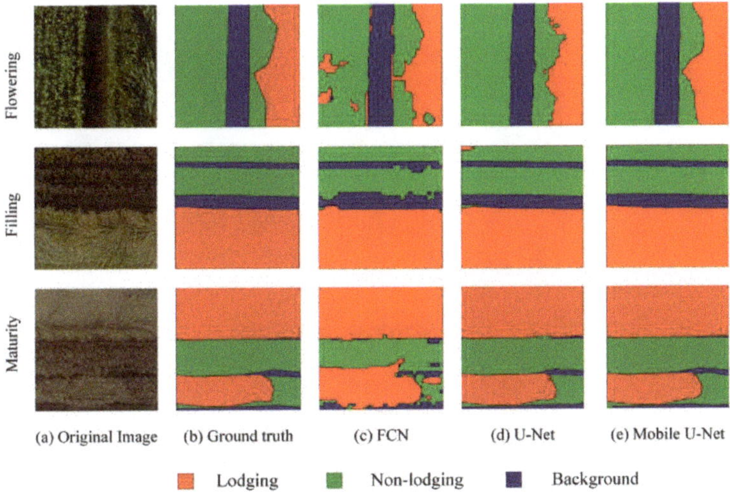

Figure 9. The results of estimating wheat lodging with different models with RGB + DSM.

4.3. Visualization of Feature Activation in Lodging Wheat

To verify the function of the depthwise separable convolution module, gradient-weighted class activation mapping (Grad-CAM) [35], which mainly uses the gradient of the target class and propagates to the final convolutional layer to generate a rough positioning map, was used to visualize the features,. The results of visualization clearly show how the network model selects important areas of the prediction class, so as to determine the impact of the depthwise separable convolution module.

As shown in Figure 10, the red area in the feature map indicates the high-weight area of the neural network to determine the lodging wheat and the blue area indicates the low-weight area of the network to determine the lodging wheat. The redder the color, the greater the influence of this area on the recognition result of the lodging wheat. It can be seen, from Figure 10b,d, that U-Net focused on the lodging area, non-lodging area and background. Figure 10c,e shows that the Mobile U-Net model paid attention to the more accurate lodging areas. Therefore, our proposed model with a depthwise separable convolution module could better learn the characteristic information of lodging wheat and improve the segmentation accuracy of lodging wheat.

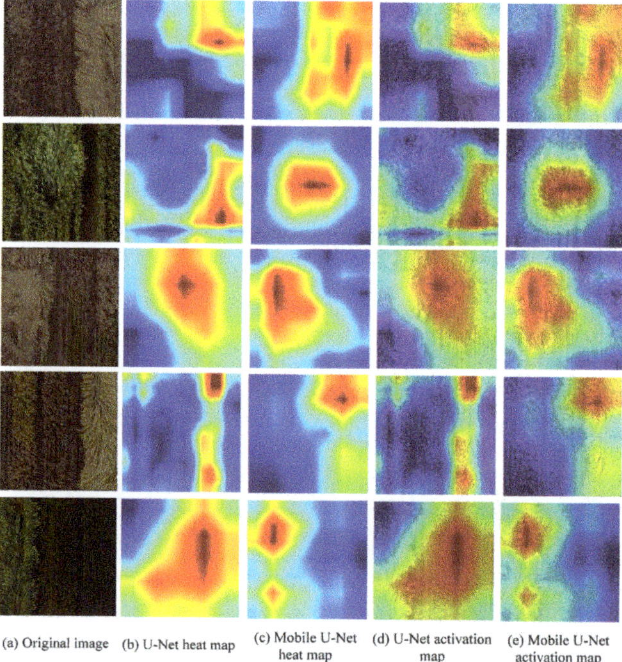

Figure 10. Visualization of feature activations using Grad-CAM.

5. Conclusions

In this study, a wheat lodging segmentation model based on a lightweight U-Net neural network with depthwise separable convolution, which was used to realize wheat lodging recognition and accurate segmentation from UAV images under field conditions, is proposed. The proposed model was trained, verified and tested with self-built wheat data sets (RGB, RGB + ExG, RGB + DSM) of different growth periods, including flowering, filling and maturity. The experiments showed that the extraction of wheat lodging effect based on the fusion image of DSM and RGB was the best; the F1-Score reached 88.99% and the mIoU reached 80.7%, indicating that the fusion image was more suitable for wheat lodging extraction. Furthermore, the parameters of the Mobile U-Net model were 9.49 million and the overall accuracy of Mobile U-Net was improved by 24.3% and 15.3% compared with FCN and U-Net, which indicate that the proposed model was suitable for the task of quickly and accurately detecting wheat lodging in the field.

Author Contributions: Methodology, B.Y. and Y.Z.; software, Y.Z.; data curation, S.Z.; writing—review and editing, B.Y. All authors have read and agreed to the published version of the manuscript.

Funding: This research project was funded by the Natural Science Foundation of Anhui Province (1808085MF195), National Natural Science Foundation of China (No.31771679), the Opening Project of Key Laboratory of Power Electronics and Motion Control of Anhui Higher Education Institutions (PEMC2001) and the Open Fund of State Key Laboratory of Tea Plant Biology and Utilization (SKLTOF20200116), Innovation and Entrepreneurship Training Program for College Students of Anhui Agricultural University (S202010364235).

Data Availability Statement: Restrictions apply to the availability of these data. Data were obtained from the Smart Agriculture Research Institute of Anhui Agricultural University and are available from the authors with the permission of the Smart Agriculture Research Institute.

Acknowledgments: We would like to thank Yuan Gao, Haiyang Xia and Zhiwei Gao for their help with field data collection. We are grateful to the reviewers for their suggestions and comments, which significantly improved the quality of this paper.

Conflicts of Interest: All the authors declare no conflict of interest.

References

1. Yang, B.; Gao, Z.; Gao, Y.; Zhu, Y. Rapid Detection and Counting of Wheat Ears in the Field Using YOLOv4 with Attention Module. *Agronomy* **2021**, *11*, 1202. [CrossRef]
2. Muhammad, A.; Hao, H.; Xue, Y.; Alam, A.; Bai, S.; Hu, W.; Sajid, M.; Hu, Z.; Samad, R.A.; Wang, L.; et al. Survey of wheat straw stem characteristics for enhanced resistance to lodging. *Cellulose* **2020**, *27*, 2469–2484. [CrossRef]
3. Islam, M.S.; Peng, S.; Visperas, R.M.; Ereful, N.; Bhuiya, M.S.U.; Julfiquar, A.W. Lodging-related morphological traits of hybrid rice in a tropical irrigated ecosystem. *Field Crop. Res.* **2007**, *101*, 240–248. [CrossRef]
4. Robertson, D.J.; Julias, M.; Gardunia, B.W.; Barten, T.; Cook, D.D. Corn Stalk Lodging: A Forensic Engineering Approach Provides Insights into Failure Patterns and Mechanisms. *Crop. Sci.* **2015**, *55*, 2833–2841. [CrossRef]
5. Weiss, M.; Jacob, F.; Duveiller, G. Remote sensing for agricultural applications: A meta-review. *Remote Sens. Environ.* **2020**, *236*, 111402. [CrossRef]
6. Zhang, J.; Gu, X.; Wang, J.; Huang, W.; Dong, Y.; Luo, J.; Yuan, L.; Li, Y. Evaluating maize grain quality by continuous wavelet analysis under normal and lodging circumstances. *Sens. Lett.* **2021**, *10*, 580–585. [CrossRef]
7. Yang, H.; Chen, E.; Li, Z.; Zhao, C.; Yang, G.; Pignatti, S.; Casa, R.; Zhao, L. Wheat lodging monitoring using polarimetric index from RADARSAT-2 data. *Int. J. Appl. Earth Obs. Geoinf.* **2015**, *34*, 157–166. [CrossRef]
8. Chauhan, S.; Darvishzadeh, R.; Lu, Y.; Boschetti, M.; Nelson, A. Understanding wheat lodging using multi-temporal Sentinel-1 and Sentinel-2 data. *Remote Sens. Environ.* **2020**, *243*, 111804. [CrossRef]
9. Chauhan, S.; Darvishzadeh, R.; Boschetti, M.; Nelson, A. Discriminant analysis for lodging severity classification in wheat using RADARSAT-2 and Sentinel-1 data. *ISPRS J. Photogramm. Remote Sens.* **2020**, *164*, 138–151. [CrossRef]
10. Murakami, T.; Yui, M.; Amaha, K. Canopy height measurement by photogrammetric analysis of aerial images: Application to buckwheat (Fagopyrum esculentum Moench) lodging evaluation. *Comput. Electron. Agric.* **2012**, *89*, 70–75. [CrossRef]
11. Zhou, X.; Zheng, H.; Xu, X.; He, J.; Ge, X.; Yao, X.; Cheng, T.; Zhu, Y.; Cao, W.; Tian, Y. Predicting grain yield in rice using multi-temporal vegetation indices from UAV-based multispectral and digital imagery. *ISPRS J. Photogramm. Remote Sens.* **2017**, *130*, 246–255. [CrossRef]
12. Näsi, R.; Viljanen, N.; Kaivosoja, J.; Alhonoja, K.; Hakala, T.; Markelin, L.; Honkavaara, E. Estimating Biomass and Nitrogen Amount of Barley and Grass Using UAV and Aircraft Based Spectral and Photogrammetric 3D Features. *Remote Sens.* **2018**, *10*, 1082. [CrossRef]
13. Jin, X.; Liu, S.; Baret, F.; Hemerlé, M.; Comar, A. Estimates of plant density of wheat crops at emergence from very low altitude UAV imagery. *Remote Sens. Environ.* **2017**, *198*, 105–114. [CrossRef]
14. Chakraborty, A.; Srikanth, P.; Murthy, C.S.; Rao, P.V.N.; Chowdhury, S. Assessing lodging damage of jute crop due to super cyclone Amphan using multi-temporal Sentinel-1 and Sentinel-2 data over parts of West Bengal, India. *Environ. Monit. Assess.* **2021**, *193*, 464. [CrossRef]
15. Guan, H.; Liu, H.; Meng, X.; Luo, C.; Bao, Y.; Ma, Y.; Yu, Z.; Zhang, X. A Quantitative Monitoring Method for Determining Maize Lodging in Different Growth Stages. *Remote Sens.* **2020**, *12*, 3149. [CrossRef]
16. Song, Z.; Zhang, Z.; Yang, S.; Ding, D.; Ning, J. Identifying sunflower lodging based on image fusion and deep semantic segmentation with UAV remote sensing imaging. *Comput. Electron. Agric.* **2020**, *179*, 105812. [CrossRef]
17. Dai, J.G.; Zhang, G.S.; Guo, P.; Zeng, T.J.; Cui, M.; Xue, J.L. Information extraction of cotton lodging based on multi-spectral image from UAV remote sensing. *Trans. Chin. Soc. Agric. Eng.* **2019**, *35*, 63–70.

18. Bai, J.; Ma, S.; Wang, F.; Xing, H.; Ma, J.; Wang, M. Performance of crop dividers with reference to harvesting lodged sugarcane. *Sugar Tech.* **2020**, *22*, 812–819. [CrossRef]

19. Cao, W.; Qiao, Z.; Gao, Z.; Lu, S.; Tian, F. Use of unmanned aerial vehicle imagery and a hybrid algorithm combining a watershed algorithm and adaptive threshold segmentation to extract wheat lodging. *Phys. Chem. Earth Parts A/B/C* **2021**, *123*, 103016. [CrossRef]

20. Mardanisamani, S.; Maleki, F.; Kassani, S.H.; Rajapaksa, S.; Duddu, H.; Wang, M.; Shirtliffe, S.; Ryu, S.; Josuttes, A.; Zhang, T.; et al. Crop Lodging Prediction From UAV-Acquired Images of Wheat and Canola Using a DCNN Augmented with Handcrafted Texture Features. In Proceedings of the IEEE/CVF Conference on Computer Vision and Pattern Recognition Workshops, Long Beach, CA, USA, 16–20 June 2019.

21. Rajapaksa, S.; Eramian, M.; Duddu, H.; Wang, M.; Shirtliffe, S.; Ryu, S.; Josuttes, A.; Zhang, T.; Vail, S.; Pozniak, C.; et al. Classification of Crop Lodging with Gray Level Co-occurrence Matrix. In Proceedings of the 2018 IEEE Winter Conference on Applications of Computer Vision (WACV), Lake Tahoe, NV, USA, 12–15 March 2018; pp. 251–258.

22. Maresma, Á.; Ariza, M.; Martínez, E.; Lloveras, J.; Casasnovas, J.A.M. Analysis of Vegetation Indices to Determine Nitrogen Application and Yield Prediction in Maize (*Zea mays* L.) from a Standard UAV Service. *Remote Sens.* **2016**, *8*, 973. [CrossRef]

23. Handique, B.K.; Khan, A.Q.; Goswami, C.; Prashnani, M.; Raju, P. Crop discrimination using multispectral sensor onboard unmanned aerial vehicle. *Proc. Natl. Acad. Sci. India-Sect. A* **2017**, *87*, 713–719. [CrossRef]

24. Feng, A.; Zhou, J.; Vories, E.D.; Sudduth, K.A.; Zhang, M. Yield estimation in cotton using UAV-based multi-sensor imagery. *Biosyst. Eng.* **2020**, *193*, 101–114. [CrossRef]

25. Yang, M.D.; Huang, K.S.; Kuo, Y.H.; Tsai, H.P.; Lin, L.M. Spatial and spectral hybrid image classification for rice lodging assessment through UAV imagery. *Remote Sens.* **2017**, *9*, 583. [CrossRef]

26. Yang, M.; Tseng, H.; Hsu, Y.; Tsai, H.P. Semantic segmentation using deep learning with vegetation indices for rice lodging identification in multi-date UAV visible images. *Remote Sens.* **2020**, *12*, 633. [CrossRef]

27. Yang, M.D.; Boubin, J.G.; Tsai, H.P.; Tseng, H.H.; Hsu, Y.C.; Stewart, C.C. Adaptive autonomous UAV scouting for rice lodging assessment using edge computing with deep learning EDANet. *Comput. Electron. Agric.* **2020**, *179*, 105817. [CrossRef]

28. Zhao, X.; Yuan, Y.; Song, M.; Ding, Y.; Lin, F.; Liang, D.; Zhang, D. Use of unmanned aerial vehicle imagery and deep learning UNet to extract rice lodging. *Sensors* **2019**, *19*, 3859. [CrossRef]

29. Li, G.; Han, W.; Huang, S.; Ma, W.; Ma, Q.; Cui, X. Extraction of Sunflower Lodging Information Based on UAV Multi-Spectral Remote Sensing and Deep Learning. *Remote Sens.* **2021**, *13*, 2721. [CrossRef]

30. Louhaichi, M.; Borman, M.; Johnson, D. Spatially Located Platform and Aerial Photography for Documentation of Grazing Impacts on Wheat. *Geocarto Int.* **2001**, *16*, 65–70. [CrossRef]

31. Ronneberger, O.; Fischer, P.; Brox, T. U-net: Convolutional networks for biomedical image segmentation. In Proceedings of the International Conference on Medical Image Computing and Computer-Assisted Intervention, Munich, Germany, 5−9 October 2015; pp. 234–241.

32. Howard, A.G.; Zhu, M.; Chen, B.; Kalenichenko, D.; Wang, W.; Weyand, T.; Andreetto, M.; Adam, H. Mobilenets: Efficient convolutional neural networks for mobile vision applications. *arXiv* **2017**, arXiv:1704.04861.

33. Yang, B.; Ma, J.; Yao, X.; Cao, W.; Zhu, Y. Estimation of Leaf Nitrogen Content in Wheat Based on Fusion of Spectral Features and Deep Features from Near Infrared Hyperspectral Imagery. *Sensors* **2021**, *21*, 613. [CrossRef]

34. Wu, W.; Wang, W.; Meadows, M.E.; Yao, X.; Peng, W. Cloud-based typhoon-derived paddy rice flooding and lodging detection using multi-temporal sentinel-1&2. *Front. Earth Sci.* **2019**, *13*, 682–694.

35. Selvaraju, R.R.; Cogswell, M.; Das, A.; Vedantam, R.; Parikh, D.; Batra, D. Grad-CAM: Visual explanations from deep networks via gradient-based localization. *Int. J. Comput. Vis.* **2020**, *128*, 336–359. [CrossRef]

sensors

MDPI

Article

Convolutional Neural Networks to Estimate Dry Matter Yield in a Guineagrass Breeding Program Using UAV Remote Sensing

Gabriel Silva de Oliveira [1], José Marcato Junior [2,*], Caio Polidoro [1], Lucas Prado Osco [2,3], Henrique Siqueira [2], Lucas Rodrigues [1], Liana Jank [4], Sanzio Barrios [4], Cacilda Valle [4], Rosângela Simeão [4], Camilo Carromeu [4], Eloise Silveira [2], Lúcio André de Castro Jorge [5], Wesley Gonçalves [1,2], Mateus Santos [4] and Edson Matsubara [1]

[1] Faculty of Computer Science, Federal University of Mato Grosso do Sul, Campo Grande 79070900, Brazil; silva.eng.gabriel@gmail.com (G.S.d.O.); cpolidoro@gmail.com (C.P.); lucas.rodrigues@ifms.edu.br (L.R.); wesley.goncalves@ufms.br (W.G.); edsontm@facom.ufms.br (E.M.)

[2] Faculty of Engineering, Architecture and Urbanism and Geography, Federal University of Mato Grosso do Sul, Campo Grande 79070900, Brazil; lucasosco@unoeste.br (L.P.O.); henrique.siqueira@ufms.br (H.S.); eloiseambiental@gmail.com (E.S.)

[3] Faculty of Engineering, Architecture and Urbanism, University of Western São Paulo, Presidente Prudente 19067175, Brazil

[4] Embrapa Beef Cattle, Brazilian Agricultural Research Corporation, Campo Grande 79106550, Brazil; liana.jank@embrapa.br (L.J.); sanzio.barrios@embrapa.br (S.B.); cbdovalle@gmail.com (C.V.); rosangela.simeao@embrapa.br (R.S.); camilo.carromeu@embrapa.br (C.C); mateus.santos@embrapa.br (M.S.)

[5] Embrapa Instrumentation, Brazilian Agricultural Research Corporation, São Carlos 13560970, Brazil; lucio.jorge@embrapa.br

* Correspondence: jose.marcato@ufms.br

Citation: de Oliveira, G.S.; Marcato Junior, J.; Polidoro, C.; Osco, L.P.; Siqueira, H.; Rodrigues, L.; Jank, L.; Barrios, S.; Valle, C.; Simeão, R.; et al. Convolutional Neural Networks to Estimate Dry Matter Yield in a Guineagrass Breeding Program Using UAV Remote Sensing. *Sensors* **2021**, *21*, 3971. https://doi.org/10.3390/s21123971

Academic Editors: Jiyul Chang and Sigfredo Fuentes

Received: 24 April 2021
Accepted: 27 May 2021
Published: 9 June 2021

Publisher's Note: MDPI stays neutral with regard to jurisdictional claims in published maps and institutional affiliations.

Abstract: Forage dry matter is the main source of nutrients in the diet of ruminant animals. Thus, this trait is evaluated in most forage breeding programs with the objective of increasing the yield. Novel solutions combining unmanned aerial vehicles (UAVs) and computer vision are crucial to increase the efficiency of forage breeding programs, to support high-throughput phenotyping (HTP), aiming to estimate parameters correlated to important traits. The main goal of this study was to propose a convolutional neural network (CNN) approach using UAV-RGB imagery to estimate dry matter yield traits in a guineagrass breeding program. For this, an experiment composed of 330 plots of full-sib families and checks conducted at Embrapa Beef Cattle, Brazil, was used. The image dataset was composed of images obtained with an RGB sensor embedded in a Phantom 4 PRO. The traits leaf dry matter yield (LDMY) and total dry matter yield (TDMY) were obtained by conventional agronomic methodology and considered as the ground-truth data. Different CNN architectures were analyzed, such as AlexNet, ResNeXt50, DarkNet53, and two networks proposed recently for related tasks named MaCNN and LF-CNN. Pretrained AlexNet and ResNeXt50 architectures were also studied. Ten-fold cross-validation was used for training and testing the model. Estimates of DMY traits by each CNN architecture were considered as new HTP traits to compare with real traits. Pearson correlation coefficient r between real and HTP traits ranged from 0.62 to 0.79 for LDMY and from 0.60 to 0.76 for TDMY; root square mean error (RSME) ranged from 286.24 to 366.93 kg·ha^{-1} for LDMY and from 413.07 to 506.56 kg·ha^{-1} for TDMY. All the CNNs generated heritable HTP traits, except LF-CNN for LDMY and AlexNet for TDMY. Genetic correlations between real and HTP traits were high but varied according to the CNN architecture. HTP trait from ResNeXt50 pretrained achieved the best results for indirect selection regardless of the dry matter trait. This demonstrates that CNNs with remote sensing data are highly promising for HTP for dry matter yield traits in forage breeding programs.

Keywords: deep learning; forage dry matter yield; high-throughput phenotyping; Brazilian pasture

1. Introduction

Pastures are the best alternative for feeding beef cattle in Brazil since they are the most economical and sustainable feed strategy available for cattle rearers. It is estimated that 86% of the beef produced in the country occurs entirely on pastures [1], which are composed mostly of perennial African species such as guineagrass (*Megathyrsus maximus*, syn. *Panicum maximum*) and brachiariagrass (*Urochloa* spp., syn. *Brachiaria* spp.) [2]. Beef cattle production in Brazilian pastures has been increased by the adoption of improved forage cultivars [3]. As an example, the first two guineagrass cultivars released by Embrapa (Brazilian Agricultural Research Corporation) in the 1990s presented 86% (cv. Tanzânia) and 136% (cv. Mombaça) higher leaf dry matter yield than the oldest cv. Colonião with a great impact on the beef and milk production systems [4]. Thus, new improved cultivars are highly recommended to increase profits and sustainability of beef cattle production.

Dry matter yield (DMY) is the most important trait in forage breeding since it contains most of the essential nutrients of the cattle's diet, such as carbohydrates, proteins, lipids, vitamins, and minerals [5]. Forage dry matter is composed of different morphological components such as leaf blades, leaf sheaths, and stems, and most of the nutrients accumulate in the leaf blades. In all the phases of cultivar development, DMY is evaluated and included in the breeder's selection criteria. However, the high labor costs and the time spent to acquire DMY phenotypes limit the genetic gains in breeding programs [6].

Currently, phenotyping for forage yield traits is performed by low-cost inaccurate visual evaluations [7] or by high-cost low throughput measurements [8]. For the latter strategy, samples are air-forced dried in a drying chamber to calculate DMY. If breeders seek to include DMY components in the selection criteria, an additional laboring step for separating different components of the forage samples is necessary prior to drying. Additionally, the multi-season x year x environment harvests increase labor costs and time of phenotyping in perennial forage breeding. In this regard, new strategies that guarantee high throughput, accuracy, low costs, and less time consumption would highly impact forage breeding programs, especially the ones with limited resources.

High-throughput phenotyping (HTP) is an emerging strategy to reduce the bottleneck of phenotyping in breeding programs [9]. HTP is performed in fully automated facilities, growth chambers, or in the field; the latter is expected to impact directly plant breeding since the information correlates better with the real environment of crop production [9]. Generally, field HTP uses unmanned aerial vehicles (UAV) platforms combined with different sensors (RGB, multispectral, hyperspectral, LiDAR) and image processing tools. Thus, features (e.g., physiological vegetation indices) are extracted from the images and correlated with ground-truth data to validate the HTP process.

Highly correlated estimates indicate that HTP may be a tool to support breeder or agricultural specialist's decisions [10–12]. Gebremedhin et al. [6] reviewed different aspects of sensor technologies applied to forage DMY and pointed out that machine learning is an important technique that can be widely applied in data analysis for HTP in large population field trials. Remote sensing in conjunction with robust and intelligent data processing methods has been an alternative used for visual inspection of agricultural landscapes in recent years [13].

Recently, machine learning methods such as deep learning have gained prominence, outperforming traditional methods. Spectral indices [14] are commonly used as input for traditional methods. Deep learning-based algorithms, using raw data, search for patterns necessary for classification or regression, with various levels of representation obtained by the composition of non-linear but straightforward models. Each one transforms the representation that is at a raw level into a higher-level representation, slightly more abstract. There are different types of deep neural networks used for image processing, speech recognition, and language interpretation. Convolutional neural networks (CNNs), which are the most used for remote sensing image processing, are deep networks with convolution and pooling layers [15]. They apply mathematical operations of convolution in vectors that represent images, thus extracting information used for training these networks [16].

Deep learning has the advantage of performing end-to-end learning, whereas conventional machine learning often requires a domain-dependent custom feature extraction process [16]. For instance, Lee et al. [17] presented a data pipeline consisting of several steps to use conventional machine learning to evaluate plant growth. The image is preprocessed and submitted to a superpixel algorithm as a feature extraction technique. Using a conventional machine learning algorithm, Random Forest, the model performed plant segmentation. Additionally, they need to compare the images to evaluate plant growth. Our proposal is much simpler and easier to tackle. We use CNNs that incorporate the feature extraction step in the lower levels of the convolutional layers and direct outputs the target information. Other similar approaches in HTP using conventional machine learning can be found in [18–21]

The applications of CNNs in agriculture are diverse [22,23]. For example, in rice [24], a set of 500 images were used to identify diseases in the field by adopting CNN inspired by AlexNet [25], and LeNet [26]. In the training part of the neural network, the 10-fold cross-validation strategy was used, and the proposed CNN achieved an accuracy of 95.48% in identifying diseases. Research that served as the inspiration for this investigation addressed the use of images and CNNs to estimate biomass in wheat [27], in which a network was proposed inspired by VGGNet [28] and RGB images. They obtained a high correlation coefficient and a low Root Mean Square Error (RMSE) compared with other techniques, such as Random Forest and Support Vector Machines.

In our previous research [29], green biomass was estimated in a tropical forage experiment applying different already known CNNs, such as AlexNet, ResNet, and VGGNet, in a bank of 330 image patches of guineagrass, taken using RGB sensor embedded in UAV. AlexNet outperformed other architectures obtaining a better correlation and intersection between the real data measured by experts and data predicted by these CNNs. However, based on literature, few studies are found in which convolutional neural networks are applied in forages, and we have not found any research that applies these techniques to estimate forage DMY in a HTP context.

Field HTP using remote sensing and CNNs generate new traits throughout the image detection, classification, and retrieval process. These traits can be used in different selection methods in the breeding program, such as indirect selection [30,31], index selection, or in prediction with genomic selection [32]. In the indirect selection, a secondary trait (e.g., HTP trait) can replace the main trait (e.g., DMY) in selection criteria if the secondary trait is highly heritable and correlated to the main trait [33]. However, if the secondary trait is more suitable to evaluate in large plant populations, as expected with HTP traits, a higher selection intensity can be applied and increase the efficiency of indirect selection [34].

To increase the efficiency of forage breeding programs, it is important to evaluate state-of-the-art methods to estimate parameters correlated to important traits. Our work hypothesis is that deep learning regression-based methods are able to estimate DMY traits using only RGB UAV-based imagery with adequate accuracy for a guineagrass breeding program. The main goal is to propose a CNN regression-based approach to estimate DMY traits using only RGB UAV-based imagery. AlexNet, ResNeXt50, DarkNet53, and two networks proposed recently for related tasks named MaCNN [27] and LF-CNN [35], which compose the state-of-the-art, were investigated in the current work. Different from previous work [29] that assessed green biomass, here we considered the dry matter. Besides, the CNN-based DMY was not assessed only in a traditional mode (RMSE, correlation coefficient, etc.) but also in the statistical genetic analysis. Also, the HTP traits from the best CNN were used to evaluate their performance in the indirect selection method.

2. Method

The proposed approach is presented in Figure 1. First, RGB images were acquired with a UAV. Orthoimages were generated using the acquired RGB images based on UAV photogrammetry technique. In the next step, the orthoimage for each plot was extracted and used as input for the CNN regression-based architectures. As ground truth, two traits were evaluated for each plot: leaf dry matter yield (LDMY) and total dry matter yield (TDMY), in kg·ha^{-1}. These traits were estimated based on field and laboratory evaluation. Finally, the estimated values from the CNNs were validated on a genetic model. Detailed descriptions of the two main steps are presented in the following sections.

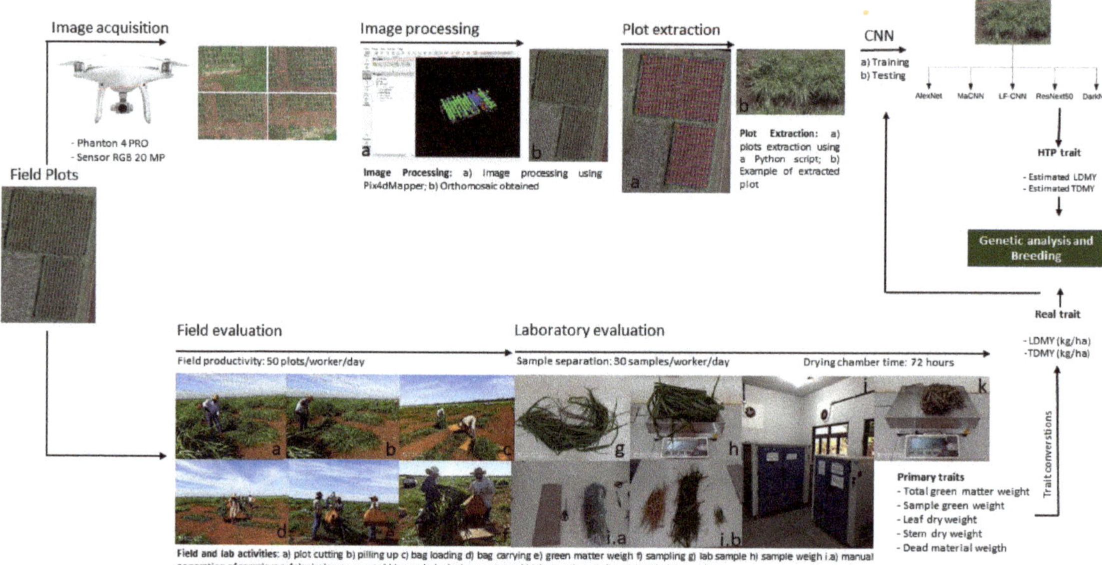

Figure 1. Workflow of phenotyping DMY traits by conventional (**bellow**) and high-throughput (**above**) processes in the guineagrass breeding program at Embrapa Beef Cattle.

2.1. Deep Learning Architectures and Evaluation

AlexNet was the architecture initially chosen for the application. According to the result of previous research focused on the estimation of green matter [29], AlexNet presented the best Pearson correlations and Root Mean Square error, compared to ResNet and VGGNet. AlexNet is a CNN proposed in [25], and is composed of eight layers in which five are convolution layers, using as activation function Rectified Linear Units (ReLU) and MaxPool between the layers, and three layers are fully connected. AlexNet and the other CNNs were adapted to the regression problem.

In addition to this network, two other CNNs containing a number of small layers were chosen for this research. Ma et al. [27] proposed in 2019 a CNN for estimating wheat biomass. Thus, it was decided to use this network and verify its performance to estimate dry matter yield traits in forages. In our study, this network was named MaCNN, and this CNN consists of four convolution layers, with layers between them with the average pooling operation, in which the image dimension is downsampled, thus containing three more levels. In addition, in each convolution layer, a batch normalization was also carried out, which contributes to the acceleration of the CNN training process. Finally, there was a fully connected dropout layer.

Another CNN used was proposed by Barbosa et al. [35] in 2020. This network was named Late Fusion (LF). It is a multi-stream network in which each input is connected to an independent convolutional layer with eight 3 × 3 filters. However, a fully connected ReLU layer with 16 neurons was added to each stream after the maximum cluster layer, followed by a single ReLu neuron. Then, the five neurons (one from each stream) were concatenated and fed to the last two layers.

Even though previous research [29] verified a higher efficiency of AlexNet in relation to deeper networks, it was decided to evaluate state-of-the-art and deeper CNNs for dry matter traits. The architecture ResNeXt [36] with 50 layers was chosen, namely ResNeXt50. In addition to ResNeXt50, the DarkNet53 proposed by Redmon et al. [37] was considered, in which a 53-layer CNN was used. In order to have a better comparison between the different architectures, Table 1 presents the architectures used in this study with the number of layers and the number of parameters.

Table 1. Comparison between the different models in terms of number of layers and parameters.

Models	Number of Layers	Number of Parameters
AlexNet	8	62 M
AlexNet pretrained	8	62 M
MaCNN	5	1.1 M
LF-CNN	10	3.6 K
ResNeXt50	50	25 M
ResNeXt50 pretrained	50	25 M
DarkNet53	53	42 M

Data augmentation related to rotation (horizontally and vertically) was also adopted to avoid overfitting, thus increasing the data set for training. For training these architectures, a 10-fold cross-validation technique was used. This procedure randomly divides the data set into 10 sets with equal size, and each set is, in turn, used to test the model trained from the other 10-1 sets [38]. In all experiments, the Adam optimization method [39] was used. The cost function used was the mean square error, described in Equation (1). The pretrained models considered the ImageNet, where the pretrained weights were loaded and adjusted to these models using the training set. Models without pretraining were fully trained using the training set.

$$\varepsilon = \frac{1}{n}\sum_{i=0}^{n}(y_i - \hat{y}_i)^2 \tag{1}$$

To evaluate the experiments, different metrics were used, as expressed in Equations (2)–(4). The MAE and RMSE metrics express the average model prediction error in units of the variable of interest. The metric r establishes a linear relationship between the real value and the predicted value. In the Equations (2) and (3), the y is the real value and the $\hat{y}$ the predicted one. In the last equation, x represents real values, and $\overline{x}$ is the average of real values. The y represents the predicted values and the $\overline{y}$ the average of the predicted values.

$$MAE = \frac{1}{n}\sum_{i=0}^{n}|y_i - \hat{y}_i| \tag{2}$$

$$RMSE = \sqrt{\frac{1}{n}\sum_{i=0}^{n}(y_i - \hat{y}_i)^2} \tag{3}$$

$$r = \frac{\sum_{i=1}^{n}(x_i - \overline{x})(y_i - \overline{y})}{\sqrt{\sum_{i=1}^{n}(x_i - \overline{x})^2(y_i - \overline{y})^2}} \tag{4}$$

Finally, to evaluate the distribution of real and estimated values, we constructed histograms. For the construction of the histograms, the number of bins was determined using the [40] elbow rule from the partitioning of the real values of the samples.

2.2. Validation of HTP Traits on a Genetic Model

Each CNN architecture generated estimates of each DMY trait from the cross-validation process for each trial plot. Here, these estimates for a given CNN were nominated HTP traits, which means a trait estimated from a UAV platform-RGB sensor-CNN architecture related to the real trait (LDMY or TDMY). Thus, DMY traits estimated by HTP and by conventional phenotyping were analyzed considering the linear mixed model shown in Equation (5).

$$y = XB + Z_1 b + Z_2 g + e \tag{5}$$

where y is the vector of observations of the real or each HTP trait; B is the vector of fixed effects of replications and checks (genitors and cultivars); b is the vector of random effects of blocks within replications where $b \sim N(0,Vb)$; g is the vector of random effects of full-sib family where $g \sim N(0,Vg)$; and e is the vector of random effects of residuals where $e \sim N(0,RVe)$. R is a matrix of (co)variances of residuals where the spacial tendencies were modeled for autocorrelation among lines and columns of the trial according to [41]. X relates y to B, whereas Z_1 and Z_2 relate y to b and g, respectively. Estimation of variance components from the data was carried out using restricted maximum likelihood (REML) by the Average Information algorithm as implemented in the ASREML-R package [42] in the R environment. Based on the most likely variance components, the fixed effects were estimated, and the random effects (family BLUPs) were predicted by solving the mixed model equations.

To compute the broad-sense heritability (H) for real and HTP traits on the basis of full-sib family means, we considered Equation (6), as suggested by [43]:

$$H = 1 - \frac{PEV}{2Vg} \tag{6}$$

where PEV is the prediction error variance, which represents the average variance of the difference between a pair of family predictions (BLUP), and Vg is the genetic variance component among full-sib families.

Correlations between family BLUPs of a given HTP trait and its related real trait (LDMY or TDMY) were estimated as an approximated genetic correlation (r) between traits.

Direct response to selection (DR) for real traits and correlated response to selection (CR) when using its related HTP trait as a secondary trait were obtained using Equations (7) and (8) [33], where i is the selection intensity, h is the square root of H, v is the square root of Vg.

$$DR = i.h_{\text{real}}.v_{g\,\text{real}} \tag{7}$$

$$CR = i.h_{\text{HTP}}.r.v_{g\,\text{real}} \tag{8}$$

For DR, we considered a selection intensity of 10%, or approximately nine selected families, which represented an $i = 1.76$; for CR we considered three scenarios that maintained the same number of selected families: (1) the same selection intensity (10%) as for the real trait, (2) increasing the selection intensity to 5% or $i = 2.06$ by simulating twice the number of full-sib families evaluated (172) by HTP and (3) increasing the selected intensity to 1% or $i = 2.67$ by simulating ten times the number of full-sib families evaluated (860) by HTP. In the last two scenarios we must consider the same experimental design as for the first one. Thus, the total number of plots considered in (2) and (3) were 660 and 3300, respectively.

3. Experiments and Results

3.1. Study Area and Data Set

For the development of this study, images from a guineagrass trial located at Embrapa Beef Cattle in Campo Grande, Mato Grosso do Sul, Brazil, were used. A total of 110 genotypes composed of 86 full-sib families, ten sexual, and ten apomictic genitors along with four commercial checks (Mombaça, MG12 Paredão, BRS Quênia, and BRS Tamani) were used as treatments. A 10×11 alpha lattice design with three replications was considered, totaling 330 plots in the trial. Each plot consisted of two rows of 2.0 m length and 0.5 m apart. Each row consisted of five plants spaced 0.5 m between plants, in a total of ten plants per plot. Plots were 1.0 m apart, representing an area of 4.5 m². This trial is described according to Figure 2.

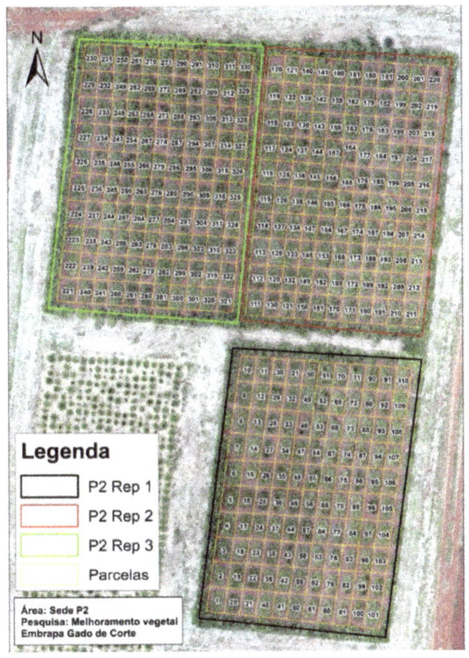

Figure 2. Sketch of the trial in the field.

The images were taken using the UAV Phantom 4 PRO, with an RGB sensor of 5427×3748 of image resolution. The flight was carried out on 23 January 2019, close to 9 am, at the height of 18 m, thus generating a resolution of 0.5 cm/pixel, with a frontal image overlap of 75% and lateral of 60%, and the flight took about 20 min.

To properly extract each plot, the images were processed using the software Pix4dMapper. An orthomosaic was generated, which is a mosaic of orthorectified and enhanced aerial images to homogenize its appearance. With the orthomosaic generated, it was possible to correctly map the area of interest and perform the extraction of images from each plot of the trial. To perform this extraction, a Python script proposed in [29] was used, in which it uses the orthomosaic as a parameter in tiff format and from the information of the trial, such as the number of blocks and the number of plots per block, as well as how the numbering of each plot is organized, generates the images of each plot identified according to the established numbers. For this experiment, 330 orthoimage patches were generated relative to the total number of plots in the trial.

Each plot was evaluated for forage yield traits on 25 January 2019. Traits were obtained by harvesting each plot 0.2 m from the soil. The harvested material (green matter) was weighed in the field using a field dynamometer to obtain the total green matter weight per plot (TGMW) in kg. After weighing, samples of 300 g to 500 g of the green matter were taken and sent to the Laboratory of Forage Sample Preparation of Embrapa Beef Cattle to obtain the sample green weight (SGW), in g, and which were then separated into leaf blades, leaf sheaths + stems and dead material. These forage sample components were then air-forced dried at 65 °C in the drying chamber for 72 h. After this period, the dried samples were weighed to obtain leaf dry matter weight (LDMW), sheaths + stem dry matter weight (SSDMW), and dead material dry matter weight (DMDMW) in kg. The Equations (9)–(12) were considered to obtain the dry matter yield per plot.

$$LDMY = \frac{TGMW * LDMW}{SGW} \tag{9}$$

$$SSDMY = \frac{TGMW * SSDMW}{SGW} \tag{10}$$

$$DMDMY = \frac{TGMW * DMDMW}{SGW} \tag{11}$$

$$TDMY = LDMY + SSDMY + DMDMY \tag{12}$$

Finally, LDMY and TDMY per plot were converted to kg·ha^{-1}.

The distribution of the values of LDMY and TDMY in kg·ha^{-1}, which for this research will be the attributes classes y, are shown in Figure 3.

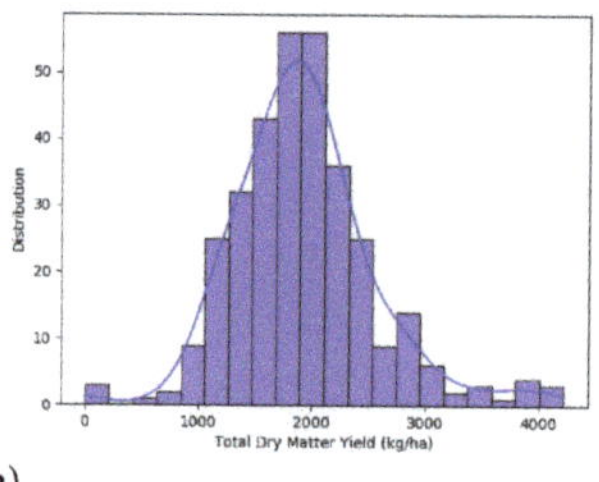

(a)

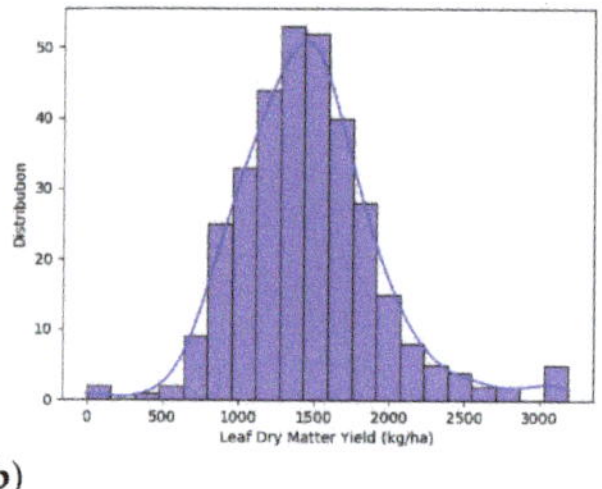

(b)

Figure 3. Data distribution. (a) TDMY; (b) LDMY.

3.2. Deep Learning Protocol

Due to the fact that only 330 image patches were used, this can become a problem when using deep neural networks, as there is a high number of parameters that are estimated. As previously mentioned, data augmentation related to rotation (horizontally and vertically) was also adopted to avoid overfitting, thus increasing the data set for training.

For training the architectures, a 10-fold cross-validation technique was used. The Adam optimization method [39] was used, with the descending gradient algorithm, and a fixed learning rate of 0.001, constant $\beta_1 = 0.9$, $\beta_2 = 0.999$ and $\epsilon = 10^{-8}$. Table 2 summarizes the CNN architectures used, the number of epochs for each experiment, and the batch size. The number of epochs for each experiment was defined empirically using early stopping evaluated every 100 epochs. The batch size sought to use the highest value that did not overflow the GPU memory. For all experiments, a server with an NVIDIA K80 (2 × 12 GB), AMD TR 1900X 3.7 GHz CPU, and 64 GB of RAM was used.

Table 2. Configurations of the experiments.

Models	Number of Epochs	Batch Size
AlexNet	500	256
AlexNet pretrained	500	256
MaCNN	500	256
LF-CNN	500	256
ResNeXt50	300	64
ResNeXt50 pretrained	300	64
DarkNet53	300	64

3.3. Results

Tables 3 and 4 present the MAE, the RMSE and the Pearson Correlation coefficient *r* for each CNN architecture with respect to the dry matter traits. From the tables presented, the MAE values ranged between 204.39 (AlexNet pretrained) and 266.77 (LF-CNN) $kg \cdot ha^{-1}$ for the LDMY trait, and ranged between 289.66 (AlexNet pretrained) and 366.93 (LF-CNN) $kg \cdot ha^{-1}$ for TDMY trait. It is important to mention that we have values, predominantly, from 500 to 4000 $kg \cdot ha^{-1}$ for these traits. Therefore, it is possible to analyze a better performance of AlexNet pretrained for LDMY trait, since it presented the lowest values for MAE and RMSE, with an absolute average error of 204.39 $kg \cdot ha^{-1}$. For TDMY trait, AlexNet pretrained presented the lowest MAE in relation to the others; however, for the other metrics, ResNeXt50 pretrained presented better results, with an RMSE of 413.07 $kg \cdot ha^{-1}$.

Table 3. Results for LDMY.

Models	Mean Absolute Error	Root Mean Square Error	Pearson Correlation (*r*)
AlexNet	248.41 ± 47.58	340.70 ± 64.85	0.70 ± 0.09
AlexNet pretrained	**204.39 ± 56.46**	**286.24 ± 80.39**	**0.79 ± 0.12**
MaCNN	240.74 ± 65.09	333.60 ± 86.93	0.71 ± 0.11
LF-CNN	266.57 ± 89.19	366.93 ± 110.33	0.62 ± 0.13
ResNeXt50	221.04 ± 54.44	319.98 ± 98.47	0.72 ± 0.12
ResNeXt50 pretrained	231.66 ± 63.41	319.58 ± 87.06	0.73 ± 0.10
DarkNet53	217.30 ± 57.09	311.76 ± 76.68	0.76 ± 0.12

Table 4. Results for TDMY

Models	Mean Absolute Error	Root Mean Squared Error	Pearson Correlation (*r*)
AlexNet	311.37 ± 88.58	441.31 ± 131.29	0.73 ± 0.17
AlexNet pretrained	**289.66 ± 96.28**	419.95 ± 136.93	0.75 ± 0.20
MaCNN	345.11 ± 97.94	477.12 ± 136.34	0.68 ± 0.20
LF-CNN	364.44 ± 145.38	506.56 ± 176.24	0.60 ± 0.17
ResNeXt50	306.09 ± 137.15	449.07 ± 175.06	0.71 ± 0.17
ResNeXt50 pretrained	294.73 ± 78.83	**413.07 ± 117.77**	**0.76 ± 0.24**
DarkNet53	291.12 ± 80.26	419.50 ± 131.87	0.75 ± 0.23

Furthermore, analyzing the coefficient *r*, we verified that there was a variation from 0.62 (LF-CNN) to 0.79 (AlexNet pretrained) for LDMY and 0.60 (LF-CNN) up to 0.76 (ResNeXt50 pretrained) for TDMY. This shows that there was a significant correlation between the data estimated by the CNN and the real data obtained in the field, realizing that it is possible to obtain a high relationship between the RGB images and the dry matter data using the CNN. Overall, the three networks with the best results were AlexNet pretrained, ResNeXt50 pretrained, and DarkNet53. LF-CNN had the worst results for both traits.

3.3.1. Graph of Predicted Versus Real

The graphs shown in Figures 4 and 5 are composed of points $(y, \hat{y})$, in which y represents the real values of LDMY and TDMY, respectively, while $\hat{y}$ represents the values predicted by CNNs for each of the traits. A perfect prediction would result in a straight line on the graph since the predicted, and real values would be the same.

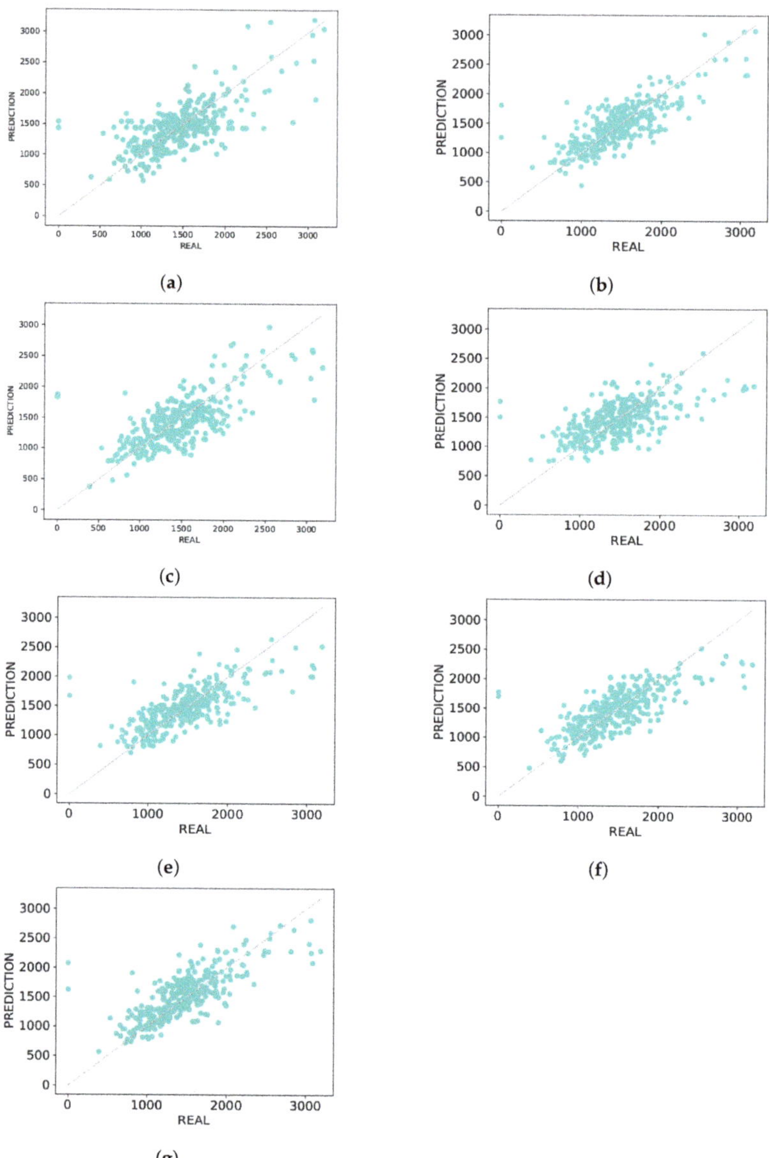

Figure 4. Predicted vs. Real plots-LDMY. (**a**) AlexNet; (**b**) AlexNet pretrained; (**c**) MaCNN; (**d**) LF-CNN; (**e**) ResNeXt50; (**f**) ResNeXt50 pretrained; (**g**) DarkNet53.

Analyzing the graphs presented in Figure 4, which presents the plot of the estimated and real values of LDMY, it is possible to notice that for high values, above 2500 kg·ha^{-1}, AlexNet pretrained (Figure 4b) provided smaller errors. The ResNeXt50 pretrained and DarkNet53 models presented more linear graphs (Figure 4f,g), providing less error mainly in the range of 1000 and 2000 kg·ha^{-1}. The other networks (Figure 4a,c–e) have some major errors with respect to values between 1000 and 2000 kg·ha^{-1}, and for higher values, they were able to estimate the values properly.

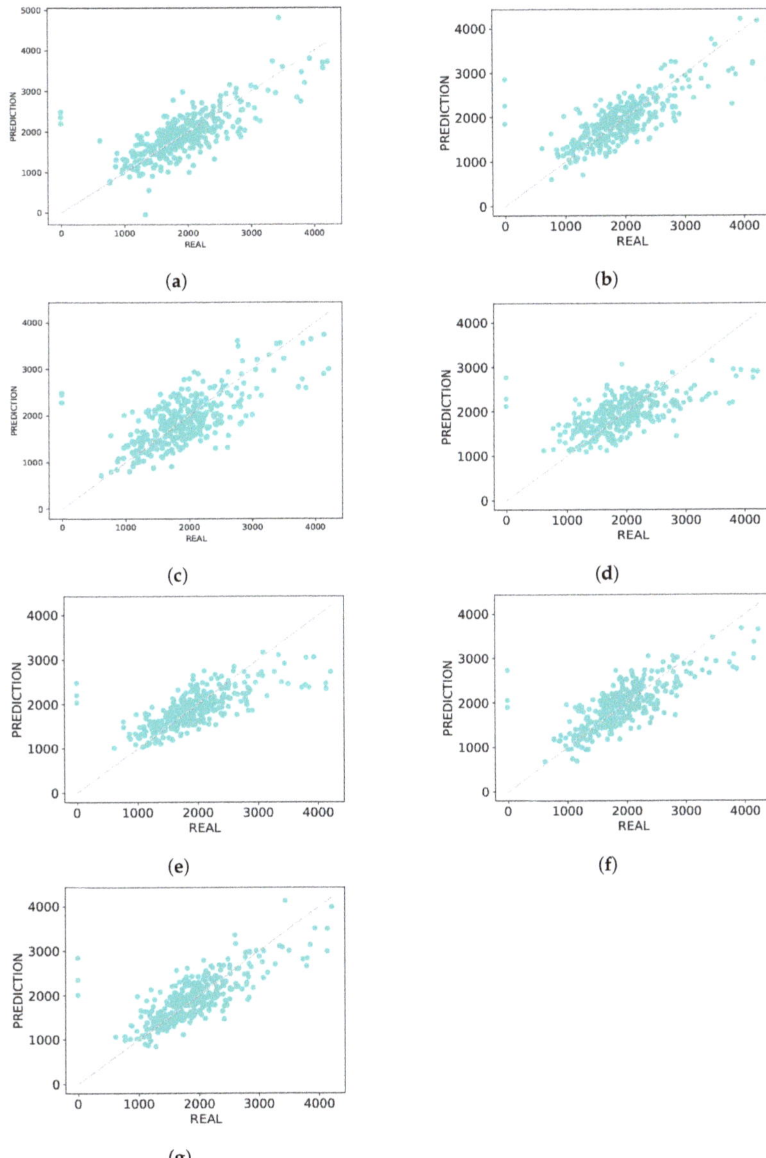

Figure 5. Predicted vs. Real plots-TDMY. (**a**) AlexNet; (**b**) AlexNet pretrained; (**c**) MaCNN; (**d**) LF-CNN; (**e**) ResNeXt50; (**f**) ResNeXt50 pretrained; (**g**) DarkNet53.

For TDMY, the analysis of the graphs presented in Figure 5 occured in a similar way. For this trait, we can notice that AlexNet pretrained (Figure 5b), ResNeXt50 pretrained (Figure 5f) and DarkNet53 (Figure 5g) provided more accurate results, mainly for values above 3000 kg·ha^{-1}. The MaCNN (Figure 5c) and LF-CNN (Figure 5d) networks presented small errors for values between 1000 and 2000 kg·ha^{-1}, but they started to disperse more as the values increased (between 2000 and 3000 kg·ha^{-1}). However, MaCNN in some data had a smaller error for values above 3000 kg·ha^{-1}, which was not verified for LDMY. Finally, we verified that AlexNet not pretrained had more spread points for values below

1500 kg·ha^{-1}. On the other hand, ResNeXt50 not pretrained presented more spread points for values above 3000 kg·ha^{-1}.

3.3.2. Histograms

The histograms (Figures 6 and 7) show the distribution of data of LDMY and TDMY, respectively, and the intersection between the predicted values $\hat{y}$ and real values y. The intersection area between the values was then calculated and is presented in Table 5. It was verified that for group values greater than 20, the addition of new groups did not significantly increase the representativeness of the data; thus, the number of bins was defined as 20.

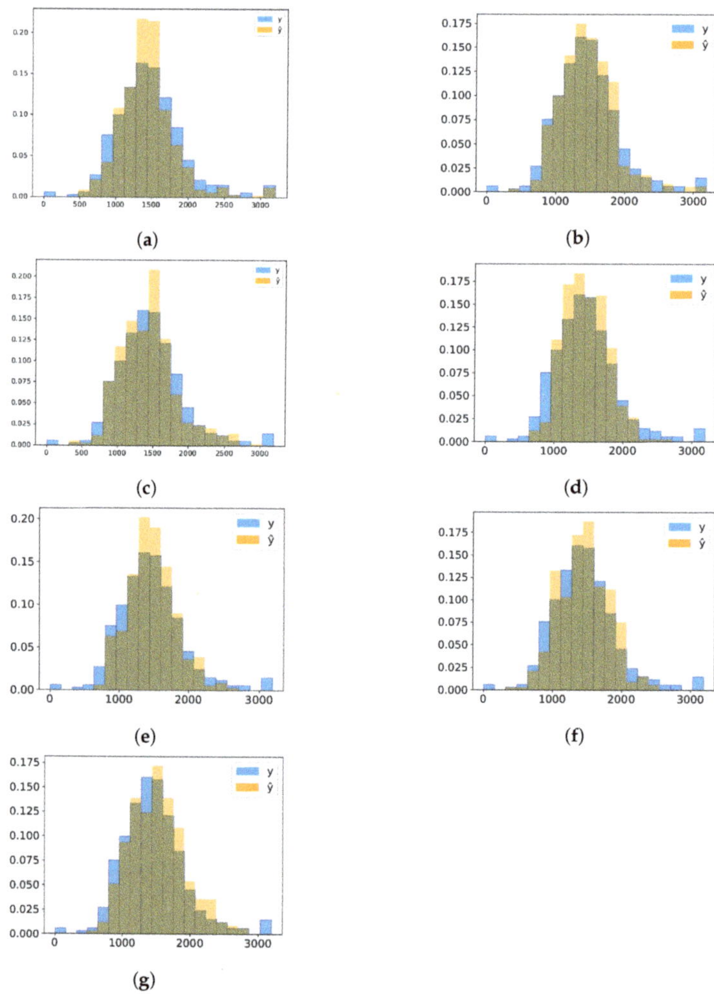

Figure 6. Comparison of predicted $\hat{y}$ vs. real y in relation to LDMY data distribution. (**a**) AlexNet; (**b**) AlexNet pretrained; (**c**) MaCNN; (**d**) LF-CNN; (**e**) ResNeXt50; (**f**) ResNeXt50 pretrained; (**g**) DarkNet53.

Table 5. Intersection area of histograms.

Experiments/Intersection	Leaf Dry Matter	Total Dry Matter
AlexNet	0.87	**0.93**
AlexNet pretrained	**0.91**	0.90
MaCNN	0.89	0.91
LF-CNN	0.86	0.81
ResNeXt50	0.88	0.85
ResNeXt50 pretrained	0.86	0.89
DarkNet53	0.89	0.90

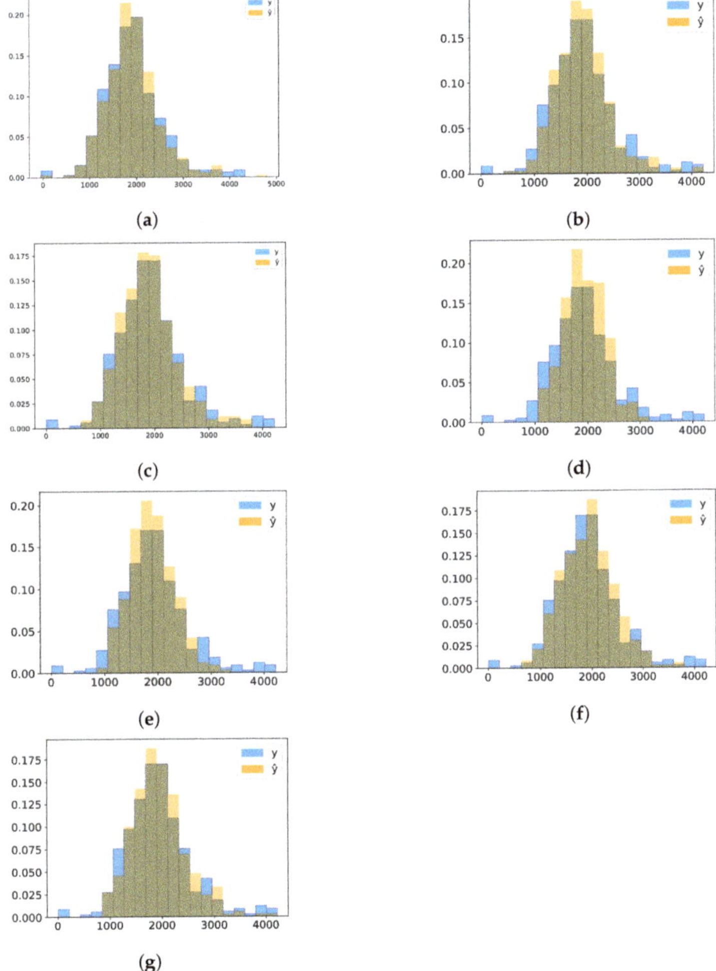

Figure 7. Comparison of predicted $\hat{y}$ vs. real y in relation to TDMY data distribution. (**a**) AlexNet; (**b**) AlexNet pretrained; (**c**) MaCNN; (**d**) LF-CNN; (**e**) ResNeXt50; (**f**) ResNeXt50 pretrained; (**g**) DarkNet53.

From the analysis of the tables and the histograms presented above, it is possible to notice a higher area of intersection for AlexNet.

3.3.3. Genetic Parameters and Indirect Selection Efficiency

Results in Table 6 showed that the V_g components changed according to the HTP trait considered, but all of them were lower than for the Real LDMY and TDMY traits. It can be observed that traits derived from ResNeXt50 pretrained showed 5.5 and 3.8 times higher Vg than those derived from LF-CNN for LDMY and AlexNet for TDMY, respectively. Compared to the Real traits, the V_g of ResNeXt50 pretrained were 0.66 and 0.73 times of the V_g for LDMY and TDMY, respectively. Interestingly, ResNeXt50 pretrained increased the estimates of Vg in comparison to ResNeXt50, a non-pretrained model, especially for TDMY. DarkNet53 was another CNN that stood out for estimating the Vg of the traits compared to the other networks.

Table 6. Genetic parameters for dry matter traits in the guineagrass breeding program using field data and their estimates by several CNNs architectures. V_g-Genetic variance among full-sib families; H-broad sense heritability; r-approximate genetic correlation between real and HTP trait.

| | Traits | | | | | |
| | **Leaf Dry Matter Yield** | | | **Total Dry Matter Yield** | | |
Data Origin	V_g	H	r	V_g	H	r
Real	44,258	0.41	1.00	87,562	0.48	1.00
AlexNet	18,299	0.08	0.71	16,956	−0.36	0.82
AlexNet pretrained	15,602	0.05	0.85	51,343	0.11	0.81
MaCNN	18,643	0.08	0.76	32,888	0.10	0.69
LF−CNN	5288	−0.49	0.71	23,975	0.23	0.75
ResNeXt50	21,188	0.36	0.78	37,580	0.47	0.80
ResNeXt50 pretrained	29,337	0.45	0.84	64,288	0.51	0.83
DarkNet53	22,347	0.44	0.84	62,378	0.45	0.85

Broad-sense heritability (H), which is related to the accuracy of phenotypes to predict genotypes, ranged from −0.49 (LF-CNN) to 0.45 (ResNeXt50 pretrained) for LDMY and from −0.36 (AlexNet) to 0.51 (ResNeXt50 pretrained) for TDMY, indicating that there were effects of the phenotyping process in the heritability of the traits. Except for HTP traits from ResNeXt50 (only for TDMY), ResNeXt50 pretrained, and DarkNet53, all the other CNN generated traits with lower H than real traits. Thus, while some CNN generated phenotypes with moderate accuracy to predict the family genotypic values, other CNN showed low or even highly inaccurate values, as in the case of LF-CNN for LDMY and AlexNet for TDMY.

Genotypic correlations (r) between HTP and real traits were positive and high, although their magnitudes varied according to the CNN architecture. The values ranged from 0.71 (AlexNet and LF-CNN) to 0.85 (AlexNet pretrained) for LDMY and from 0.69 (MaCNN) to 0.85 (DarkNet53) for TDMY. The highest correlations were achieved by HTP traits generated by AlexNet pretrained, DarkNet53 and ResNeXt50 pretrained for LDMY and DarkNet53, ResNext50 pretrained, and AlexNet for TDMY.

ResNeXt50 pretrained was chosen to estimate the efficiency of indirect selection using HTP secondary traits, since it showed a generally better performance for Vg, H and r for both dry matter traits. Table 7 shows that the CR/DR ratios were 0.88 and 0.86 for LDMY and TDMY, respectively, when considering the same number of plots (330), population size (86) and selection intensity (10%). This means that the CR was 88% and 86% of the efficiency of DR for LDMY and TDMY, respectively.

Simulated scenarios shown in Table 7 compare the selection response using the HTP process in larger trial and population sizes with conventional process maintaining the current trial and population size. It is important to mention that the estimated selection efficiency would only be achieved if the same experimental design were to be repeated, since the genetic parameters are considered the same regardless of the population size. If selection intensity increased to 5% by evaluating 660 plots (172 full-sib families) in the field, the CR/DR ratio would be 1.03 for LDMY and 1.00 for TDMY, indicating that CR would be 103% of the DR for LDMY and similar for TDMY. The CR might be efficient if we applied a selection intensity of 1% by evaluating 3300 plots (860 full-sib families) instead. In this latter scenario, the CR/DR ratio would be 1.34 for LDMY and 1.30 for TDMY, which means that the CR would be 134 and 130% of the DR for these DMY traits.

Table 7. Direct (DR) and correlated (CR) response to selection for dry matter traits in the guineagrass breeding program using field and data estimated by the ResNeXt50 pretrained CNN architecture. SI—selection intensity.

Responses to Selection	Trait	
	Leaf Dry Matter Yield (kg·ha^{-1})	Total Dry Matter Yield (kg·ha^{-1})
	330 plots	
DR (SI = 10%)	237	361
CR (SI = 10%)	209	309
CR/DR	0.88	0.86
	660 plots	
CR (SI = 5%)	244	361
CR/DR	1.03	1.00
	3300 plots	
CR (SI = 1%)	317	468
CR/DR	1.34	1.30

4. Discussion

The results showed that the CNNs with the best performance for the estimate of the LDMY and TDMY were the AlexNet pretrained, ResNeXt50 pretrained, and DarkNet53. Regarding the statistical-genetic analysis, it was noticed that ResNeXt50 pretrained and DarkNet53 were the best CNNs to estimate genetic parameters. The literature usually reports deeper networks with better performance than shallower networks [44,45]. For our final aim, which is related to the genetic analysis, we also verified that deeper networks provided more accurate solutions.

The results for the LDMY and TDMY traits showed a high correlation with the ground truth, in general, higher than 0.75 for the three best CNNs. This indicates that the high-throughput and the conventional phenotyping processes were highly correlated in this study. The result presented in Ma et al. [27] to estimate the above-ground biomass in winter wheat showed an R^2 correlation of 0.80, which is slightly greater than that obtained in our investigation. To compare the results with their study, we used the same network to estimate dry matter in guineagrass, but it did not achieve the same performance presented in the paper. In the research of [46], using RGB images and crop surface models to estimate dry matter in barley, they obtained an RMSE between 97 and 234 g·m^{-2}, while this experiment had an RMSE between 41.3 to 50.65 g·m^{-2}.

Regarding the correlation between predicted and real data, it was possible to verify that, to estimate dry matter yield traits, the RGB images obtained with UAV and the CNNs can have a high correlation. However, as presented in the previous research [29], the correlation was better for estimating green than dry biomass. A possible explanation is that the image presented to the neural network, which represents the plot before harvest, is more related to the green matter yield. Thus, the dry matter obtained in the laboratory process was not fully represented in the RGB images, thus making extracting characteristics through the network more laborious. Nevertheless, the performance obtained was encouraging, even when compared with other methodologies for estimating biomass.

From the statistical-genetic analysis, it was noticed that ResNeXt50 pretrained and DarkNet53 were the best CNNs to estimate genetic variance and heritability. Estimates from AlexNet pretrained failed to exploit these parameters, especially the heritability, thus showing the importance of the validation of the CNN in a genetic model. Our results showed that the HTP traits generated by these networks presented a larger Vg and higher

H when compared to the other CNN. When compared to the Real traits, it showed lower *Vg* and a similar *H*. Thus, while the accuracy of the high-throughput process was comparable to the conventional process, it failed to exploit the available genotypic variance among families. This situation is expected due to the small number of examples to train the model, mainly for data in the extreme of the normal distribution, as shown in Figure 3. Thus, to overcome this limitation, we expect that increasing the size as well as the variability of examples of the data set will increase the power of ResNeXt50 and DarkNet53 to estimate phenotypes more representative of the genetic variability.

Theoretically, selection for one trait will cause a correlated response to selection in a second trait if a genetic correlation exists between the two traits [33]. This is an important concept in applying HTP traits in indirect selecion in plant breeding. Our results indicated high genetic correlation ($r > 0.69$) between the HTP and Real traits even for CNN with lower performance for the other genetic parameters. This indicates the high potential of HTP process (UAV-RGB-CNN) used in this research to be considered in different strategies in guineagrass breeding for DMY traits. The CNN, AlexNet pretrained, ResNeXt50 pretrained and DarkNet53 stood out among the other CNN showing the highest values of r (>0.81). Thus, the results of genotypic correlations are in accordance with the results shown by the standard evaluations of the algorithms. From a selection point of view, high correlations indicate that the ranking of the family genotypic values were highly coincident between HTP and Real traits, and the best-selected families would also be highly coincident.

Indirect selection is one of the main applications of HTP in breeding programs [32]. The potential of HTP in improving the efficiency of early generation selection has been studied in sugarcane [30], and wheat [31]. In both studies, the results were encouraging to use HTP traits and indirect selection. In our results with guineagrass, considering the same selection intensity for DR and CR ($i = 10\%$), the efficiency of indirect selection (CR/DR ratio) was 88 and 86% of the direct selection for LDMY and TDMY traits, respectively. These lower indirect selection efficiencies for the DMY traits were due to the lower magnitudes of the product between the heritability of the HTP traits (h_{HTP}) and the genetic correlation between the HTP and the real traits (r) compared to the heritability of the real trait (h_{real}). Natarajan et al. [30], using indirect selection based on NDVI as HTP trait, reached 73% of the efficiency of direct selection for yield in sugarcane. Also, using NDVI, Krause et al. [31] reported a higher efficiency of indirect selection when compared to visual selection for grain yield in lines of wheat in early generations. Thus, these results in different crops show that HTP is a promising strategy to improve the efficiency of breeding programs.

Increasing the size of the breeding program to enable higher selection intensity is one of the main benefits of using HTP [34]. This important aspect of HTP stimulated us to simulate scenarios where the only parameter to be changed is the selection intensity. This is achieved by increasing the number of plots and families evaluated and maintain the heritability and genotypic variance of HTP traits unchanged. Our results showed a high increase in the indirect selection efficiencies for DMY traits were achieved when multiplying by two (100 to 103%) and by ten (130 to 134%) times the size of the population evaluated in the field by the high throughput process using ResNeXt50 pretrained. Here the CR/DR ratio increased due to the higher values of i (2.06 for $i = 5\%$ and 2.67 for $i = 1\%$) applied in larger populations evaluated by the proposed process. This means that families with superior DMY values are more likely to be selected in a larger population evaluated with the HTP process. Although these results are promising, they are not yet useful to extend to different populations or environments (regions/climates). In further researches, we will include data from a wider range of environments (seasons, locations) to investigate the generalization of the methodology to a diversity of scenarios. Investigation regarding domain adaptation will also be performed.

Resources aspects as labor, time and cost are important phenotyping issues for breeding programs, since evaluations are performed in a range of environments, seasons and populations. Thus, we compared these issues between conventional and high-throughput phenotyping processes (Figure 1) used in guineagrass breeding program. We used LDMY trait for comparison since it requires all steps of phenotyping in the field and in the laboratory. Conventional phenotyping took about eight days to obtain phenotypes: one day for field evaluations (seven workers), three and a half days for sample separation in the laboratory (three workers), three days for samples drying in the dry chamber, and a half-day for dried samples weighing and preparing the data for genetic analysis (one worker). With the HTP process, it took us about four and a half hours as following: 13 min to UAV flight covering an area of 1.5 ha, four hours for the process to generate the orthomosaic, approximately 10 min, based on the period of organizing the data to use in the algorithm and the execution of it for extracting plots, which is very fast, approximately 15 s to apply the ResNeXt50 pretrained algorithm, considering it was just trained. Therefore, we see here that the HTP process with ResNeXt50 pretrained can produce phenotypes with the same accuracy with less time (a half-day) labor and cost, when compared to the conventional process.

If the size of the program were increased by ten times, as suggested here to increase until 34% the response to selection, the conventional phenotyping process for DMY traits would be impracticable for the current resources available for the guineagrass breeding program. But it would be viable using the HTP phenotyping process, due to the advantages just presented. Resources saved by using the high-throughput phenotyping would be reallocated to the management of activities related to the increased breeding program size and the infrastructure necessary for HTP, as UAV, sensor, and computation services.

HTP based on UAV brings some technical challenges related to working conditions (e.g., sun angle, wind, clouds) that alter the image quality. Also, there are some UAV-related problems such as flight safety, flight time, and limited payload [15,47]. Additionally, the image overlap associated with the pasture features on images can cause errors during the orthomosaic generation. Forage breeding images obtained from UAV are very similar, making the matching process difficult. An overlap higher than 80% (front and side) is recommended to reduce gaps in the generated orthoimages.

5. Conclusions

The high-throughput phenotyping has been increasingly used in research to improve different species, as well as tropical forages. Thus, it is possible to conclude that remote sensing with low cost unmanned aerial vehicles embedded with high-resolution RGB sensors, together with convolutional neural networks, is a promising technique to be used to estimate dry matter yield in the guineagrass breeding program. Moreover, the ResNeXt50 with pretraining shows the best results since this network is able to estimate more accurately the genetic parameters. In future investigations, we expect to increase the data set and its variability by evaluating other experimental fields with other environmental characteristics. Finally, when performing these procedures, it is expected that CNN's robustness will be optimized and, with this, it will be applied as a tool for increase the efficiency of selection in forage breeding programs.

Author Contributions: Conceptualization, J.M.J., L.P.O., M.S. and E.M.; Data curation, G.S.d.O., H.S., L.J., S.B., C.V., R.S., C.C., E.S., L.A.d.C.J., M.S. and E.M.; Formal analysis, M.S. and E.M.; Funding acquisition, M.S.; Investigation, L.A.d.C.J.; Methodology, G.S.d.O., J.M.J., C.P., L.J., S.B., C.V., R.S., C.C. and E.M.; Project administration, M.S.; Software, G.S.d.O., C.P., H.S., L.R., E.S., W.G. and E.M.; Supervision, J.M.J., M.S. and E.M.; Writing—original draft, G.S.d.O., J.M.J., L.P.O., M.S. and E.M.; Writing—review and editing, J.M.J., H.S., L.R., L.J., S.B., C.V., R.S., C.C., L.A.d.C.J. and W.G. All authors have read and agreed to the published version of the manuscript.

Funding: This research was funded by Fundação de Apoio ao Desenvolvimento do Ensino, Ciência e Tecnologia do Estado do Mato Grasso do Sul (FUNDECT-MS) grant numbers 59/300.075/2015 and 59/300.066/2015, Empresa Brasileira de Pesquisa Agropecuária (EMBRAPA), and Associaçao para o Fomento à Pesquisa de Melhormento de Forrageiras (UNIPASTO), CNPq (433783/2018-4, and 303559/2019-5). The authors acknowledge the support of the UFMS (Federal University of Mato Grosso do Sul) and CAPES (Finance Code 001).

Institutional Review Board Statement: Not applicable.

Informed Consent Statement: Not applicable.

Data Availability Statement: Not applicable.

Conflicts of Interest: The authors declare no conflict of interest.

References

1. Perfil da Pecuária no Brasil. Beef Report. Abiec. Available online: http://abiec.com.br/publicacoes/beef-report-2020/ (accessed on 28 May 2021).
2. do Valle, C.B.; Jank, L.; Resende, R.M.S. Melhoramento de Forrgaeiras Tropicais no Brasil. *Rev. Ceres* **2009**, *56*, 460–472.
3. Jank, L.; Barrios, S.C.; do Valle, C.B.; Simeão, R.M.; Alves, G.F. The value of improved pastures to Brazilian beef production. *Crop Pasture Sci.* **2014**, *65*, 1132–1137. [CrossRef]
4. Jank, L.; Resende, R.M.S.; do Valle, C.B.; de Resende, M.D.V.; Chiari, L.; Cançado, L.J.; Simioni, C. Melhoramento genético de Panicum maximum. *Melhor. Forrageiras Trop.* **2008**, *56*, 55–87.
5. Da Silva, G.M.; da Silva, F.F.; Viana, P.T.; de Oliveira Rodrigues, E.S.; Moreira, C.N.; de Almeida Meneses, M.; de Souza Abreu Júnior, J.; de Araújo Rufino, C.; da Silva Barreto, L. Avaliação de forrageiras tropicais: Revisão. *Publicações em Medicina Veterinária e Zootecnia* **2016**, *10*, 190–196. [CrossRef]
6. Gebremedhin, A.; Badenhorst, P.E.; Wang, J.; Spangenberg, G.C.; Smith, K.F. Prospects for Measurement of Dry Matter Yield in Forage Breeding Programs Using Sensor Technologies. *Agronomy* **2019**, *9*, 65. [CrossRef]
7. Teixeira, D.H.L.; Gonçalves, F.M.A.; Nunes, J.A.R.; Souza Sobrinho, F.; Benites, F.R.G.; Dias, K.O.D.G. Visual selection of Urochloa ruziziensis genotypes for green biomass yield. *Acta Sci. Agron.* **2019**, *42*, e42444. [CrossRef]
8. Gouveia, B.T.; Barrios, S.C.L.; do Valle, C.B.G.; da Costa Machado, R.; Filho, W.K.R.B.; de Sousa Nunes, J.S.; Rodrigues, J.A. Selection strategies for increasing the yield of high nutritional value leaf mass in Urochloa hybrids. *Euphytica* **2020**, *216*, 38. [CrossRef]
9. Araus, J.L.; Cairns, J.E. Field high-throughput phenotyping: The new crop breeding frontier. *Trends Plant Sci.* **2014**, *19*, 52–61. [CrossRef]
10. Shi, Y.; Thomasson, J.A.; Murray, S.C.; Pugh, N.A.; Rooney, W.L.; Shafian, S.; Rajan, N.; Rouze, G.; Morgan, C.L.S.; Neely, H.L.; et al. Unmanned Aerial Vehicles for High-Throughput Phenotyping and Agronomic Research. *PLoS ONE* **2016**, *11*, e0159781. [CrossRef] [PubMed]
11. Watanabe, K.; Guo, W.; Arai, K.; Takanashi, H.; Kajiya-Kanegae, H.; Kobayashi, M.; Yano, K.; Tokunaga, T.; Fujiwara, T.; Tsutsumi, N.; et al. High-Throughput Phenotyping of Sorghum Plant Height Using an Unmanned Aerial Vehicle and Its Application to Genomic Prediction Modeling. *Front. Plant Sci.* **2017**, *8*, 421. [CrossRef] [PubMed]
12. Tattaris, M.; Reynolds, M.P.; Chapman, S.C. A Direct Comparison of Remote Sensing Approaches for High-Throughput Phenotyping in Plant Breeding. *Front. Plant Sci.* **2016**, *7*, 1131. [CrossRef] [PubMed]
13. Leiva, J.N.; Robbins, J.; Saraswat, D.; She, Y.; Ehsani, R.J. Evaluating remotely sensed plant count accuracy with differing unmanned aircraft system altitudes, physical canopy separations, and ground covers. *J. Appl. Remote Sens.* **2017**, *11*, 1–15. [CrossRef]
14. Pathak, C.; Chandra, S.; Maurya, G.; Rathore, A.; Sarif, M.O.; Gupta, R.D. The Effects of Land Indices on Thermal State in Surface Urban Heat Island Formation: A Case Study on Agra City in India Using Remote Sensing Data (1992–2019). *Earth Syst. Environ.* **2021**, *5*, 135–154. [CrossRef]
15. Osco, L.P.; Junior, J.M.; Ramos, A.P.M.; de Castro Jorge, L.A.; Fatholahi, S.N.; de Andrade Silva, J.; Matsubara, E.T.; Pistori, H.; Gonçalves, W.N.; Li, J. A Review on Deep Learning in UAV Remote Sensing. *arXiv* **2021**, arXiv:2101.10861.
16. LeCun, Y.; Bengio, Y.; Hinton, G. Deep Learning. *Nature* **2015**, *521*, 436–444. [CrossRef] [PubMed]
17. Lee, U.; Chang, S.; Putra, G.A.; Kim, H.; Kim, D.H. An automated, high-throughput plant phenotyping system using machine learning-based plant segmentation and image analysis. *PLoS ONE* **2018**, *13*, e0196615. [CrossRef] [PubMed]
18. Šulc, M.; Matas, J. Fine-grained recognition of plants from images. *Plant Methods* **2017**, *13*, 1–14. [CrossRef]
19. Zhang, S.; Wang, H.; Huang, W. Two-stage plant species recognition by local mean clustering and Weighted sparse representation classification. *Clust. Comput.* **2017**, *20*, 1517–1525. [CrossRef]
20. Piiroinen, R.; Heiskanen, J.; Maeda, E.; Viinikka, A.; Pellikka, P. Classification of tree species in a diverse African agroforestry landscape using imaging spectroscopy and laser scanning. *Remote Sens.* **2017**, *9*, 875. [CrossRef]
21. Mochida, K.; Koda, S.; Inoue, K.; Hirayama, T.; Tanaka, S.; Nishii, R.; Melgani, F. Computer vision-based phenotyping for improvement of plant productivity: A machine learning perspective. *GigaScience* **2019**, *8*, giy153. [CrossRef]
22. Jiang, Y.; Li, C. Convolutional Neural Networks for Image-Based High-Throughput Plant Phenotyping: A Review. *Plant Phenomics* **2020**, *2020*, 4152816. [CrossRef]
23. Kattenborn, T.; Leitloff, J.; Schiefer, F.; Hinz, S. Review on Convolutional Neural Networks (CNN) in vegetation remote sensing. *ISPRS J. Photogramm. Remote Sens.* **2021**, *173*, 24–49. [CrossRef]
24. Lu, Y.; Yi, S.; Zeng, N.; Liu, Y.; Zhang, Y. Identification of Rice Diseases using Deep Convolutional Neural Networks. *Neurocomputing* **2017**, *267*, 378–384. [CrossRef]

25. Krizhevsky, A.; Sutskever, I.; Hinton, G.E. Imagenet classification with deep convolutional neural networks. In *Advances in Neural Information Processing Systems*; Curran Associates Inc.: New York, NY, USA, 2012; pp. 1097–1105.

26. Lecun, Y.; Bottou, L.; Bengio, Y.; Haffner, P. Gradient-based learning applied to document recognition. *Proc. IEEE* **1998**, *86*, 2278–2324. [CrossRef]

27. Ma, J.; Li, Y.; Chen, Y.; Du, K.; Zheng, F.; Zhang, L.; Sun, Z. Estimating above ground biomass of winter wheat at early growth stages using digital images and deep convolutional neural network. *Eur. J. Agron.* **2019**, *103*, 117–129. [CrossRef]

28. Simonyan, K.; Zisserman, A. Very Deep Convolutional Networks for Large-Scale Image Recognition. *arXiv* **2014**, arXiv:1409.1556.

29. Castro, W.; Marcato Junior, J.; Polidoro, C.; Osco, L.; Gonçalves, W.; Rodrigues, L.; Santos, M.; Jank, L.; Barrios, S.; Valle, C.; et al. Deep Learning Applied to Phenotyping of Biomass in Forages with UAV-Based RGB Imagery. *Sensors* **2020**, *20*, 4802. [CrossRef]

30. Natarajan, S.; Basnayake, J.; Wei, X.; Lakshmanan, P. High-Throughput Phenotyping of Indirect Traits for Early-Stage Selection in Sugarcane Breeding. *Remote Sens.* **2019**, *11*, 2952. [CrossRef]

31. Krause, M.R.; Mondal, S.; Crossa, J.; Singh, R.P.; Pinto, F.; Haghighattalab, A.; Shrestha, S.; Rutkoski, J.; Gore, M.A.; Sorrells, M.E.; et al. Aerial high-throughput phenotyping enables indirect selection for grain yield at the early generation, seed-limited stages in breeding programs. *Crop Sci.* **2020**, *60*, 3096–3114. [CrossRef]

32. Morota, G.; Jarquin, D.; Campbell, M.T.; Iwata, H. Statistical methods for the quantitative genetic analysis of high-throughput phenotyping data. *arXiv* **2019**, arXiv:1904.12341.

33. Bernardo, R. *Breeding for Quantitative Traits in Plants*, 3rd ed.; Stemma Press: Woodbury, MN, USA, 2020.

34. Araus, J.L.; Kefauver, S.C.; Zaman-Allah, M.; Olsen, M.S.; Cairns, J.E. Translating High-Throughput Phenotyping into Genetic Gain. *Trends Plant Sci.* **2018**, *23*, 451–466. [CrossRef]

35. Barbosa, A.; Trevisan, R.; Hovakimyan, N.; Martin, N.F. Modeling yield response to crop management using convolutional neural networks. *Comput. Electron. Agric.* **2020**, *170*, 105197. [CrossRef]

36. Xie, S.; Girshick, R.; Dollár, P.; Tu, Z.; He, K. Aggregated Residual Transformations for Deep Neural Networks. In Proceedings of the IEEE Conference on Computer Vision and Pattern Recognition (CVPR), Honolulu, HI, USA, 21–26 June 2017.

37. Redmon, J.; Farhadi, A. Yolov3: An incremental improvement. *arXiv* **2018**, arXiv:1804.02767.

38. Wong, T.T. Performance evaluation of classification algorithms by k-fold and leave-one-out cross validation. *Pattern Recognit.* **2015**, *48*, 2839–2846. [CrossRef]

39. Kingma, D.P.; Ba, J. Adam: A method for stochastic optimization. *arXiv* **2014**, arXiv:1412.6980.

40. Thorndike, R.L. Who belongs in the family. In *Psychometrika*; Citeseer: Princeton, NJ, USA, 1953.

41. Gilmour, A.R.; Cullis, B.R.; Verbyla, A.P. Accounting for Natural and Extraneous Variation in the Analysis of Field Experiments. *J. Agric. Biol. Environ. Stat.* **1997**, *2*, 269–293. [CrossRef]

42. Butler, D.G.; Cullis, B.R.; Gilmour, A.R.; Gogel, B.J.; Thompson, R. *ASReml-R Reference Manual Version 4*; VSN International Ltd.: Hemel Hempstead, UK, 2017.

43. Cullis, B.; Smith, A.; Coombes, N. On the design of early generation variety trials with corrected data. *J. Agric. Biol. Environ. Stat.* **2006**, *11*, 381–393. [CrossRef]

44. Osco, L.P.; Ramos, A.P.M.; Pinheiro, M.M.F.; Moriya, É.A.S.; Imai, N.N.; Estrabis, N.; Ianczyk, F.; de Araújo, F.F.; Liesenberg, V.; de Castro Jorge, L.A.; et al. A machine learning framework to predict nutrient content in valencia-orange leaf hyperspectral measurements. *Remote Sens.* **2020**, *12*, 906. [CrossRef]

45. Osco, L.P.; dos Santos de Arruda, M.; Gonçalves, D.N.; Dias, A.; Batistoti, J.; de Souza, M.; Gomes, F.D.G.; Ramos, A.P.M.; de Castro Jorge, L.A.; Liesenberg, V.; et al. A CNN Approach to Simultaneously Count Plants and Detect Plantation-Rows from UAV Imagery. *arXiv* **2021**, arXiv:2012.15827.

46. Brocks, S.; Bareth, G. Estimating Barley Biomass with Crop Surface Models from Oblique RGB Imagery. *Remote Sens.* **2018**, *10*, 268. [CrossRef]

47. Xie, C.; Yang, C. A review on plant high-throughput phenotyping traits using UAV-based sensors. *Comput. Electron. Agric.* **2020**, *178*, 105731. [CrossRef]

Article

Evaluation of Random Forests (RF) for Regional and Local-Scale Wheat Yield Prediction in Southeast Australia

Alexis Pang [1,*], Melissa W L Chang [1,2] and Yang Chen [1,3]

1   School of Agriculture and Food, Faculty of Veterinary and Agricultural Sciences, The University of Melbourne, Parkville 3010, Australia; Melissa_CHANG@sfa.gov.sg (M.W.L.C.); y.chen@csiro.au (Y.C.)
2   Singapore Food Agency, JEM Office Tower, 52 Jurong Gateway Road, #14-01, Singapore 608550, Singapore
3   CSIRO Data61, Goods Shed North, 34 Village St., Docklands 3008, Australia
*   Correspondence: alexis.pang@unimelb.edu.au

Abstract: Wheat accounts for more than 50% of Australia's total grain production. The capability to generate accurate in-season yield predictions is important across all components of the agricultural value chain. The literature on wheat yield prediction has motivated the need for more novel works evaluating machine learning techniques such as random forests (RF) at multiple scales. This research applied a Random Forest Regression (RFR) technique to build regional and local-scale yield prediction models at the pixel level for three southeast Australian wheat-growing paddocks, each located in Victoria (VIC), New South Wales (NSW) and South Australia (SA) using 2018 yield maps from data supplied by collaborating farmers. Time-series Normalized Difference Vegetation Index (NDVI) data derived from Planet's high spatio-temporal resolution imagery, meteorological variables and yield data were used to train, test and validate the models at pixel level using Python libraries for (a) regional-scale three-paddock composite and (b) individual paddocks. The composite region-wide RF model prediction for the three paddocks performed well ($R^2 = 0.86$, $RMSE = 0.18\,\mathrm{t\,ha^{-1}}$). RF models for individual paddocks in VIC ($R^2 = 0.89$, $RMSE = 0.15\,\mathrm{t\,ha^{-1}}$) and NSW ($R^2 = 0.87$, $RMSE = 0.07\,\mathrm{t\,ha^{-1}}$) performed well, but moderate performance was seen for SA ($R^2 = 0.45$, $RMSE = 0.25\,\mathrm{t\,ha^{-1}}$). Generally, high values were underpredicted and low values overpredicted. This study demonstrated the feasibility of applying RF modeling on satellite imagery and yielded 'big data' for regional as well as local-scale yield prediction.

Keywords: wheat; yield prediction; random forests; satellite imagery; Normalized Difference Vegetation Index (NDVI)

Citation: Pang, A.; Chang, M.W.L.; Chen, Y. Evaluation of Random Forests (RF) for Regional and Local-Scale Wheat Yield Prediction in Southeast Australia. *Sensors* **2022**, *22*, 717. https://doi.org/10.3390/s22030717

Academic Editors: Jiyul Chang and Sigfredo Fuentes

Received: 10 December 2021
Accepted: 14 January 2022
Published: 18 January 2022

Publisher's Note: MDPI stays neutral with regard to jurisdictional claims in published maps and institutional affiliations.

1. Introduction

Wheat is a key component of the Australian grain industry. Regional and national-scale wheat yield forecasting and prediction provide essential information to all parts of the value chain from farm production, aggregation, processing, distribution and through to the commodity markets, as well as governmental agricultural and economic departments. At the farm scale, this is the ability to monitor and predict crop health and, by extension, yields, in a spatially-variable manner within a farm paddock using NDVI facilitates precision variable-rate nitrogen application to achieve high production efficiencies and profitability [1]. The mainland southeast Australian wheat belt accounts for 53% of all wheat production regions [2], but is particularly vulnerable to significant volatility in yields due to climactic variability [3,4]. Therefore, this is a region that would benefit greatly from accurate yield prediction. Comprehensive and up-to-date reviews of crop yield prediction methods have been reported by [5,6].

High and ultra-high-resolution imagery using aerial platforms such as UAVs and manned aircraft can now provide high-precision quantitative information for crop monitoring of crop health and stresses at the sub-meter scale [7]. However, these techniques tend to be beyond the capabilities of normal producers or regional assessors and can also be limited by the spatial coverage and revisitation frequency (cadence) meaning that satellite-based data remain a critical component of regional and local-scale yield predictions. Cloud cover is a persistent problem [8] but this can be largely addressed with high-cadence imagery. Planet's (www.planet.com; last accessed 17 July 2019) [9] constellation of Dove satellites offers an unprecedented observing potential of daily land surface imagery increasing the chances of acquiring cloud-free images for analysis, with an orthorectified spatial resolution of 3 m, enabling the detection of reflectance variations over very small distances and matching them with yield data [10]. This allows investigation of within-field yield variation which aids farmers in precision agriculture decisions. While somewhat limited in spectral resolution and range, PlanetScope imagery can bridge the spatio-temporal and spectral characteristics of MODIS (36 bands; 250 to 1000 m spatial resolution; daily revisit), Landsat 8 (9 bands; 30 m spatial resolution with 15 m for Band 8; 16-day revisit) and Sentinel 2 MSI (10 to 60 m spatial resolution; 5-day effective revisit) platforms that have recent multisensory data fusion strategies [11–13].

Machine Learning (ML)-driven approaches show much potential for the retrieval of key parameters such as biomass and soil moisture from satellite imagery [14]. While much previous work has focused on using Artificial Neural Networks (ANNs); the potential of random forests (RF) [15], being quicker and requiring fewer training dataset volumes, have yet to be comprehensively evaluated [14], particularly for dynamic, in-season wheat yield prediction at multiple scales. RF is a supervised ML algorithm based on decision-tree procedures to predict output classes based on patterns learnt in the training datasets. These involve building tree ensembles whose growth are controlled by randomized selection of (input-output) vectors from the training dataset; which are then assembled as classification or regression models to predict the most likely output class (or values) from the inputs of the test dataset with good accuracy and robustness to outliers with lower likelihood of generalization errors [15]. RF have the potential to generate better models compared to single decision-tree models [16], are more efficient computationally and therefore suitable for regional and global applications in agriculture [17] where Big Data dominates [18,19]. For instance, RF-driven yield prediction for sugar cane in Australia has been found to be more accurate and reliable than traditional approaches such as multiple linear regression [20,21]. For wheat yield prediction, methods ranging from a traditional crop-weather analysis model relating crop yield to stress (water, temperature) indices [22], to computationally-driven crop model simulation tools such as DSSAT and APSIM [23–25] have been used to varying degrees of success but require substantial calibration to reduce uncertainty. Recently, Feng et al. [26] adopted a hybrid approach combining a biophysical model and RF to improve dynamic yield forecasts for 29 sites across the New South Wales wheat belt and achieved good yield forecasting results (r = 0.87, *RMSE* = 0.64 t ha^{-1}) based on the end of milk development stage. However, this study used NDVI derived from MODIS/MOD09GA surface reflectance composites at 500 m spatial resolution, precluding the assessment of intra-paddock variability.

Recent examples of RF-driven yield prediction include evaluating the effective use of RF at the global and country (USA) scale using wheat, maize and potato yield, climate, soil and fertilizer management datasets [27]; wheat biomass estimation in Jiangsu province of southern China using experimental plots and vegetation indices (VIs) from 30 m resolution multispectral imagery from HJ-1A/B satellites [28]; broad-scale wheat yield prediction over nine agricultural divisions in north China using Terra MODIS MOD13Q1 data, where RF was found to be one of the top best-performing ML algorithms [29]. These studies demonstrated the higher performance, robustness and accuracy of RF compared to statistical models, artificial neural networks (ANNs) and support vector regressions (SVRs). Furthermore, work on the use of ML techniques for within-farm wheat yield forecasting has been found to be still in their early stages [30,31] and therefore can provide novel and accurate information to aid farmers' precision agriculture decision-making such as variable-rate nitrogen or phosphorus application for improved production efficiency and sustainability [32–34] as well as downstream stakeholders in the grain industry.

The main objective of this research was to evaluate the integration of ML (RFR) algorithms, high-resolution satellite imagery with multiple field and weather data to develop advanced, data-driven yet generalizable models for wheat yield prediction for wheat-growing paddocks in different parts of southeast Australia. This would therefore develop a foundation for developing region-centric algorithms for national-scale yield prediction. A key enabling objective was to build a parsimonious model (i.e., having a maximum predictive power using a minimum number of parameters) to predict yield in-season prior to, and up to harvest at various phenological stages while minimizing costs and complexity, and maximizing applicability to potential users (e.g., growers and agronomists).

2. Materials and Methods

The project process workflow is summarized in Figure 1 and elaborated in the following sections.

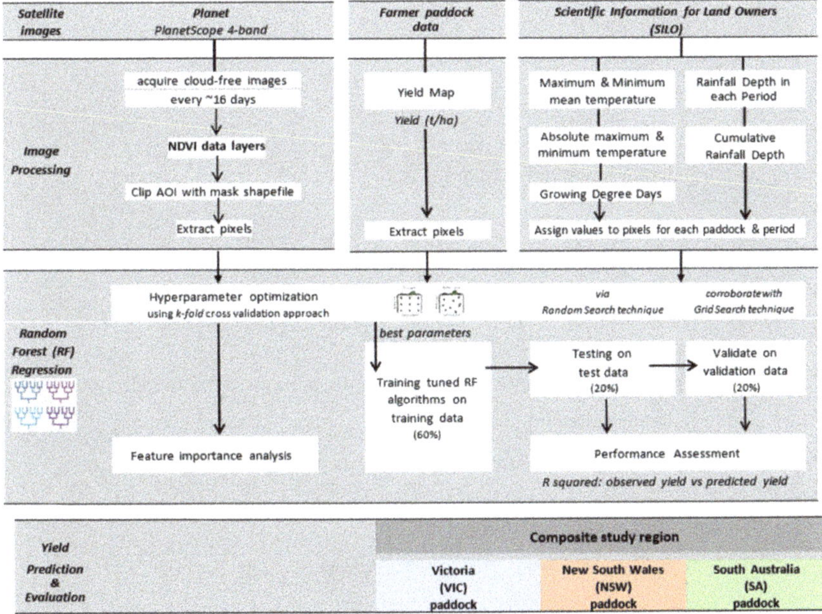

Figure 1. Summary of workflow processes and datasets used for building, testing and evaluating RF model wheat yield prediction method.

2.1. Study Region Paddocks

Spatially-distributed and referenced wheat yield values (t ha^{-1}) were the pixel-level target variable for the RF prediction model. Three paddocks in southeast Australia viz. the states of Victoria (VIC), New South Wales (NSW) and South Australia (SA), that grew wheat in 2018 (Figure 2), were selected from a pool of private yield data collected from collaborating farmers; 5 m grid resolution yield maps were generated using a semi-automated procedure involving block kriging of yield monitor data, detailed in [35]. The verified yield maps were resampled to 3 m resolution to match with the PlanetScope imagery detailed below. The paddocks varied in hydroclimatic conditions, and soil characteristics and the preceding 3 years' cropping/fallow sequences were likely to have affected fertility, water availability and crop residue cover leading into the 2018 season [36]. Different wheat varieties were also grown, adding another layer of complexity with which to test the robustness of the present technique. For instance, Kord is a mid-maturing variety that is robust to drought stresses, though not necessarily with the highest potential yields. Lancer is a mid to late-maturing variety suitable for early sowing with good resistance to lodging. Scepter is an early-mid season maturing type that has moderate resistance to lodging and one of the highest average yields of up to 3.0 t ha^{-1} in the SA wheat National Variety Trials (NVTs). These yield maps were used as training, testing and validation datasets for the RF model development [37].

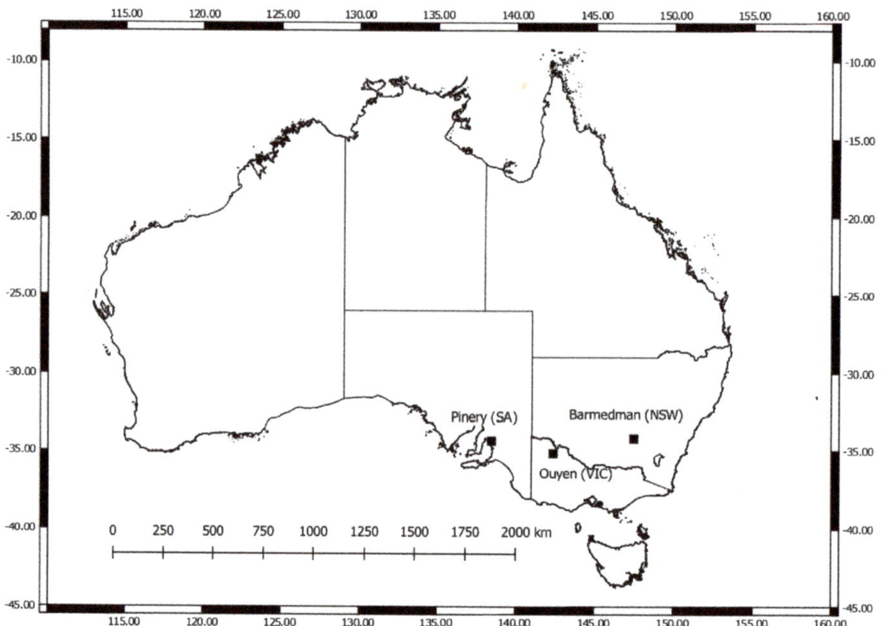

Figure 2. Location of study paddocks in southeast Australia, covering the states of Victoria (VIC), New South Wales (NSW) and South Australia (SA).

According to study [38], 2018 was a particularly difficult growing season for southeast Australia cropping with the region experiencing rainfall in decile one range and temperatures in decile ten.

2.2. Planet^{TM} Satellite Imagery

NDVI data was used as one of the predictor variables (features); 16-day Periods spanning sowing to harvest dates for all three paddocks were created to constrain the temporal variability of the wide range of data and imagery, and also enable foreseen later work to compare with LANDSAT-based studies and imagery [39,40] (Table 1). In total, 41 PlanetScope Analytic Ortho Scene (Level 3B), cloud-free BGRN imagery (VIC: 13, NSW: 15, SA: 13) for the target paddocks were selected from available datasets, spanning the southeast Australia winter wheat-growing season, ~April to December 2018, from sowing to harvest. Ground Sample Distance (GSD) was 3.7 m and pixel dimensions were 3 m × 3 m. This spatial resolution was relevant to practical precision agronomic management by farmers (e.g., variable-rate fertilization), and harvesting header swath width varying between approx. 5 to 12 m. Normalized Difference Vegetation Index (NDVI) layers were generated for each scene using the Red (R) and Near Infra-Red (NIR) bands following [41]; see also [42,43] in QGIS 3.4 [44], before cropping to paddock boundaries.

Table 1. Location, cropping, climate and soil characteristics of study paddocks.

Location	2018 Wheat Crop Information	Paddock Area & Cropping Sequence 2015–2016–2017	Climate	Soil Description
Ouyen, VIC 142.37 E 35.12 S	Variety: *Kord* Sowing: 15 May Harvest: 30 Nov Growing days: 199 Mean yield: 1.53 t ha^{-1}	181.2 ha Barley–Wheat–*Fallow*	Mean Max Temp: 23.8 °C Mean Min Temp: 9.8 °C Mean Annual Rainfall: 331.2 mm	Calcarosol (dune systems with series of alkaline sandy/loamy duplex, and sandy clay soils).
Barmedman, NSW 147.46 E 34.15 S	Variety: *Lancer* Sowing: 4 April Harvest: 27 Jan Growing days: 298 Mean yield: 1.06 t ha^{-1}	67.6 ha Canola–Wheat–Canola	Mean Max Temp: 24.0 °C Mean Min Temp: 9.9 °C Mean Annual Rainfall: 470.9 mm	Brown Vertosol (heavy clay soil, alkaline with strongly sodic subsoil).
Pinery, SA 138.46 E 34.32 S	Variety: *Scepter* Sowing: 9 May Harvest: 11 Dec Growing days: 216 Mean yield: 1.95 t ha^{-1}	120.1 ha Wheat–Wheat–Lentils	Mean Max Temp: 23.6 °C Mean Min Temp: 9.7 °C Mean Annual Rainfall: 408.9 mm	Calcarosol (alkaline silty clay loam to medium-heavy clay) variable soil profiles on dune systems.

In total, there were 377,475 pixels (3 m resolution; total area: 400 ha) across the VIC (188,865 pixels; 170 ha), NSW (67,830 pixels; 61 ha) and SA (120,780 pixels; 109 ha) paddocks. Areas covered by pixels analyzed were lower than actual paddock areas (Table 1) because the data were cropped internally from paddock boundaries to mitigate edge effects.

The main dataset comprising all three paddocks was split into individual paddock datasets, giving two levels: regional-scale (three-paddock composite) and local-scale (individual paddock). All datasets were randomly divided into 60% training, 20% testing and 20% validation.

2.3. Weather Data

Location-specific daily weather data were compiled for each paddock from 5 km grid resolution values interpolated from local and regional networks of the Bureau of Meteorology and affiliated contractors' weather station measurements, extracted from the Scientific Information for Land Owners (SILO) database (http://www.longpaddock.qld.gov.au/silo, last accessed 20 June 2019) [45], and assembled into the individual Periods (Table 2). For each Period, mean maximum and minimum, absolute maximum and minimum temperatures were prepared as predictor variables (features) that would help indicate heat or frost occurrence that could impact yield negatively; particularly pertinent at critical growth stages such as anthesis [46]. Growing degree days (GDD) corresponding to the imagery dates were also calculated and included as predictor variable [47]. Two rainfall datasets were prepared: rainfall depth (mm) in the preceding Period and cumulative rainfall depth (mm) since sowing date. Because of the coarse spatial resolution of the weather data, they were applied uniformly at the paddock scale for each Period by assigning the same value for all individual pixels within each paddock.

Table 2. PlanetScope imagery fortnightly Periods, dates and corresponding Days After Sowing (DAS) in year 2018 for each location in the states of Victoria (VIC), New South Wales (NSW) and South Australia (SA), Australia.

Location	Ouyen, VIC		Barmedman, NSW		Pinery, SA	
Period	2018 Date	DAS	2018 Date	DAS	2018 Date	DAS
1	-	-	19 April	15	-	-
2	-	-	30 April	26	-	-
3	25 May	10	14 May	40	16 May	7
4	31 May	16	29 May	55	31 May	22
5	14 June	30	22 June	79	13 June	35
6	30 June	46	30 June	87	29 June	51
7	14 July	60	14 July	101	14 July	66
8	29 July	75	12 August	130	29 July	81
9	13 August	90	27 August	145	26 August	109
10	7 September	115	4 September	153	4 September	118
11	20 September	128	21 September	170	17 September	131
12	4 October	142	30 September	179	1 October	145
13	19 October	157	18 October	197	19 October	163
14	4 November	173	11 November	221	2 November	177
15	18 November	187	26 November	236	17 November	192
16	-	-	12 December	252	-	-

2.4. RF Model Development

Pandas software library functions for Python [48] were used for data preparation, manipulation and analysis. Time-series NDVI and weather data were used together as predictor variables. The NDVI data layers were parsed into CSV format with each cell value representing an individual pixel value. Weather variables were assembled as individual pixel values homogenous for each Period. Yield data (t ha^{-1}) for individual spatially-referenced pixels were used as the target values for the prediction algorithms. All input and target values were indexed to retain their individual geographic locations to enable their reassembly for examination of their spatial distributions.

The RF approach is an ensemble learning technique that makes predictions by combining decisions from a sequence of base models, with individual base models known as trees [49]. Hyper-parameters (e.g., weather and NDVIs) are tuned using the best cross-validation (CV) results. Random Forest Regression (RFR) was performed using the Scikit-learn machine learning module for Python [50]. Each tree in the RFR was built by using randomly selected variable sets from the training dataset with the final prediction for the testing datasets derived by averaging the tree outputs. Cross-validation was conducted to check the accuracy of the model on the independent validation dataset [51].

Calibration of each RFR model was done by hyperparameter tuning to obtain the optimal combination of: (i) number of trees in ensemble (*n_estimators*); (ii) maximum number of levels in each decision tree; (iii) maximum number of features considered for splitting a node, and (iv) method for sampling data points (with or without replacement). Random Grid Search was to incorporate a wide range of possible values and hyperparameter combinations in an unbiased manner, with superior computation times [52], an important consideration for mining large volumes of agricultural data. Twenty iterations of five-fold cross validation, with different model settings each time, were performed to facilitate model optimization and generalizability, while avoiding overfitting on the test dataset [50,53].

2.5. Feature Importance Analysis

Identifying and ranking the importance of individual features used in the RFR models we built, was conducted via Scikit-learn toolkit RF feature importance function, in order to understand the underlying dynamics contributing to model accuracy in yield prediction and ascertain their generalizability and meaningfulness [15,54,55]. To improve model performance while reducing the risk of overfitting, a forward-selection process was implemented following [21,56]. The optimum parameter combination giving the highest mean validation score was selected for model training. There was a need to balance performance against computational costs, even though model accuracy would expectedly increase with number of trees. To quantify and evaluate the tradeoffs made with different hyperparameter combinations, mean validation score was compared against number of trees, with the latter changed one at a time. Grid Search was then used for the selected numbers of trees to corroborate the optimality of the tuned settings, thus giving converged parameter settings of practical value.

3. Results

3.1. Regional (Composite) Yield Prediction

The RFR model developed for predicting yield of the three paddocks combined, i.e., at the regional scale, performed well with good generalizability across the VIC, NSW and SA locations. Table 3 compares the descriptive statistics of the observed and predicted yield datasets; the independent validation dataset. They were very similar, albeit with predicted minimum yield slightly higher, and maximum yield, slightly lower than the observed yield. Performance metrics shown in Table 4 demonstrate the good accuracy of the developed model. Notably, the adjusted R^2 value and validated regression metric scores were similar, indicating good model generalization ability and absence of overfitting, performing well on unseen data.

Table 3. Descriptive statistics for regional-scale observed and RF model predicted yield.

	Observed Yield	Predicted Yield
sample size, n	75,495	75,495
minimum (t ha^{-1})	0.35	0.38
maximum (t ha^{-1})	2.79	2.67
mean (t ha^{-1})	1.60	1.60
standard deviation (t ha^{-1})	0.47	0.44

Table 4. Statistical performance of regional-scale RF yield prediction model.

Metric	Test Dataset	Validation Dataset
R Squared (R^2)	0.858	0.860
Adjusted R Squared (R^2)	0.858	0.860
Mean Absolute Error (*MAE*)	0.126	0.126
Mean Squared Error (*MSE*)	0.032	0.031
Root Mean Squared Error (*RMSE*) (t ha^{-1})	0.179	0.177

As seen in Figure 3, the datapoints were mostly closely clustered around the reference line, particularly for yield values between 0.8 to 1.3 t ha^{-1}. However, they were more dispersed between the 1.3 to 2.8 t ha^{-1} yield. While the VIC paddock (blue) yield values were broadly distributed, NSW paddock (orange) yield values tended towards the lower, and for SA paddock (green), the higher ranges.

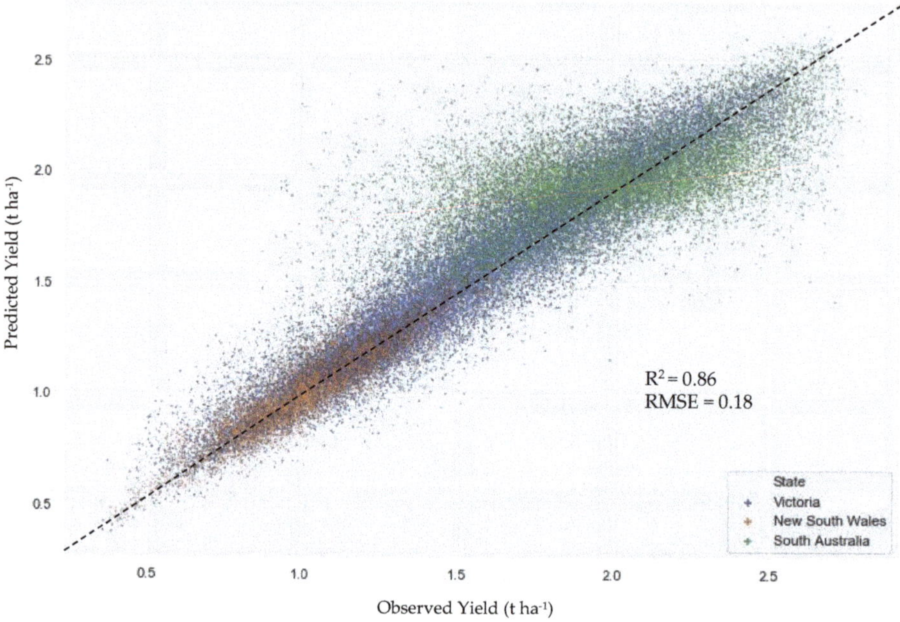

Figure 3. Scatterplot of observed and predicted yield of VIC, NSW and SA paddocks combined.

Feature importance analysis found that NDVI data acquired in late September/early October were most important to the prediction accuracy of the RF model developed for the 3-paddock composite (Figure 4; Table 2). This corresponded to 142, 179 and 145 DAS for VIC, NSW and SA paddocks, respectively. If the NDVI data for Period 12 (P12) were excluded as input to the model, a mean decrease in prediction accuracy of 53% occurred. In contrast, excluding NDVI data from later or earlier time Periods led to only 2% to 6% mean decrease in prediction accuracy. Notably, only NDVI images from P5 to P14 featured in the top 10 most important features.

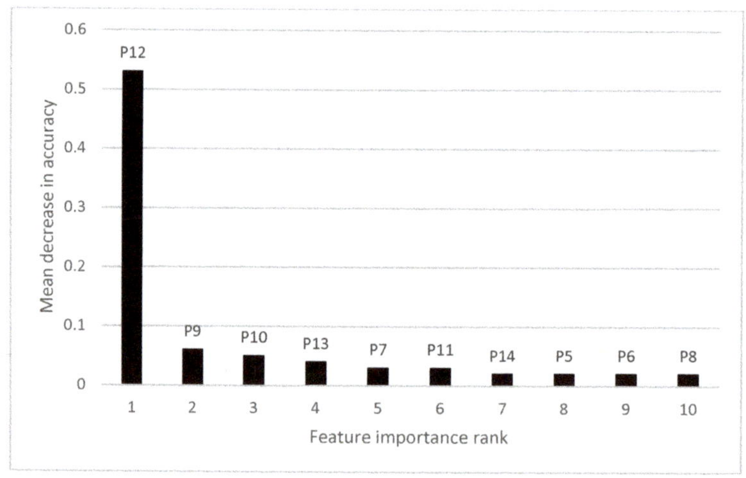

Figure 4. Top 10 features of importance for regional RF yield prediction model. Note: Data labels e.g., P12 refer to NDVI in Periods described in Table 2.

We also found low feature importance of weather (temperature and rainfall) datasets, being ranked outside of the top 10; this also applied to the individual paddock RF prediction models discussed below.

3.2. Individual Paddock Yield Prediction Models

For all three paddocks, predicted mean yields were very close to the observed mean yield with less than 1% difference (Table 5). Standard deviation values showed that RF model predictions resulted in lower variations around the mean compared to observed yield, with the worst performance for SA paddock and best performance for NSW paddock. This was also shown in the overprediction of minimum yields by up to 0.06 t ha-1 for NSW paddock, and underprediction of maximum yields by up to 0.14 t ha^{-1} for SA paddock.

Table 5. Descriptive statistics for predicted yields from individual RF models compared with observed yields for VIC, NSW and SA paddocks.

Yield Statistic (t ha^{-1})	VIC (n = 37,773)		NSW (n = 13,566)		SA (n = 24,156)	
	Observed	Predicted	Observed	Predicted	Observed	Predicted
mean	1.55	1.56	1.08	1.08	1.95	1.94
standard deviation	0.44	0.41	0.20	0.19	0.33	0.22
minimum	0.36	0.38	0.34	0.40	0.91	0.96
maximum	2.72	2.66	1.67	1.59	2.80	2.66

The individual paddock RF model performance metrics are presented in Table 6. RF prediction models for VIC and NSW paddocks performed well with high R^2 values, although with only moderate performance for the SA RF prediction model with R^2 at 0.447. Nevertheless, all adjusted R^2 values indicated the absence of overfitting. *MAE*, *MSE* and *RMSE* values were generally good, with lowest values for the NSW paddock but for the SA paddock, relatively higher error values.

Table 6. Statistical performance of VIC, NSW and SA RF yield prediction models.

Metric	VIC		NSW		SA	
	Test Dataset	Validation Dataset	Test Dataset	Validation Dataset	Test Dataset	Validation Dataset
R^2	0.890	0.887	0.870	0.878	0.447	0.443
Adjusted R^2	0.890	0.887	0.869	0.877	0.445	0.441
Mean Absolute Error (*MAE*)	0.110	0.111	0.056	0.054	0.186	0.185
Mean Squared Error (*MSE*)	0.021	0.022	0.005	0.005	0.061	0.060
Root Mean Squared Error (*RMSE*) (t ha^{-1})	0.146	0.147	0.073	0.071	0.246	0.246

Figure 5a–c compare the RF predicted and observed yield for VIC, NSW and SA paddocks, respectively. There was a close clustering of data around the reference line for VIC paddock for yield values between 1.0 to 1.3 t ha^{-1}, while this was seen for the NSW paddock between 0.8 to 1.3 t ha^{-1}. SA paddock displayed quite widely-dispersed values around the reference line with clear underprediction 2.0 t ha^{-1} and overprediction below it.

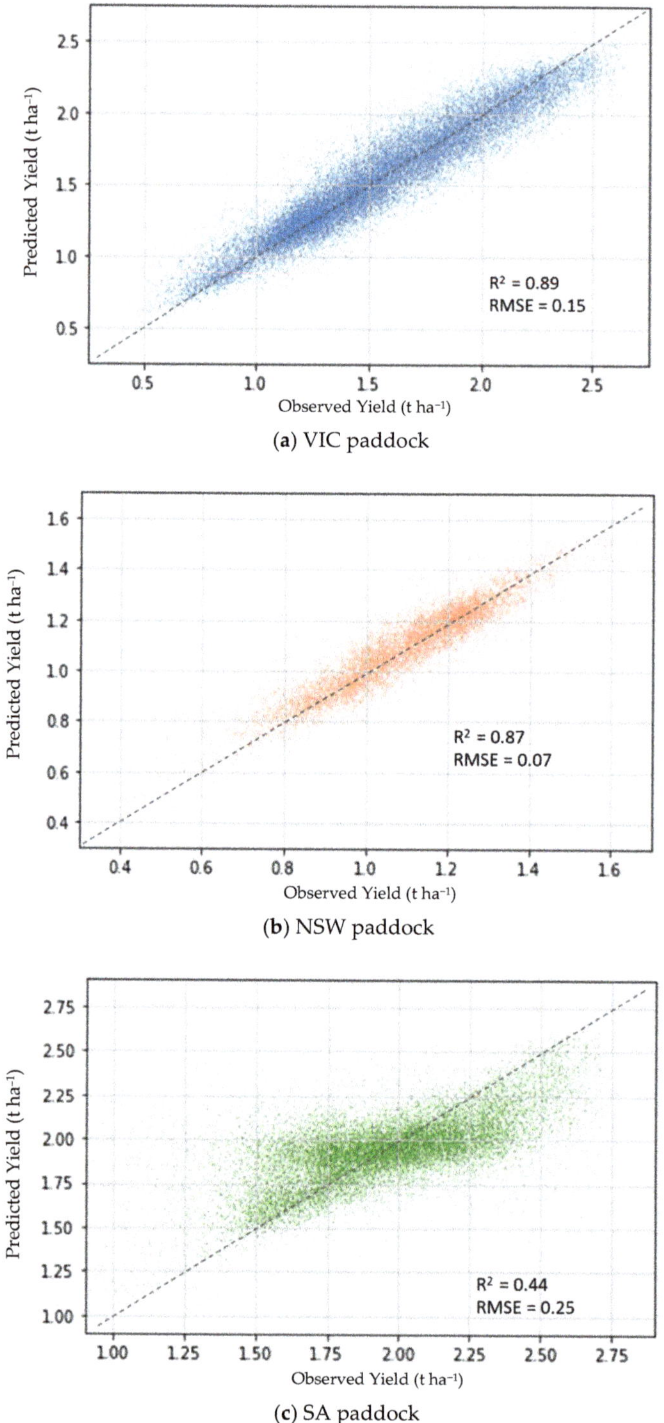

(**a**) VIC paddock

(**b**) NSW paddock

(**c**) SA paddock

Figure 5. Comparison of predicted vs. observed yield for (**a**) VIC; (**b**) NSW and (**c**) SA paddocks.

Figure 6a–c present the yield map and histogram for VIC, NSW and SA paddocks, respectively. For the VIC paddock, we saw from the yield map, good spatial correspondence between the observed and predicted values. The histograms showed a higher number of high-yield values being predicted compared to the observed yield values, quite apparent for the yield values above 2.0 t ha^{-1}. The NSW paddock yield map also showed good spatial correspondence between observed and predicted values. The NSW yield histograms also showed good similarities in the general distribution of values, although the prediction was not able to replicate the bimodal pattern of the observed yield with peaks at 1.0 and 1.3 t ha^{-1}. The prediction gave a single high peak around the 1.25 t ha^{-1} yield value. The SA paddock yield map had comparatively poorer spatial correspondence between the observed and predicted values. The predicted yield histogram had a higher peak of average values around 1.95 t ha^{-1} compared to the observed yield histogram, which had gentler peaks around 1.75 t ha^{-1} and 2.15 t ha^{-1}. This corroborated with the lower standard deviation of 0.22 t ha^{-1} for predicted yield compared to 0.33 t ha^{-1} for observed yield in Table 5.

3.3. Feature Importance Analysis for Individual Paddocks

Table 7 shows the mean decrease in accuracy (MDA)—the arithmetic averaged loss of prediction accuracy for all individual pixels comparing predicted output with target output values, if one of the features were excluded as predictor input for the RF model, for the top 10 most important features, and the corresponding Period (P) (Table 2) of the NDVI data.

Table 7. Top ten most important features for VIC, NSW and SA paddock RF models, and corresponding NDVI Period and mean decrease in accuracy (MDA) if excluded.

Feature Importance Rank	VIC		NSW		SA	
	NDVI Period	MDA	NDVI Period	MDA	NDVI Period	MDA
1	18	0.68	16	0.68	13	0.22
2	17	0.11	17	0.14	16	0.12
3	20	0.04	7	0.02	18	0.09
4	13	0.03	21	0.02	15	0.08
5	12	0.03	18	0.02	14	0.07
6	16	0.02	19	0.02	17	0.06
7	15	0.02	20	0.02	12	0.06
8	19	0.02	13	0.02	22	0.06
9	14	0.02	14	0.01	19	0.04
10	11	0.01	9	0.01	11	0.04

For the VIC paddock, NDVI data for Period 12 (30 September to 15 October; 138 to 153 DAS), with the imagery on 4 October (142 DAS) used for the VIC RF yield prediction model. This image contributed 68% to the prediction accuracy. The second most important NDVI map in the Period 11 (20 September, 128 DAS) contributed 11% to prediction accuracy.

For the NSW paddock, NDVI data for Period 10 (29 August to 13 September; 147 to 162 DAS), with imagery obtained on 4 September (153 DAS) used for the NSW RF yield prediction model. This image contributed 68% to prediction accuracy.

For the SA paddock, NDVI data for Period 7 (12 July to 27 July; 64 to 79 DAS) with imagery obtained on 14 July (66 DAS) used for the SA RF yield prediction model. In contrast to the results for VIC and NSW paddocks, this image contributed only 22% to prediction accuracy. The second most important NDVI map was obtained in Period 10 on 4 September (118 DAS) contributing 12% to prediction accuracy. Distribution of RF yield prediction model feature importances of NDVI data for SA paddock were hence more evenly distributed across the growing period, albeit with lower prediction accuracy.

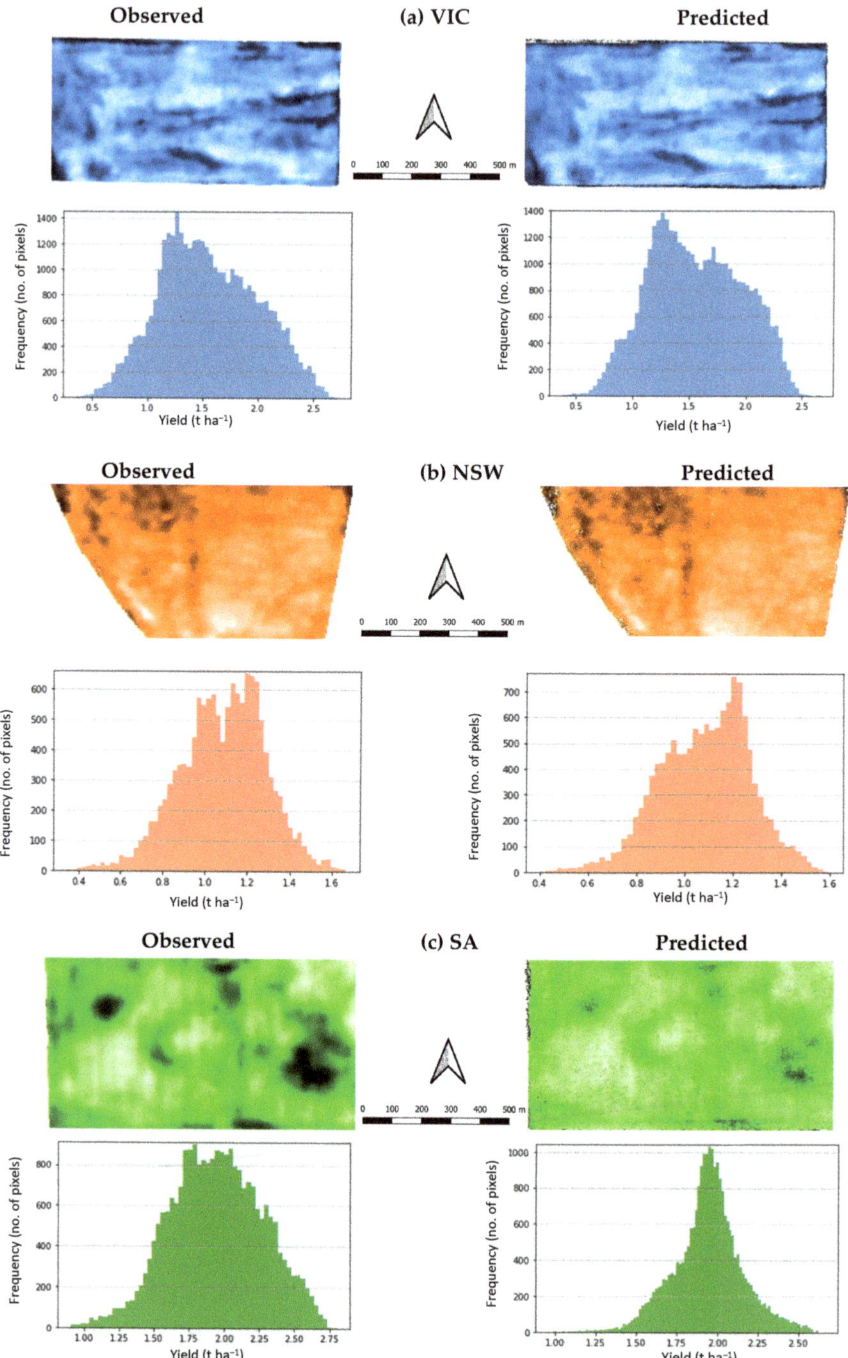

Figure 6. Yield maps and histograms for (**a**) VIC; (**b**) NSW and (**c**) SA paddocks. Notes: Yield maps—darker colors indicate higher yield values; yield histogram y-axes differ in range for NSW and SA paddocks.

4. Discussion

The regional-scale RF regression model was able to provide accurate wheat yield prediction at a high R^2 value of 0.86 and low *RMSE* of 0.18 t ha^{-1}. The results show that the model is robust at prediction across the three different paddocks with distinct conditions. Despite its limitations, NDVI continues to be a useful Vegetation Index (VI) for yield prediction, and the results from this study concurs with previous work using UAV-mounted cameras [57], LANDSAT [58] and MODIS imagery [59]. The present work further demonstrates the ability for spatially-explicit predictions by using high-resolution imagery and machine learning (RF) approach. Furthermore, the high-cadence of Planet imagery enabled the acquisition of cloud-free images of our target paddocks within a constrained time period, an important consideration for operational applications at the regional and local scales.

Interestingly, we found that the weather data were not significant features for all developed RFR yield prediction models, even though it is indubitable that these are important factors affecting crop health and growth [60], and their inclusion have improved accuracy of various yield prediction techniques [61–63]. None of the weather data layers were found in the top ten features of importance. The key explanation could be that while NDVI is able to indicate plant health, including their responses to varying weather and climatic conditions [64], high spatial resolution 3 m NDVI used in this study (and indeed other VIs), the precision with which plant growth conditions are reflected, and the fidelity with which the data can be extrapolated via RFR to reasonably accurate yield predictions, render near-term weather data unnecessary. Hence, RFR could enable parsimonious wheat yield prediction models to be built by possibly precluding the requirement for accessing and assembling large weather datasets to aid the prediction process.

While good agreement was found between predicted and observed yield, the reported differences can be attributed to several factors. Firstly, NDVI estimates live vegetative biomass [65] which has good, but not perfect correspondence with yield. This is especially so for grain crops, such as wheat, where the yield comprises grains in storage organs in contrast with pasture or forage crops. Secondly, temperature extremes such as frost damage to foliage, particularly during winter, can initiate leaf senescence and lower vegetation greenness (higher red reflectance, lower NDVI value) but lag in time for these to manifest (i.e., in later images normally of lower importance). The yield impacts of frost, particularly during critical periods during the reproductive and grain development phases, strongly determine wheat grain number and size [46]. Thirdly, index value saturation, and obscuring of the biomass beneath the closed canopy can lead to high uncertainty in biomass estimates [66,67] and, consequently, wide variation in accuracy of yield predictions.

Model calibration required few tunable model parameters, similar to how Houborg and McCabe [42] found good accuracy by simply optimizing the number of trees (*n_estimators*). This study concurs with other studies across different crops including wheat (biomass) [28], sugarcane [21] mango [68] and corn [69]. Thus, RFR has been found to be a suitable and parsimonious technique for regional-scale wheat yield prediction.

Examination of RFR yield predictions for individual paddocks found good accuracy for VIC and NSW paddocks. The most important NDVI data for these two paddocks correspond well to the start of anthesis where peak biomass (and NDVI values) are likely to translate predictably to grain yield [70], barring any unpredictable perturbations in the intervening time to harvest, such as temperature stresses (heat/frost) or strong winds causing lodging. These demonstrate the viability of RFR for aspatial paddock-level prediction of mean yields, as well as the good accuracy of spatially-explicit pixel-level yield predictions in the given conditions. However, SA paddock RFR prediction model outputs lowered the overall regional prediction accuracy, and had only moderate accuracy at the individual paddock level illustrated by substantial statistical and spatial differences between the predicted and observed yields. The SA RF model feature importances' lower values, more even distribution, and higher importance of earlier Periods (Table 7) indicate that some unpredicted factors were not comprehensively accounted for, when compared to the VIC and NSW RF model prediction accuracies and feature importance analyses. Such inter-regional variability could be mainly attributed to the inter-paddock differences in wheat variety, sowing density, soil, topography, local weather conditions and farmer management practices, not all of which can be pragmatically quantified. The earlier time of most important NDVI map in P7 was likely to have coincided with tillering stage [70], but was probably confounded by paddock-level variabilities and perturbations affecting crop health and yield later in the season.

We noted that only SA paddock had legume residue from the previous crop, which has been shown to supply additional N to wheat crop via fixed soil N as well as organic N that mineralized as it decomposed [71] and therefore enhanced yield outcomes. However, there would likely to have been substantial spatial heterogeneity in these decomposition processes [72]. Furthermore, Scepter was one of the highest yielding and drought tolerant varieties (trial mean yield: 3 t ha^{-1}) during the 2018 NVT [73]. Hence, this could have led to ample overall biomass growth, canopy closure and NDVI value saturation which could have obfuscated the predicted yield, leading to the overprediction of lower- performing areas and underprediction of high-performing areas. This may not have happened on the VIC and NSW paddocks which grew comparatively lower-yielding and less drought-tolerant varieties. This uncertainty would have been compounded by the variable topography and soil characteristics for the SA paddock. For instance, we observed high-frequency microtopographical variations over SA paddock compared to the more regular undulating terrain for VIC and flat terrain for NSW paddocks. For SA paddock, the high intra-paddock soil variability, and corresponding soil moisture and fertility variations, could have contributed to substantial uncertainties in yield outcomes. This corresponds to how study [74] found close relationships between yield and mean surface curvature due to correlations with soil productivity (e.g., moisture).

Furthermore, haying-off [75], leading to reduced yields due to post-anthesis drought and heat stress despite vigorous growth through the season (detected as high NDVI values) aided by ample N supply, can be quite unpredictable at both the regional and paddock-level scale. This could have contributed to the overprediction of yield, particularly in SA paddock. For instance, SA paddock recorded maximum daily temperatures above 35 °C for three consecutive Periods prior to harvest, compared to 1 each for VIC and NSW. These numerically small occurrences may not have been adequately accounted for amongst all the other feature datasets used in the RFR. Altogether, the high spatio-temporal variability in crop phenology throughout the season with drought and heat stresses led to only moderate prediction accuracy for the SA paddock at Pinery. These results highlight nuances in crop phenology, and their variable presentation via satellite imagery and NDVI that are difficult to capture even using RFR.

The results for the composite dataset, and exemplified by the SA paddock results, show that poor prediction accuracies occurred at the lower and higher ends of the yield values. Similar outcomes were also found by [27] who found that while overall accuracy of RF yield predictions were excellent, poor accuracy was found at extreme values or for values that were outside the range of the training dataset. Nevertheless, similar to study [59], the developed RF yield prediction models were able to predict yields up to two months before harvest, a timeline that is useful for farmers and other wheat crop stakeholders further along the value chain.

The results of the present study for wheat yield prediction (regional RF model: $R^2 = 0.86$, $RMSE = 0.18$ t ha^{-1}, n = 75,495) compare favorably with similar studies such as [28], who reported for wheat biomass prediction using RF and HJ-1/2 30 m satellite imagery, $R^2 = 0.79$, $RMSE = 1.81$ t ha^{-1} (n = 49); [76] applied RF yield prediction methods to wheat, barley and canola using MODIS 250 m derived Enhanced Vegetation Index and reported Lin's Concordance Correlation Coefficient (LCCC) of 0.89 to 0.92 at the field resolution ($RMSE = 0.36$ to 0.42 t ha^{-1}) at 10 m spatial resolution. Relatedly, study [30] evaluated convolutional neural networks (CNN) with bootstrapped regression trees (BRR), and the effects of different data quality and resolution (Landsat 8, Sentinel 2 and proximal sensing) at 5 m spatial resolution; they found optimal wheat yield predictions at LCCC = 0.63 (*RMSE* = 0.08) for three selected fields using BRR with Sentinel 2 data. The results from this study, particularly at the paddock scale for VIC and NSW paddocks at 3 m spatial resolution, demonstrate the viability of RF modeling and, more broadly, data and ML-driven techniques for wheat prediction. The spatial variations in predicted as well as observed yield are particularly helpful in the era of precision agriculture where farmers are able to make better spatial accuracy scouting or fertilizer management decisions (e.g., via management zoning) [76].

Key challenges involved in this work include the inability to evaluate various other Vegetation Indices (VIs) that could enable even higher prediction accuracies. Although the broad spectral resolution limited the range and precision of vegetation indices that could be harnessed, this study showed that good yield prediction results were possible by using RF algorithms with NDVI data. This also points to the high potential for further work using other VIs such as chlorophyll content index (CCCI) or Photochemical Reflectance Index (PRI) [77], as air and spaceborne platforms with more spectral bands become available as sensor technologies advance [78].

RF algorithms have some limitations which the present research encountered and researchers should be aware of. Dang et al. [79] highlighted that the lower performance of RFR autumn crop yield prediction compared to Support Vector Regression (SVR) and Deep Neural Network (DNN). This was attributed to its inability to make predictions beyond the range of values of the training set data, the tendency of overfitting when modeling noisy data, and discreteness of output values defined by categories (however narrowly defined), which would otherwise give continuous range of output values provided by, e.g., SVR.

This research also did not integrate data reflecting field management practices such as fertilization and pest management. Although these are important factors affecting crop health and yield [80], it is typically very difficult to obtain such information in a timely way from farmers at the individual level, as well as prepare and input them into the model. It is also likely the effects of these practices manifest in the crop performance and health for which the spatially-distributed NDVI and yield values reflect to a reasonable extent, although with some time lag. Thus, excluding management practices data is not critical to the yield prediction objectives while allowing the RF modeling process to stay as parsimonious as possible.

Beyond the present research, further work can include (i) increasing or decreasing temporal resolution of predictor variables (e.g., NDVI) to optimize modeling and data processing times and higher accuracy; (ii) evaluation of other VIs or the use of different VIs at different growth stages [74]; (iii) increased number of paddocks distributed throughout the region to increase size of training datasets and to capture greater variability for better model generalizability; (iv) evaluating RFR yield prediction models for other areas such as the western Australian wheat belt; (v) comparative evaluation of RFR with other ML algorithms such as SVR, DNN, Least Absolute Shrinkage and Selection Operator (LASSO) and Sequential Forward Selection (SFS) [81]. At the time of writing in late 2021, southeast Australia and much of the rest of the country is estimated to record harvests at least 10% above the 10-year average [77]. Application of the RF modeling method to this "good" growing season in contrast to the "difficult" season examined in this research would help to further test its robustness and viability for operational use, as well as reexamine the importance of various features such as weather parameters, and the integration of spatially-explicit soil data [28].

5. Conclusions

This study evaluated the use of RFR to perform in-season wheat yield prediction at regional and paddock-level scales in southeast Australia using (3 m) NDVI data derived from high-cadence, high-resolution (3 m) PlanetScope satellite imagery and weather data through the winter crop-growing season with actual yield data as the reference. Evaluation of the RFR models found that good yield prediction results were possible by using

NDVI data, even though the broad spectral resolution limited the range and precision of vegetation indices that could be harnessed.

With high accuracy at the regional scale and for two out of three paddocks at the paddock scale, this research shows how RFR-driven yield prediction could be successfully performed in data-rich, information-poor (lack of information on soil, topography, farmer management actions) contexts. Hence, RFR methods have much potential for regional-scale surveillance and monitoring of wheat crop that can benefit various business stakeholders, while paddock-level yield predictions can aid spatially-explicit tactical crop management, harvest and post-harvest decision-making by farmers. When fully or partially automated, the modeling outputs can be generated efficiently, accurately and communicated effectively to various stakeholders for timely decision-making. Where yields with significant departures from the mean in terms of amount (t ha^{-1}) or quality (protein, grain size), further investigations of the contributing factors (soil, pests, microclimate) can be done. Additionally, the high spatio-temporal resolution of Planet CubeSatCubeSat data exploited by RFR modeling can also be particularly relevant in smallholder farm contexts (e.g., economically less-developed countries) where plot sizes are modest compared to industrial-scale paddocks in countries such as Australia.

Author Contributions: Conceptualization, A.P. and Y.C.; Methodology, Y.C. and M.W.L.C.; formal analysis, M.W.L.C. and Y.C.; writing—initial version, M.W.L.C.; writing—manuscript draft preparation, A.P.; writing—review and editing, Y.C.; visualization, M.W.L.C. and Y.C.; supervision, A.P. and Y.C. All authors have read and agreed to the published version of the manuscript.

Funding: This research received no external funding.

Data Availability Statement: Data is currently withheld due to farmer privacy concerns.

Acknowledgments: We acknowledge the Planet Education and Research Program for access to the satellite imagery used in this research. Appreciation is extended to numerous industry participants who provided training data in the form of yield maps, accessed by the project team. We thank the three anonymous reviewers and the Editor of *Sensors* for comments which helped us improve this paper.

Conflicts of Interest: The authors declare no conflict of interest.

References

1. Nordblom, T.L.; Hutchings, T.R.; Godfrey, S.S.; Schefe, C.R. Precision variable rate nitrogen for dryland farming on waterlogging riverine plains of southeast Australia? *Agric. Syst.* **2021**, *186*, 102962. [CrossRef]
2. Kath, J.; Mushtaq, S.; Henry, R.; Adeyinka, A.A.; Stone, R.; Marcussen, T.; Kouadio, L. Spatial variability in regional scale drought index insurance viability across Australia's wheat growing regions. *Clim. Risk Manag.* **2019**, *24*, 13–29. [CrossRef]
3. Feng, P.; Liu, D.L.; Wang, B.; Waters, C.; Zhang, M.; Yu, Q. Projected changes in drought across the wheat belt of southeastern Australia using a downscaled climate ensemble. *Int. J. Climatol.* **2019**, *39*, 1041–1053. [CrossRef]
4. Ray, D.K.; Gerber, J.S.; MacDonald, G.K.; West, P.C. Climate variation explains a third of global crop yield variability. *Nat. Commun.* **2015**, *6*, 5989. [CrossRef]
5. Van Klompenburg, T.; Kassahun, A.; Catal, C. Crop yield prediction using machine learning: A systematic literature review. *Comput. Electron. Agric.* **2020**, *177*, 105709. [CrossRef]
6. Hao, S.; Ryu, D.; Western, A.; Perry, E.; Bogena, H.; Franssen, H.J.H. Performance of a wheat yield prediction model and factors influencing the performance: A review and meta-analysis. *Agric. Syst.* **2021**, *194*, 103278. [CrossRef]
7. Aasen, H.; Honkavaara, E.; Lucieer, A.; Zarco-Tejada, P.J. Quantitative remote sensing at ultra-high resolution with UAV spectroscopy: A review of sensor technology, measurement procedures, and data correction workflows. *Remote Sens.* **2018**, *10*, 1091. [CrossRef]
8. Whitcraft, A.K.; Vermote, E.F.; Becker-Reshef, I.; Justice, C.O. Cloud cover throughout the agricultural growing season: Impacts on passive optical earth observations. *Remote Sens. Environ.* **2015**, *156*, 438–447. [CrossRef]
9. Planet Team. Planet Application Program Interface: In Space for Life on Earth. San Francisco, CA. Available online: https://api.planet.com (accessed on 17 July 2019).
10. Houborg, R.; McCabe, M.F. High-resolution NDVI from planet's constellation of earth observing nano-satellites: A new data source for precision agriculture. *Remote Sens.* **2016**, *8*, 768. [CrossRef]
11. Zhu, X.; Cai, F.; Tian, J.; Williams, T.K.-A. Spatiotemporal fusion of multisource remote sensing data: Literature survey, taxonomy, principles, applications, and future directions. *Remote Sens.* **2018**, *10*, 527. [CrossRef]

12. Gao, F.; Hilker, T.; Zhu, X.; Anderson, M.; Masek, J.; Wang, P.; Yang, Y. Fusing Landsat and MODIS data for vegetation monitoring. *IEEE Geosci. Remote Sens. Mag.* **2015**, *3*, 47–60. [CrossRef]

13. Liao, C.; Wang, J.; Dong, T.; Shang, J.; Liu, J.; Song, Y. Using spatio-temporal fusion of Landsat-8 and MODIS data to derive phenology, biomass and yield estimates for corn and soybean. *Sci. Total Environ.* **2019**, *650*, 1707–1721. [CrossRef] [PubMed]

14. Ali, I.; Greifeneder, F.; Stamenkovic, J.; Neumann, M.; Notarnicola, C. Review of machine learning approaches for biomass and soil moisture retrievals from remote sensing data. *Remote Sens.* **2015**, *7*, 16398–16421. [CrossRef]

15. Breiman, L. Random forests. *Mach. Learn.* **2001**, *45*, 5–32. [CrossRef]

16. Mutanga, O.; Adam, E.; Cho, M.A. High density biomass estimation for wetland vegetation using Worldview-2 imagery and random forest regression algorithm. *Int. J. Appl. Earth Obs. Geoinf.* **2012**, *18*, 399–406. [CrossRef]

17. Wolanin, A.; Camps-Valls, G.; Gómez-Chova, L.; Mateo-García, G.; van der Tol, C.; Zhang, Y.; Guanter, L. Estimating crop primary productivity with Sentinel-2 and Landsat 8 using machine learning methods trained with radiative transfer simulations. *Remote Sens. Environ.* **2019**, *225*, 441–457. [CrossRef]

18. Liakos, K.G.; Busato, P.; Moshou, D.; Pearson, S.; Bochtis, D. Machine learning in agriculture: A review. *Sensors* **2018**, *18*, 2674. [CrossRef] [PubMed]

19. Cravero, A.; Sepúlveda, S. Use and adaptations of machine learning in big data—Applications in real cases in agriculture. *Electronics* **2021**, *10*, 552. [CrossRef]

20. Everingham, Y.; Sexton, J.; Robson, A. In A statistical approach for identifying important climatic influences on sugarcane yields. In Proceedings of the 37th Annual Conference of the Australian Society of Sugar Cane Technologists, Bundaberg, Australia, 28–30 April 2015; pp. 8–15.

21. Everingham, Y.; Sexton, J.; Skocaj, D.; Inman-Bamber, G. Accurate prediction of sugarcane yield using a random forest algorithm. *Agron. Sustain. Dev.* **2016**, *36*, 27. [CrossRef]

22. Stephens, D.; Lyons, T.; Lamond, M. A simple model to forecast wheat yield in Western Australia. *J. R. Soc. West. Aust.* **1989**, *71*, 77–81.

23. Marletto, V.; Ventura, F.; Fontana, G.; Tomei, F. Wheat growth simulation and yield prediction with seasonal forecasts and a numerical model. *Agric. Forest Meteorol.* **2007**, *147*, 71–79. [CrossRef]

24. Ahmed, M.; Akram, M.N.; Asim, M.; Aslam, M.; Hassan, F.-U.; Higgins, S.; Stöckle, C.O.; Hoogenboom, G. Calibration and validation of APSIM-Wheat and CERES-Wheat for spring wheat under rainfed conditions: Models evaluation and application. *Comput. Electron. Agric.* **2016**, *123*, 384–401. [CrossRef]

25. Mehrabi, F.; Sepaskhah, A.R. Winter wheat yield and DSSAT model evaluation in a diverse semi-arid climate and agronomic practices. *Int. J. Plant Prod.* **2020**, *14*, 221–243. [CrossRef]

26. Feng, P.; Wang, B.; Liu, D.L.; Waters, C.; Xiao, D.; Shi, L.; Yu, Q. Dynamic wheat yield forecasts are improved by a hybrid approach using a biophysical model and machine learning technique. *Agric. Forest Meteorol.* **2020**, *285–286*, 107922. [CrossRef]

27. Jeong, J.H.; Resop, J.P.; Mueller, N.D.; Fleisher, D.H.; Yun, K.; Butler, E.E.; Timlin, D.J.; Shim, K.-M.; Gerber, J.S.; Reddy, V.R.; et al. Random forests for global and regional crop yield predictions. *PLoS ONE* **2016**, *11*, e0156571. [CrossRef] [PubMed]

28. Wang, L.; Zhou, X.; Zhu, X.; Dong, Z.; Guo, W. Estimation of biomass in wheat using random forest regression algorithm and remote sensing data. *Crop J.* **2016**, *4*, 212–219. [CrossRef]

29. Han, J.; Zhang, Z.; Cao, J.; Luo, Y.; Zhang, L.; Li, Z.; Zhang, J. Prediction of winter wheat yield based on multi-source data and machine learning in China. *Remote Sens.* **2020**, *12*, 236. [CrossRef]

30. Fajardo, M.; Whelan, B.M. Within-farm wheat yield forecasting incorporating off-farm information. *Precis. Agric.* **2021**, *22*, 569–585. [CrossRef]

31. Kamilaris, A.; Prenafeta-Boldú, F.X. Deep learning in agriculture: A survey. *Comput. Electron. Agric.* **2018**, *147*, 70–90. [CrossRef]

32. Bramley, R.G.V.; Ouzman, J. Farmer attitudes to the use of sensors and automation in fertilizer decision-making: Nitrogen fertilization in the Australian grains sector. *Precis. Agric.* **2019**, *20*, 157–175. [CrossRef]

33. Guerrero, A.; Mouazen, A.M. Evaluation of variable rate nitrogen fertilization scenarios in cereal crops from economic, environmental and technical perspective. *Soil Tillage Res.* **2021**, *213*, 105110. [CrossRef]

34. Maleki, M.R.; Mouazen, A.M.; De Ketelaere, B.; Ramon, H.; De Baerdemaeker, J. On-the-go variable-rate phosphorus fertilisation based on a visible and near-infrared soil sensor. *Biosyst. Eng.* **2008**, *99*, 35–46. [CrossRef]

35. Gobbett, D.; Ouzman, J.; Ratcliff, C.; Bramley, R. Yield map workflow for precision agriculture: Challenges of real-world data. In Proceedings of the Collaborative Conference on Computational and Data Intensive Science (C3DIS), Online, 6–8 July 2021; Available online: http://www.c3dis.com/wp-content/uploads/2020/05/A-Precision-Agriculture-workflow-v1g.pdf (accessed on 10 December 2021).

36. Angus, J.F.; Kirkegaard, J.A.; Hunt, J.R.; Ryan, M.H.; Ohlander, L.; Peoples, M.B. Break crops and rotations for wheat. *Crop Pasture Sci.* **2015**, *66*, 523–552. [CrossRef]

37. GRDC. *Grdc Grownotes—Wheat, Southern*; Grains Research and Development Corporation: Canberra, Australia, 2018.

38. Bureau of Meteorology Australia. Annual Climate Statement 2018. Available online: http://www.bom.gov.au/climate/current/annual/aus/2018/ (accessed on 1 November 2021).

39. Fan, X.; Liu, Y.; Wu, G.; Zhao, X. Compositing the minimum NDVI for daily water surface mapping. *Remote Sens.* **2020**, *12*, 700. [CrossRef]

40. Chen, Y.; Donohue, R.J.; McVicar, T.R.; Waldner, F.; Mata, G.; Ota, N.; Houshmandfar, A.; Dayal, K.; Lawes, R.A. Nationwide crop yield estimation based on photosynthesis and meteorological stress indices. *Agric. For. Meteorol.* **2020**, *284*, 107872. [CrossRef]

41. Rouse, J.W.; Haas, R.H.; Schell, J.A.; Deering, D.W. *Monitoring Vegetation Systems in the Great Plains with Erts*; Freden, S.C., Mercanti, E.P., Becker, M., Eds.; NASA: Washington, DC, USA, 1974; Volume I, pp. 309–317.

42. Houborg, R.; McCabe, M.F. Daily retrieval of NDVI and LAI at 3 m resolution via the fusion of CubeSat, Landsat, and MODIS data. *Remote Sens.* **2018**, *10*, 890. [CrossRef]

43. Guilherme Teixeira Crusiol, L.; Sun, L.; Chen, R.; Sun, Z.; Zhang, D.; Chen, Z.; Wuyun, D.; Rafael Nanni, M.; Lima Nepomuceno, A.; Bouças Farias, J.R. Assessing the potential of using high spatial resolution daily NDVI-time-series from Planet CubeSat images for crop monitoring. *Int. J. Remote Sens.* **2021**, *42*, 7114–7142. [CrossRef]

44. QGIS Development Team. QGIS Geographic Information System. Open Source Geospatial Foundation Project. Available online: http://qgis.osgeo.org (accessed on 1 November 2021).

45. Queensland Department of Environment and Science. Scientific Information for Land Owners (SILO), Queensland Government. Available online: https://www.longpaddock.qld.gov.au/silo/ (accessed on 20 June 2019).

46. Barlow, K.M.; Christy, B.P.; O'Leary, G.J.; Riffkin, P.A.; Nuttall, J.G. Simulating the impact of extreme heat and frost events on wheat crop production: A review. *Field Crops Res.* **2015**, *171*, 109–119. [CrossRef]

47. Figueiredo, B.; Dhillon, J.; Eickhoff, E.; Nambi, E.; Raun, W. Value of composite Normalized Difference Vegetative Index and growing degree days data in Oklahoma, 1999 to 2018. *Agrosystems Geosci. Environ.* **2020**, *3*, e20013. [CrossRef]

48. McKinney, W. *Python for Data analysis: Data Wrangling with Pandas, Numpy, and Ipython*; O'Reilly Media, Inc.: Newton, MA, USA, 2012.

49. Breiman, L. Bagging predictors. *Mach. Learn.* **1996**, *24*, 123–140. [CrossRef]

50. Pedregosa, F.; Varoquaux, G.; Gramfort, A.; Michel, V.; Thirion, B.; Grisel, O.; Blondel, M.; Prettenhofer, P.; Weiss, R.; Dubourg, V. Scikit-learn: Machine learning in Python. *J. Mach. Learn. Res.* **2011**, *12*, 2825–2830.

51. Fawagreh, K.; Gaber, M.M.; Elyan, E. Random forests: From early developments to recent advancements. *Syst. Sci. Control Eng.* **2014**, *2*, 602–609. [CrossRef]

52. Bergstra, J.; Bengio, Y. Random search for hyper-parameter optimization. *J. Mach. Learn. Res.* **2012**, *13*, 281–305.

53. Kohavi, R. A study of cross-validation and bootstrap for accuracy estimation and model selection. In Proceedings of the 14th International Joint Conference on Artificial Intelligence—Volume 2; Morgan Kaufmann Publishers Inc.: Montreal, QC, Canada, 1995; pp. 1137–1143.

54. Rogers, J.; Gunn, S. *Identifying Feature Relevance Using a Random Forest*; Springer: Berlin/Heidelberg, Germany, 2006; pp. 173–184.

55. Menze, B.H.; Kelm, B.M.; Masuch, R.; Himmelreich, U.; Bachert, P.; Petrich, W.; Hamprecht, F.A. A comparison of random forest and its Gini importance with standard chemometric methods for the feature selection and classification of spectral data. *BMC Bioinform.* **2009**, *10*, 213. [CrossRef]

56. Abdel-Rahman, E.M.; Ahmed, F.B.; Ismail, R. Random forest regression and spectral band selection for estimating sugarcane leaf nitrogen concentration using EO-1 Hyperion hyperspectral data. *Int. J. Remote Sens.* **2013**, *34*, 712–728. [CrossRef]

57. Hassan, M.A.; Yang, M.; Rasheed, A.; Yang, G.; Reynolds, M.; Xia, X.; Xiao, Y.; He, Z. A rapid monitoring of NDVI across the wheat growth cycle for grain yield prediction using a multi-spectral UAV platform. *Plant Sci.* **2019**, *282*, 95–103. [CrossRef] [PubMed]

58. Lai, Y.R.; Pringle, M.J.; Kopittke, P.M.; Menzies, N.W.; Orton, T.G.; Dang, Y.P. An empirical model for prediction of wheat yield, using time-integrated Landsat NDVI. *Int. J. Appl. Earth Obs. Geoinf.* **2018**, *72*, 99–108. [CrossRef]

59. Nagy, A.; Fehér, J.; Tamás, J. Wheat and maize yield forecasting for the Tisza river catchment using MODIS NDVI time series and reported crop statistics. *Comput. Electron. Agric.* **2018**, *151*, 41–49. [CrossRef]

60. Xiao, D.; Moiwo, J.P.; Tao, F.; Yang, Y.; Shen, Y.; Xu, Q.; Liu, J.; Zhang, H.; Liu, F. Spatiotemporal variability of winter wheat phenology in response to weather and climate variability in China. *Mitig. Adapt. Strateg. Glob. Change* **2015**, *20*, 1191–1202. [CrossRef]

61. Johnson, D.M. An assessment of pre- and within-season remotely sensed variables for forecasting corn and soybean yields in the united states. *Remote Sens. Environ.* **2014**, *141*, 116–128. [CrossRef]

62. French, R.; Schultz, J. Water use efficiency of wheat in a Mediterranean-type environment. I. The relation between yield, water use and climate. *Aust. J. Agric. Res.* **1984**, *35*, 743–764. [CrossRef]

63. Saeed, U.; Dempewolf, J.; Becker-Reshef, I.; Khan, A.; Ahmad, A.; Wajid, S.A. Forecasting wheat yield from weather data and MODIS NDVI using random forests for Punjab province, Pakistan. *Int. J. Remote Sens.* **2017**, *38*, 4831–4854. [CrossRef]

64. Li, C.; Li, H.; Li, J.; Lei, Y.; Li, C.; Manevski, K.; Shen, Y. Using NDVI percentiles to monitor real-time crop growth. *Comput. Electron. Agric.* **2019**, *162*, 357–363. [CrossRef]

65. Fu, Y.; Yang, G.; Wang, J.; Song, X.; Feng, H. Winter wheat biomass estimation based on spectral indices, band depth analysis and partial least squares regression using hyperspectral measurements. *Comput. Electron. Agric.* **2014**, *100*, 51–59. [CrossRef]

66. Tan, C.-W.; Zhang, P.-P.; Zhou, X.-X.; Wang, Z.-X.; Xu, Z.-Q.; Mao, W.; Li, W.-X.; Huo, Z.-Y.; Guo, W.-S.; Yun, F. Quantitative monitoring of leaf area index in wheat of different plant types by integrating NDVI and Beer-Lambert law. *Sci. Rep.* **2020**, *10*, 929. [CrossRef] [PubMed]

67. Prabhakara, K.; Hively, W.D.; McCarty, G.W. Evaluating the relationship between biomass, percent groundcover and remote sensing indices across six winter cover crop fields in Maryland, United States. *Int. J. Appl. Earth Obs. Geoinf.* **2015**, *39*, 88–102. [CrossRef]

68. Fukuda, S.; Spreer, W.; Yasunaga, E.; Yuge, K.; Sardsud, V.; Müller, J. Random forests modelling for the estimation of mango (*Mangifera indica* L. cv. Chok Anan) fruit yields under different irrigation regimes. *Agric. Water Manag.* **2013**, *116*, 142–150. [CrossRef]

69. Khanal, S.; Fulton, J.; Klopfenstein, A.; Douridas, N.; Shearer, S. Integration of high resolution remotely sensed data and machine learning techniques for spatial prediction of soil properties and corn yield. *Comput. Electron. Agric.* **2018**, *153*, 213–225. [CrossRef]

70. Vannoppen, A.; Gobin, A. Estimating farm wheat yields from NDVI and meteorological data. *Agronomy* **2021**, *11*, 946. [CrossRef]

71. Muschietti-Piana, P.; McBeath, T.M.; McNeill, A.M.; Cipriotti, P.A.; Gupta, V.V.S.R. Combined nitrogen input from legume residues and fertilizer improves early nitrogen supply and uptake by wheat. *J. Plant Nutr. Soil Sci.* **2020**, *183*, 355–366. [CrossRef]

72. Thilakarathna, M.S.; Raizada, M.N. Challenges in using precision agriculture to optimize symbiotic nitrogen fixation in legumes: Progress, limitations, and future improvements needed in diagnostic testing. *Agronomy* **2018**, *8*, 78. [CrossRef]

73. Wheeler, R.; Grains Research and Development Corporation (GRDC). Wheat and Barley Variety Update 2018. Available online: https://grdc.com.au/resources-and-publications/grdc-update-papers/tab-content/grdc-update-papers/2019/02/wheat-and-barley-variety-update-2018 (accessed on 12 November 2021).

74. Drăguţ, L.; Schauppenlehner, T.; Muhar, A.; Strobl, J.; Blaschke, T. Optimization of scale and parametrization for terrain segmentation: An application to soil-landscape modeling. *Comput. Geosci.* **2009**, *35*, 1875–1883. [CrossRef]

75. Nuttall, J.G.; O'Leary, G.J.; Khimashia, N.; Asseng, S.; Fitzgerald, G.; Norton, R. 'Haying-off' in wheat is predicted to increase under a future climate in south-eastern Australia. *Crop Pasture Sci.* **2012**, *63*, 593–605. [CrossRef]

76. Filippi, P.; Jones, E.J.; Wimalathunge, N.S.; Somarathna, P.D.S.N.; Pozza, L.E.; Ugbaje, S.U.; Jephcott, T.G.; Paterson, S.E.; Whelan, B.M.; Bishop, T.F.A. An approach to forecast grain crop yield using multi-layered, multi-farm data sets and machine learning. *Precis. Agric.* **2019**, *20*, 1015–1029. [CrossRef]

77. Patel, M.K.; Ryu, D.; Western, A.W.; Suter, H.; Young, I.M. Which multispectral indices robustly measure canopy nitrogen across seasons: Lessons from an irrigated pasture crop. *Comput. Electron. Agric.* **2021**, *182*, 106000. [CrossRef]

78. Lu, B.; Dao, P.D.; Liu, J.; He, Y.; Shang, J. Recent advances of hyperspectral imaging technology and applications in agriculture. *Remote Sens.* **2020**, *12*, 2659. [CrossRef]

79. Dang, C.; Liu, Y.; Yue, H.; Qian, J.; Zhu, R. Autumn crop yield prediction using data-driven approaches:- support vector machines, random forest, and deep neural network methods. *Can. J. Remote Sens.* **2021**, *47*, 162–181. [CrossRef]

80. Liu, Y.-N.; Li, Y.-C.; Peng, Z.-P.; Wang, Y.-Q.; Ma, S.-Y.; Guo, L.-P.; Lin, E.-D.; Han, X. Effects of different nitrogen fertilizer management practices on wheat yields and N_2O emissions from wheat fields in north China. *J. Integr. Agric.* **2015**, *14*, 1184–1191. [CrossRef]

81. Shafiee, S.; Lied, L.M.; Burud, I.; Dieseth, J.A.; Alsheikh, M.; Lillemo, M. Sequential forward selection and support vector regression in comparison to lasso regression for spring wheat yield prediction based on UAV imagery. *Comput. Electron. Agric.* **2021**, *183*, 106036. [CrossRef]

MDPI

St. Alban-Anlage 66

4052 Basel

Switzerland

Tel. +41 61 683 77 34

Fax +41 61 302 89 18

www.mdpi.com

Sensors Editorial Office

E-mail: sensors@mdpi.com

www.mdpi.com/journal/sensors